ROBOTICS AGE

In THE Beginning

ROBOTICS AGE

In The Beginning

Selected from ROBOTICS AGE magazine

Edited by
CARL HELMERS

HAYDEN BOOK COMPANY, INC.
Hasbrouck Heights, New Jersey

Acquisitions Editor: PRIJONO HARDJOWIROGO
Production Editor: RONNIE GROFF
Art Director: JIM BERNARD
Printed and bound by: MAPLE-VAIL BOOK MANUFACTURING GROUP
Cover illustration by Mike Hagelberg, courtesy of the MACK Corporation,
Flagstaff, Arizona, manufacturer of robotic components.

Library of Congress Cataloging in Publication Data
Main entry under title:

Robotics age, in the beginning

 Includes bibliographical references.
 1. Automata—Addresses, essays, lectures. I. Helmers,
Carl. II. Robotics age.
TJ211.R564 1983   629.8'92   83-8600
ISBN 0-9104-6325-3

3 4 5 6 7 8 9 PRINTING

84 85 86 87 88 89 90 91 YEAR

PREFACE

We live in interesting times. The times are made all the more interesting by the technologies we humans invent. Robotics is one of the latest in a series of such technologies, and *Robotics Age* is a vehicle for spreading the word about the intelligent machine technologies we call robotics.

Let me introduce myself. My name is Carl Helmers. I am a long-time believer in the inspirational value of science fiction. I am a thinker and experimenter by nature. I occasionally write about the technological ferment that is producing the exciting advancements in electronics, computers, robotics, genetics, and space travel that we all see.

I read my first science fiction book at the age of eleven. I spent my youth as "... the kid in your high school who always went around with a slide rule dangling from his belt— the one who called it a 'slip stick.'"* I earned an undergraduate degree in physics, paid for by summer programming jobs, and went to work on a project to produce software tools for development of Space Shuttle flight programs in the early 1970s. But I had not found my calling. I spent much of my time tinkering with electronics and computer-related applications, writing about the experiments and self-publishing the results.

In 1975, I found my "true" calling ... putting together that mysterious combination of readers and writers that defines the editorial end of a special purpose magazine devoted to technological experimentation with contemporary engineering concepts. With much help from hundreds of writers who were also readers, I created and set the directions for an idea forum called *BYTE*, which became the bible of the personal computer field. In that previous context I often wrote about and encouraged work in that classic of science fiction: robotics and artificial intelligence.

Robotics Age the magazine began in 1979 when robotics researcher Alan Thompson and his friend Phil Flora brainstormed the idea of starting a new journal of intelligent machines. At that time, Phil ran a print shop in Houston, Texas, and Alan worked for NASA's Jet Propulsion Laboratory in Pasadena, California. So, they put their heads together and started what would become a quarterly magazine. The first issue came out in the summer of 1979.

I first met Phil and Alan at about the same time, while they were putting together the first issue and talking about the prospects of the new magazine. I knew Phil through my good friend Robert Tinney, the artist responsible for so many vivid and imaginative covers of *BYTE* and other contemporary computer magazines. At the time, Robert was living in Houston, where he and Phil frequented the libertarian circles where I had originally met Robert.

Phil, Alan, and I had a pleasant meeting discussing topics ranging from artificial intelligence to the physics of information, from solar eclipses to the engineering of prosthetics. Having established a friendly relation, we parted. Alan and Phil proceeded to devote much time and energy to publication of the magazine, and I went back to what I had been doing as editor of *BYTE* magazine.

Time passes. After another year, I was no longer involved with *BYTE*, which had been sold to McGraw-Hill shortly before I met Phil and Alan. A dozen issues or so of *Robotics Age* had been published. Phil came to me at the end of 1981 and asked if I'd get involved with *Robotics Age*. It came to pass that ownership of the magazine was transferred to a new company owned by my company as well as Phil and Alan—and the magazine moved to my home of Peterborough, New Hampshire, a publishing center in the southwestern part of the state.

Robotics Age has a continuing emphasis on the technology and techniques for experimentation and implementation of robotic systems. The tools and products we talk about range from inexpensive, personally owned computers, robot arms, and machine shop equipment to the sophisticated computer-aided design, graphic design, and artificial intelligence equipment now coming to market. Whether the designing is done by professionals, educators teaching new professionals, or simply by amateurs with personal computers and conventional home workshop tools, it contributes to the growth of the field. We've seen this same phenomenon earlier in the eras of automotive and aviation pioneering, radio pioneering, and recently in personal computing.

The amateur in robotics is an important and vital link in the chain of events needed to train competent robotics engineers for the coming decades. The technologist who builds to a personal budget while exploring the problems of the field has been shown to be a vital force in previous eras of innovation. Whether it is a Steve Jobs or Steve Wozniac of Apple Computer fame, or in a different era, a Henry Ford, the spark of innovative applications of technology is an individual matter. A truly professional technological innovator does not turn off his or her mind upon leaving the official place of work. As in past eras, new companies

---

From The Nine Nations of North America by Joel Garreau, page 42. Published by Houghton Mifflin Company, Boston, 1981.

starting with new ideas often spring from the workshop of the amateur—who has already become the true professional with a love of his work.

One characteristic of serious technical magazines such as *Robotics Age* is that their technology discussions do not soon pass into obsolescence. In an expanding field such as the intelligent machines field of *Robotics Age*, there is a continual influx of new people with a great thirst for knowledge. The writers of articles are in many respects simply the first people through the doors of opportunity and enlightenment represented by the technology. In order to bring about a wider understanding of the field, the technical articles deserve—even demand—reprinting for a wider audience than the original readers. Hence this book.

Hayden Book Co. was selected to produce and distribute the book because of their excellent library of books on the subject of experimentation with robotics concepts.

As preparation for the reprint, we sent letters to all the authors of the articles asking for any fixes of faux pas and other miscellany in the original production and publication. These fixes and a few we've found here as well were incorporated into the texts by the good people at Hayden following a checklist I provided for each article. (The number of changes was actually quite small considering the volume of material involved. The typical change was a spelling error, a transposition, a change of a reference, or other evidence of the great god Murphy in his publishing form.)

The articles contained in this book are selected reprints of technical articles first published in *Robotics Age*, Volumes 1 to 3. The articles were originally obtained and then produced under the editorial direction of Alan Thompson.

I organized the articles selected for this reprint into three major categories of content: *Power and Control* articles refer to the lower level electronic details of hardware for robotics. *Interactions: Senses, Vision, and Voice* provides a place for articles, largely on vision input, concerning the interface of intelligent systems with the real world of humans and objects. The final section is entitled *Applications and Development*. This introduction, along with an introduction for each section, are the only parts of this book that have not been published in the first three volumes of *Robotics Age*.

This is just the first book of reprints of *Robotics Age* articles. The magazine is a continuing, lively forum for people participating in the *Robotics Age*. If you'd like to get current information *before* waiting for the next reprint book, drop us a line.

Carl Helmers
Editor/Publisher
Robotics Age Magazine
174 Concord Street
Peterborough, NH 03458
(603) 924-7136

CONTENTS

─────────────── **Section 3** ───────────────
APPLICATIONS AND DEVELOPMENT

ROBOTICS AGE

In The Beginning

SECTION 1

POWER AND CONTROL

A robotic device is fundamentally a computer closely bound to the real world of space and time. In order to apply robotic principles, we have to explore the electronic and electromechanical world of design. We start this reprint book with a grouping of several articles dealing with concepts of power and control.

John Craig provides the opening article in this group, "Digital Speed Control of DC Motors." On a similar theme, Wesley E. Snyder and Jorg Schött provide us with an article on "Using Optical Shaft Encoders" to find out where the rotating shafts are. The principles are simple and the applications are universal.

Then Slobodan Ćuk and R. D. Middlebrook provide a two-part article on "Advances in Switched-Mode Power Conversion." This article provides a tutorial background on the principles of switching power supplies. This is done by way of introducing a specific account of the authors'

newly patented development of a highly efficient switcher.

Two articles explore the problems of moving a suitable mechanism (such as, for example, a robot arm) through desired paths in three-dimensional space. Wesley E. Snyder discusses a method of "Microcomputer Based Path Control." Then Shafi Motiwalla and Richard Tseng discuss the problems and strategies needed for "Continuous Path Control with Stepping Motors" in a two-part article.

We conclude the section on *Power and Control* with a tutorial background article on the efficient computation of trigonometric functions. Calculation of transcendental functions like sines and cosines is an inherently slow process. In order to use trigonometric functions in a real-time robotic control application we must pay attention to the speed of such calculations. Carl Ruoff provides us with practical mathematics to implement "Fast Trig Functions for Robot Control."

Inexpensive and effective servo techniques suitable for microcomputer-controlled robots are one of our major interests. This article describes a hardware servo method that is both cheap and highly accurate.

Digital Speed Control of DC Motors

John Craig
NASA Jet Propulsion Laboratory
California Institute of Technology
Pasadena, CA

Most home robotics hobbyists would agree that immobile robots just aren't fun. Many robots in research labs and in industry are just mechanical arms sticking out of a table or mounted along an assembly line. In contrast, most robots built by hobbyists for fun are capable of locomotion. Although the captive robots may be capable of impressive feats within their limited world, we usually want our robots to wander about the house looking for an outlet to recharge themselves, seeking light, fetching the evening paper, and a host of other tasks, all requiring the ability to move. We may strive to give our robot eyes, ears, or an arm, but the first step is to make it MOVE! Many home robots are barely more than moving boxes, but from that base new features can be added as fast as the inventor can dream them up. The great majority of today's robots that move do so with wheels. Robots that walk on articulated legs are still rare, and are complex, expensive, and, so far, limited in capability. Although nature has found it easier to evolve creatures with legs (for movement on dry land), with today's technology wheels are the best alternative. With the exception of staircases, wheels will allow a robot to rove almost anywhere in the house!

I. Introduction to Feedback Control.

The methods of controlling a motor in a robot can range from the most simple on/off power switch to the complex servo control circuits used in industrial manipulators. If variations in the speed of the motor are not important, but different speed ranges are desired, then a simple approach is just to set the input voltage to the motor with a variable power supply. But suppose you want your robot to roll at the same speed regardless of variations in the load. If you have a fixed voltage setting on the motor, the robot will move faster going down inclines and slower when it is climbing or if it is pushing or pulling a load. More important, if more than one motor is used to drive the wheels, as in the system discussed here, accurate speed control of each motor is critical. When commanded to walk a straight line, each motor must run at the proper speed or the robot will follow an embarassingly crooked path. If the speed of each wheel is accurately known, then a robot with a microcomputer can also "navigate" by keeping track of its position relative to some initial starting point or coordinate frame. This can be done by a numerical integration of the robot's velocity. Put more simply, the formula "distance = rate x time" is used, for which the "rate" or speed must be accurately known during the elapsed "time" of the movement for the calculation to be meaningful.

To insure accurate speed it is essential to "close the loop" by using feedback. The "loop" is the feedback loop, and "closing" it refers to using a "feedback" signal from a sensor to automatically regulate the input to the drive system. In this case, feedback indicating wheel speed must be used to control the input voltage to the motor. In the general case, however, feedback may be used to control almost any dynamic quality of a system, subject to stability restrictions that can be determined by a mathematical analysis of the problem. A generalized feedback loop is shown in Figure 1. The commanded value of the controlled parameter is set from outside the loop or else could be an intrinsic value of the system, as often occurs in living systems. The commanded value is compared against the measured value of the parameter as determined by the feedback sensor. If there is no difference, then no action need be taken, but if there is, then that difference is "amplified", or otherwise processed, by the "gain" element of the loop to produce the input to the system's effector. The effector is called such because it "effects" reality in some way to accomplish the desired change in the controlled parameter, which is being measured by the sensor.

In complex systems, there may be loops within loops, or measurements from different types of sensors may be used. The gain process may range from simple amplification to involved combination of several feedback terms. For example, suppose it is desired to control the force that a motor-driven effector applies to an object. This might occur if an assembly-line robot has to insert a fragile part in an object under construction, or if you want your personal robot to properly use a sponge! A suitable sensor would be a pressure transducer such as a strain guage, which produces a signal that can be processed to measure the pressure applied by the effector. The difference between the measured pressure and the desired pressure can be integrated, e.g., by an operational amplifier circuit, to determine the current to a DC permanent magnet motor driving the effector (mechanical manipulator, etc.)

Feedback loops of one form or another are among the building blocks of life and are the cornerstones of Robotics. In living systems, feedback loops can be found regulating heartbeat, respiration, blood chemistry, intra-cellular pressure, and many forms of motion. Industrial robots depend on many types of feedback loops, typically position, rate, or force control of manipulators. In advanced robots, visual feedback may be employed to aid in grasping, parts positioning, and other tasks. Feedback loops applied to controlling motors are called servo systems, or servos. For a good introduction to servo design, consult Reference [4] (listed at the end of the article) or one of the many other college texts on control system design.

II. Robot Locomotion Using a Phase Locked Servo

Figure 2 shows a basic system for robot locomotion. The front wheels are motor driven and the two rear wheels are small casters. If the motors are reversible this robot can roll forward, backward, turn right or left, and even do fairly graceful piroettes (one motor driven forward and one backward). Photo 1 shows an example of the two motor drive system. Using a microcomputer to control the robot is not a necessity, but doing so will certainly make any robot more talented through the flexibility and power of software. Microcomputers and robotics are two hobbies that mix excellently! Single-board microcomputers may now be bought for a few hundred dollars, putting them within the economic reach of most robotics

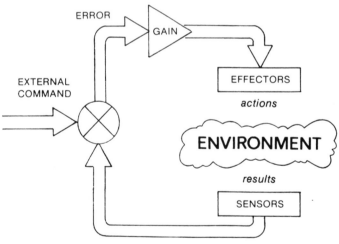

Figure 1. Generalized Feedback Loop: Sensors measure some parameter in the environment. The processed signal is compared with the commanded value, producing an error signal if there is a difference. The error signal is processed by the gain element, resulting in the control signal to the effectors which then alter the environment.

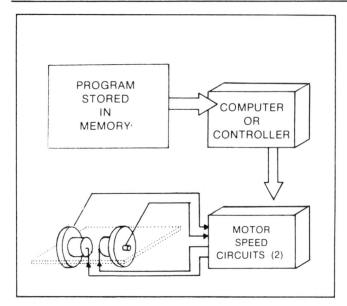

Figure 2. Basic system for robot locomotion: robot is built on a wheeled base which has two powered wheels in front and two free rolling casters in the back. A speed control circuit is needed for each motor to ensure that it turns at the desired speed. A microcomputer can give "high-level" commands to the speed control circuits.

experimenters. Today's microcomputers can, after all, do more than play TV games! The addition of even the smallest microcomputer opens up a whole new world of autonomous operation for your robot when properly interfaced and programmed. But don't be discouraged if your robot doesn't have one, many home built robots haven't had that advantage and still have served their masters well!

The motor speed control circuit is a feedback loop which is given a desired speed command (from an output port of a microcomputer perhaps) and, by comparing it to feedback from the motor, outputs the voltage necessary to drive the motor at that speed. This kind of servo can be realized in many forms. Servos often use feedback from

analog tachometers and are designed with analog circuitry which may drift as it ages or with temperature changes and needs to be "tweaked" regularly to be correct. A digital circuit offers many advantages. First, it is easy to connect to a microcomputer, no digital to analog converter is necessary. Digital circuits are much less susceptible to electrical noise than analog circuits, which is an important consideration when working near motors. The accuracy of a digital circuit may be made far superior to that of an analog loop, since it may be driven by a crystal-controlled clock signal.

This article describes a digital motor speed control system called a Phase Locked Servo (PLS). This is a fairly new technique, but the circuitry can be amazingly simple, inexpensive, and extremely accurate. The circuit requires no digital to analog or analog to digital converters. It also can be run off a single voltage power supply, so the need for a multiple-level power supply on board the robot may be avoided. The motor can be made reversible with the addition of a simple relay (mechanical or solid-state). It is a complete hardware servo system; it can be used without a microcomputer, or, if a micro is used, the computer only issues speed commands and is free to do other more interesting tasks while the PLS controls the motor speed.

The only drawback of a PLS system is that it requires a pulse rate tachometer compatible with digital circuitry. Commercially available digital tachs are still fairly expensive (usually over $100). One way around this problem is to build one yourself! An optical encoder is a device which outputs pulses at a rate proportional to the rotational speed of its shaft. It is used as the feedback sensor in the PLS to measure the motor (or wheel) speed. Figure 3 shows a possible way of building your own

Photo 1. Example of the 2-wheel, 2-caster system as implemented on the author's home robot. The powered wheels were bought as units, originally designed for use on electric-powered wheelchairs.

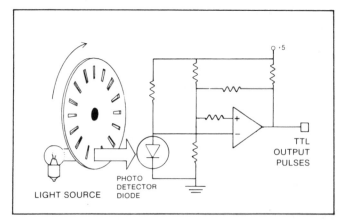

Figure 3. Simple Encoder Circuit: Possible method for converting light pulses to TTL level pulses.

encoder. A disk with evenly spaced holes around its perimeter is mounted on the motor shaft or wheel of the robot. A light source illuminates it from one side so that a photo-detector is illuminated by light pulses as the shaft is turned. The circuit shown is one of many possibilities; it simply converts the light pulses to TTL level digital pulses. You may want to experiment with the circuit to get the best results with your light source and photo-detector.

For a stable system, the number of holes in the disk should be sufficiently large and they should be evenly spaced. If the disk is mounted on a high speed shaft (before a gear reduction to the wheel) the number of holes can be fewer. The number of holes times the disk's rotational speed gives the pulse rate output by the encoder. At the minimum desired shaft speed this value should not be less than about 100 Hz. for good control. If you don't have access to a drill press with a dividing gear to accurately measure the hole spacing, you can geometrically define equally spaced holes by constructing regular polygons on the disk or a template. Start with a perfect square inscribed in your disk and find the midpoint of each chord. The radius through the midpoint of each chord gives the location of four new points on the perimeter, from which eight new chords may be drawn to form an octagon. (See Figure 4.) Continue constructing new chords until the number of perimeter points is the power of two that is near to the number of holes determined by dividing 100 (pulses per second) by your minimum desired shaft rate in revolutions per second. The encoder for the circuit described here has 256 holes. If the holes are too close together for you to drill them accurately, either use a larger disk or put the encoder on a higher speed shaft. Use your construction as a template for drilling or punching uniformly sized holes at a constant radius of your disk (sheet metal, painted plexiglass, etc.) For a given hole spacing, the hole size determines the duty cycle of the output square wave; use small holes with clean edges for a good signal.

Another important element in the PLS is the phase-frequency detector. The one used here is the Motorola MC4044. This circuit accepts two input waveforms and generates an error voltage that is proportional to the frequency and/or phase difference of the input signals. The error voltage is zero when the feedback signal matches the reference signal in phase and frequency but grows as the feedback deviates from the reference. If only phase difference alone was measured, the system could lock onto multiples of the reference frequency. Using the MC4044, this cannot occur, so that the motor may be started from zero and will always reach the right speed,

provided, that is, that your power supply is sufficient. The MC4044 also contains an amplifier which will serve in the gain element of the PLS system.

Figure 5 shows a block diagram of a PLS system. The speed command is a 3 bit binary number. The first bit specifies the motor's direction and the other 2 bits select one of four speeds. A clock circuit generates a reference signal which is divided down appropriately in accordance with the command. The resulting square wave is used as the reference input to the phase-frequency detector. The frequency at which the master clock runs must be selected by/from the encoder's characteristics and the desired speed of the motor. If the encoder produces N pulses per revolution, and the maximum (encoder) shaft speed desired is R revolutions per second, then the clock frequency should be N*R cycles per second. By dividing this frequency down, the lower reference frequencies are obtained. In the circuit shown, the minimum shaft speed (used above to determine the proper number of holes in the encoder disk) is one fourth the maximum speed. The reference signal is compared with the output of the encoder, generating an error signal at the output of the phase-frequency detector. The error signal must be filtered to remove the high frequency components. Its average DC level is used as the input to an inverting amplifier which drives the motor. The system is self-regulating; when the error signal grows in magnitude, the motor's voltage is raised or lowered to reduce the error. If the circuit is operating properly there will be a small offset error in phase, giving an average DC level to the amplifier, but almost no error in velocity. PLS systems can achieve **speed accuracies as good as 0.02%!**

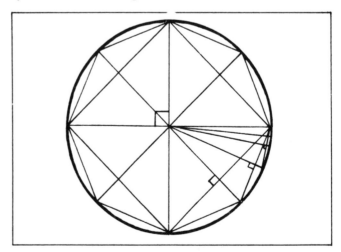

Figure 4. A method of constructing regular polygons of 2^n sides. Used to locate positions of holes in encoder disk.

A working circuit is shown in Figure 6. The master clock is implemented using the familiar NE555 timer integrated circuit. The values of R_a, R_b, and C will set its frequency according to the formula:

$$f_c = 1.44/[(R_a + 2R_b) C]$$

(with R in Ohms, C in Farads)

For example, with $R_a = R_b = 33k$ ohms, and C = .015 microfarads, the master clock frequency will be about 1000 Hz. The values shown in Figure 4 were used for the encoder and motor speed shown, they will have to be changed for a different encoder or different motor speeds.

Reference [1] gives complete information on the NE555. The divider circuit uses two JK flip flops in the "toggle" mode, connected as shown in Figure 4. The 1-of-4 selector allows one of the four different speeds to be selected by the two input bits. It will select the master clock frequency (f_c), $f_c/2$, $f_c/4$, or zero. I C's 2, 3, and 5 are all common TTL chips, described fully in References [1] and [2].

The MC4044, described in Reference [3], is used as the phase-frequency detector. An active low-pass filter is constructed using the amplifier in the MC4044 package and the resistors R_1, R_2, and capacitor C. Their values will set the break frequency of the filter and were found experimentally. They should be selected so your motor runs smoothly and doesn't oscillate. The values shown will probably be fine for many small motors. When experimenting with different values, try to keep the ratio R_1/R_2 about equal to 10. If the ratio is less than 10, the filter begins to pass high frequencies. The value of C will have an effect on the "damping" of the system. In general a large C will tend to make the system sluggish and take too long to lock, while too small a C will make it oscillate. The voltage at the output of the filter should vary between 0.75 and 2.25 volts. For a 12 volt motor, a gain in the driver amplifier of about 5 should allow for a good range in motor speed. The voltage gain of the driver amplifier can be changed by varying resistors R_c and R_e. The gain is approximately equal to $-R_c/R_e$. A control theorist would have a good theoretical understanding of the influences of changing the filter's characteristics and varying the loop gain, but all that's really needed is a little experimenting to find the values which best suit your motor/encoder combination. Readers interested in learning more about the control theory involved should pursue References [4], [5], and [6].

III. Operating Limitations

The circuit described here was designed to demonstrate the PLS principles and to give you a working circuit with a low chip count that can be easily added to your robot. It is capable of operating at only three different speeds in each direction, and the speeds must be 0, 1, 2, or 4 times the minimum speed. You can increase the range of possible speeds by building a more elaborate programmable reference clock circuit. Without going into complete details, several options are available that you might want to try. If you just want particular speeds other than the ones

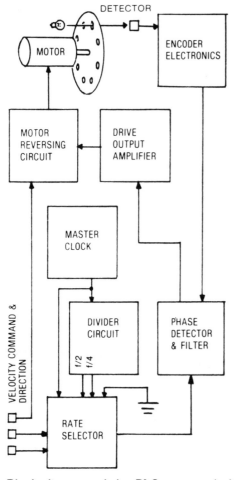

Figure 5. Block diagram of the PLS system: A three-bit velocity command is issued to the PLS system to choose one of four speeds in either the forward or reverse direction.

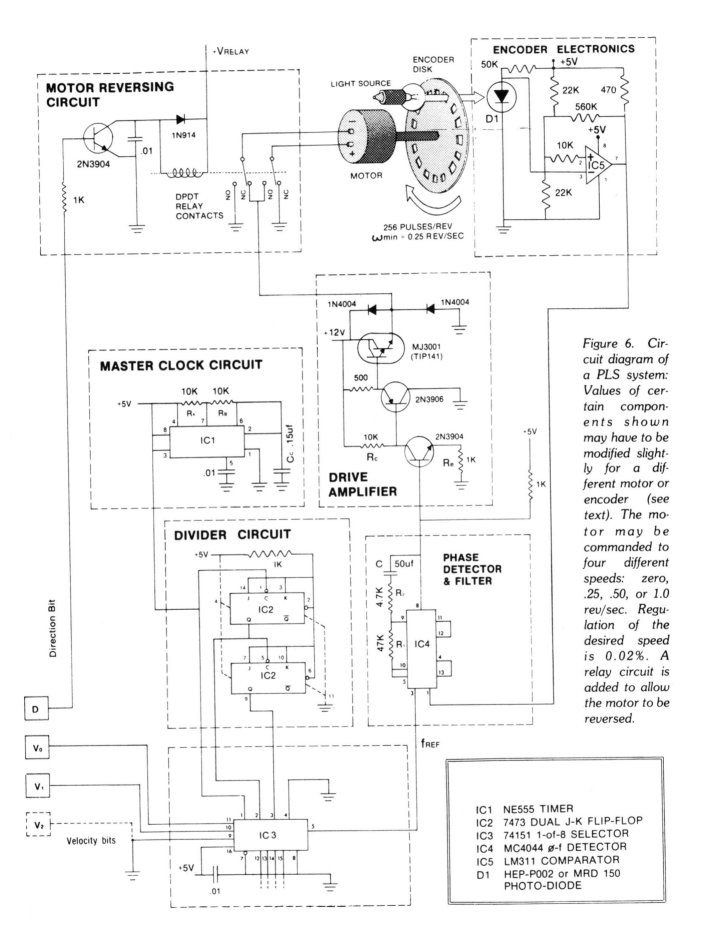

Figure 6. Circuit diagram of a PLS system: Values of certain components shown may have to be modified slightly for a different motor or encoder (see text). The motor may be commanded to four different speeds: zero, .25, .50, or 1.0 rev/sec. Regulation of the desired speed is 0.02%. A relay circuit is added to allow the motor to be reversed.

IC1   NE555 TIMER
IC2   7473 DUAL J-K FLIP-FLOP
IC3   74151 1-of-8 SELECTOR
IC4   MC4044 ø-f DETECTOR
IC5   LM311 COMPARATOR
D1   HEP-P002 or MRD 150
      PHOTO-DIODE

available here, the outputs of different clocks could be selected by the input commands. The 74151 shown in the circuit is capable of selecting one of eight different inputs; for simplicity, only four inputs were used in this circuit. A circuit with the same chip count could provide eight possible speeds by using the outputs of a 74160 binary counter to divide down the clock, as shown in Figure 7(a).

You can obtain a continuous range of possible speeds by using a variable frequency oscillator for the reference signal. Options here would be to use a pot to control the frequency of the 555 (Reference [1]), or to use a Voltage Controlled Oscillator (VCO) chip (Intersil 8038, or others). A disadvantage of going to an analog circuit for the clock, however, is that you lose the accuracy of the PLS; if you don't know the accuracy of the reference signal, the fact that the PLS can lock to it within 0.02% doesn't mean much. This may be fine if your robot isn't doing its own navigation. Also, if your PLS is to be interfaced to a microcomputer, you have to use a D/A converter to set the control voltage.

An accurate programmable digital clock with a practically continuous output frequency range can be built by using a crystal-controlled oscillator as the input to an N-bit binary up-counter (Figure 7(b)). Use the counter's carry output to produce the output signal and to reload the counter from the speed command latch register. This forms a "Divide-by-N" circuit with a minimum frequency of $f_o/2^r$, where f_o is the crystal clock rate and r is the number of bits in the counter. If needed, gate some zeros into the most significant bits to limit the maximum rate and define the useful range of the clock.

When given a new speed command, the reference rate suddenly becomes different from the feedback signal, and the PLS compensates by changing the motor voltage appropriately. The problem is that the motor cannot change speeds instantaneously, and during that time the velocity is not accurately controlled. The accelleration/deaccelleration curve of the system is influenced by the characteristics of the low-pass filter as determined by R_1, R_2, and C. Finding the lowest value of C that will work without oscillating gives a "critically damped" system that will lock to the reference the quickest. This is especially important when the robot is navigating; the velocity errors result in position uncertainty. Angular error in a turn causes position error proportional to the distance traveled! A solution is to make speed changes in small steps, so that the PLS only has to make small adjustments. Depending on the resolution of the speed command, a microcomputer-controlled PLS can have a programmed accelleration curve for each motor that minimizes the errors.

With a proper heatsink, the motor output transistor shown in the circuit is good for small motors requiring only a few amps at 12V. If you try to drive a more powerful motor, this drive amplifier may overload and die, since the output transistor operates in the linear region and may

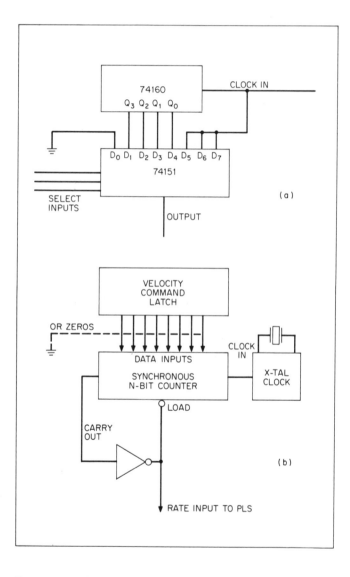

Figure 7. Options for multiple-speed selection: (a): Outputs of binary counter are used as rate outputs. (b): Carry output of counter is used as rate output and counter reload, forming a divide-by-N circuit. Rate output is clock frequency divided by $(2^r \cdot V)$, where r is the number of bits in the counter and V the number loaded into the counter from the velocity command latch.

dissipate too much power. The best way to control large motors is with a Pulse Width Modulated (PWM) switching power circuit, which turns the output transistor completely off and on at a controlled rate, supplying the desired average power to the motor and dissipating the minimum possible power in the transistor. ®

REFERENCES:

[1] Linear Data Book, National Semiconductor, 1976
[2] The TTL Data Book, Second Edition, Texas Instruments Inc., 1976
[3] Motorola Data Sheet, MC4344/MC4044 Phase Frequency Detector, MTTL Complex Functions, Issue A, Motorola
[4] Dorf, Modern Control Systems, Addison Wesley, 1974
[5] More, A. W., "Phase Locked Loops for Motor Speed Control", IEEE Spectrum, April 1973
[6] Tal, J.,"Speed Control by Phase Locked Servo Systems   New Possibilities and Limitations", IEEE Transactions on Industrial Electronics and Control Instrumentation, February 1977

The author, John Craig, has experimented with robots since high school. Formerly with the Robotics Project at Rensselaer Polytechnic Institute, where he received his Master's degree, he is now a member of the Jet Propulsion Laboratory Robotics Research Program, which is developing advanced robot systems for NASA. John is one of those fortunate people whose vocation is also his avocation, and his microcomputer-controlled home robot employs many cost-saving techniques such as the PLS.

USING OPTICAL SHAFT ENCODERS

Wesley E. Snyder and
Jorg Schött

Control of a robot requires knowledge of the position and velocity of each joint. There are a number of ways to get this information. Among the more recent and more accurate techniques are those which involve a particular kind of transducer, a calibrated rotary shaft called an *optical shaft encoder*, and special purpose hardware to interface the encoder to the computer.

We will describe the two major types of optical shaft encoders and present a design for a system which will translate pulses from one of these to position increment signals which may be processed by a simple counter circuit for use by a microprocessor, in this case a TI 9900.

The cost of building and maintaining solid-state hardware depends to a large extent on the number of interconnections between chips which are required. Therefore, our design has been optimized to reduce the number of chips rather than according to some other optimizing factor such as the total number of gates. Also, since most engineers can seldom resist the temptation to innovate and improve on "canned" designs, we discuss a few of the crucial design considerations and caveats.

Optical Shaft Encoders

There are two basic types of optical shaft encoders, absolute and incremental. In our design we chose the incremental shaft encoder because it can provide a high degree of resolution at a relatively low cost. However, in some applications the absolute optical shaft encoder is easier to use and can avoid certain problems in information transmission, which we shall explain.

Absolute Optical Shaft Encoders

This type of encoder consists of a circular glass disc imprinted with rows of broken concentric arcs. A light source is assigned to each row with a corresponding detector on the opposite side of the disc. The arcs and sensors are arranged so that, as the light shines through the disc, the position of the shaft can be uniquely identified by the pattern of activated sensors (to within a given angular resolution).

The pattern of activated sensors is, of course, in machine-readable code. However, instead of representing the angular shaft position in the standard set of sequential binary members, as given by the encoder disc in Figure 1, absolute encoders usually employ a special *grey code*, as shown in Figure 2.

In a grey code, the configuration of ones and zeros is chosen such that only one bit changes between any two consecutive numbers. Grey codes are used to help alleviate errors which inevitably occur when constantly changing numbers are being counted and stored. When a binary encoder is turning or a counter is counting binary numbers, any or all of the bits involved may change from one reading to the next. However, since these changes do not occur absolutely simultaneously in the real world, several transition states may occur between the true readings. In the worst cases, such as the transition from 7 (0111) to 8 (1000), for example, every bit changes. For a few nanoseconds the system is unstable. If by ill luck the computer reads the counter or the encoder during this moment of flux, any manner of garbage may be recorded.

Since the duration of the instability is quite short, the probability of a bad reading is quite small, and one erroneous reading out of every thousand does not matter for most applications in which numerous readings are averaged. A motor, for instance, is much slower in its response than an electronic controller, and it could not switch directions in response to one false signal. However, some optimal control applications may depend on near infallibility on the part of the computer (if not on the engineer), and in these applications erroneous readings are intolerable.

Of course, digital counting circuits may be designed to avoid this problem by synchronizing their operation with a clock and reading the count only during stable intervals, but the addition of such synchronization to an absolute binary-coded encoder represents an additional expense which may be unnecessary if a grey code is used.

Since the absolute encoder assigns to every location a unique coded number, it may be read directly by the computer, and much of the interfacing hardware which we describe here for an incremental encoder can be eliminated. With the grey code, when erroneous readings do occur, they are much less serious, since the error is at most one count.

On the other hand, an absolute encoder requires 12 or more separate sensors in order to attain good resolution, and is therefore rather expensive. An incremental shaft encoder which can identify locations within 0.04 degrees costs about $275. Others are available with much higher resolution, as well as less expensive ones. Incremental encoders require more interfacing hardware, but such designs are reasonably straightforward, as we shall explain.

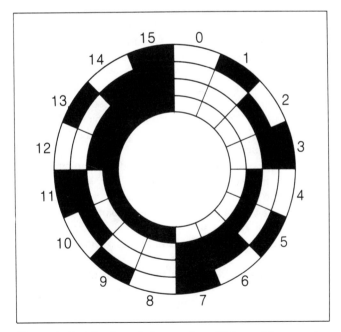

Figure 1. An absolute shaft encoder utilizing a binary code.

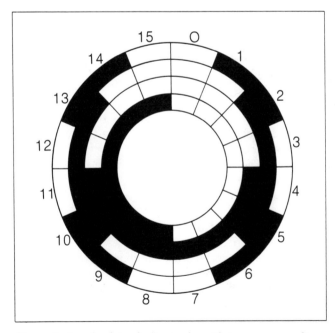

Figure 2. An absolute shaft encoder utilizing a grey code.

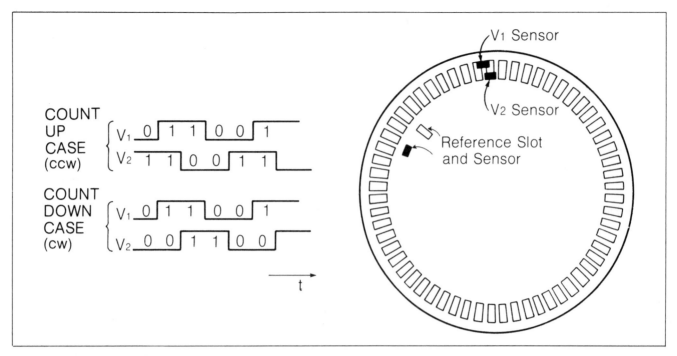

Figure 3. Output of incremental encoder.

Incremental Optical Shaft Encoders

Like the absolute encoder, the incremental encoder is a glass disc. However, it is imprinted with only one circular row of slots, all the same size and distance apart. One additional slot serves as a reference point.

Only three sensors are used with the incremental encoder, although the number of slots in the circular row increases depending on the resolution desired. Two of the sensors are focused on the row of slots. These are placed exactly one-half slot-width apart so that, as the disk rotates, a light shining through the slots produces detector signals that are 90° out of phase with each other. Figure 3 shows that if the disc is moving in a clockwise direction, one sensor (V_1) is always activated first; if the disc is moving counterclockwise, the other sensor (V_2) is always activated first. Thus, the pattern of the phase shift indicates direction.

The third sensor is focused on the reference point. Location can be determined by counting the number of pulses that occur from the time the reference point sensor is activated. The length of time required for a single pulse to be completed indicates velocity.

Position-Counting Hardware

While all necessary information about direction, velocity and position can be obtained directly from an incremental encoder, this information is imbedded in a continuous stream of pulses. It is too obscure for most simple counters, which are designed to rely on separate up or down counting signals to indicate direction. Therefore, we must provide special hardware to translate the pulses from the encoder into up and down count pulses.

It is possible to derive position simply by counting pulses directly from the encoder. One could count pulses from a single input, which, for the encoder used in our experiments, would allow us to attain an accuracy of 1/2500 of a revolution. Even so, it is necessary to consider both inputs, since phase information is needed to determine direction and hence to decide whether to count up or down.

Figure 3 shows the four unique states of the encoder, and the unique sequences of possible encoder outputs for continuous motion in each direction.

One could design a sequential machine to decode this complex of signals and output count up (CU) or count down (CD) pulses at appropriate changes of the encoder. Figure 4 shows such a design, where the state of the machine has been given the same designation as the "current" input. When the input changes, so that the input does not match the current state, the machine will change into a state which does match the input, and will output a CU or CD pulse as shown.

The design of a finite-state automaton to implement the state diagram of Figure 4 is reasonably straightforward. It requires two flip-flops and a few gates.

Alternatively, one could sacrifice resolution for simplicity and get a design which counts every cycle of the encoder rather than every state. Figure 5 shows one such design.

We have used a different design which utilizes the full resolution of the encoder but requires only one chip, a read-only memory (ROM), connected in a feedback structure.

The ROM used is a 32x8-bit programmable ROM, which

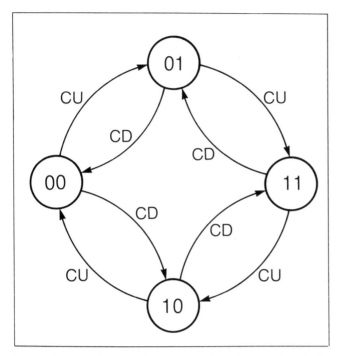

Figure 4. State diagram of a sequential machine to decode the state transitions of an incremental encoder.

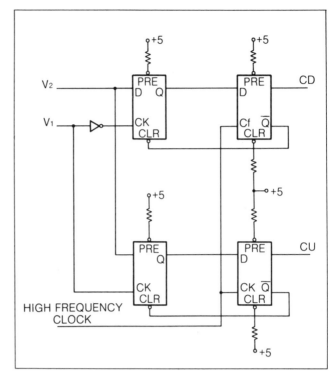

Figure 5. A low resolution position decoder (1 part in 2500 for a 2500 count encoder). This design is free of race conditions even for direction reversal. All flip flops are 7474s. The clock frequency must be much higher than the highest possible pulse rate on V_1. A rising edge indicates count up and a falling edge indicates count down.

has five address lines and eight data lines, connected as shown in Figure 6. Address lines A_1 and A_2 receive signals from the two encoder sensors. Data lines D_1 and D_2 output "up" or "down" signals to the counter. Address lines A_3 and A_4 are connected to data lines D_3 and D_4 so that they must contain identical information to be stable. The ROM is initially loaded with a stable configuration, that is, $A_3 = D_3$ and $A_4 = D_4$.

It is important to understand the internal operation of the ROM in this feedback structure. In this application, the ROM may be regarded as having 16 memory locations, each storing 4 bits. An example of how the data might be stored in the ROM is shown in Table 1 (page 8). Let us consider only one example.

Suppose the inputs V_1 and V_2 are in state $V_1 = 1$, $V_2 = 0$. Further suppose the output of the ROM is a 0010. (Don't be concerned at this point with *how* the output of the ROM came to be 0010, just assume that it is.) Then, since the lower two bits of the output are connected to the input, the address at the input to the ROM is 1010. Looking at address 1010 in Table 1, we see that the data stored there is 0010. The 10's in the lower two bits of input and output match, and, sure enough, the system is stable.

Now, suppose encoder line V_2 undergoes a transition from 0 to 1. We can see from Figure 3 that this change from 10 to 11 indicates clockwise motion and should result in a count down pulse. Consequently, the address inputs to the ROM change from 1010 to 1110. A few nanoseconds later, the ROM responds to this new input, by outputting the contents of location 1110, a 1011.

Since the low order two bits are different, address 1110 is not stable. As soon as the 1011 appears at the output,

the address changes, since input *must* equal output on the bottom two bits, and the ROM sees an address of 1111. Again, a lookup is done, and a few nanoseconds later, the contents of location 1111 appear at the output, 0011. This time, the low order two bits of output equal input, and the state is stable. For as long as V_1 and V_2 both remain 1's, the output will remain 0011.

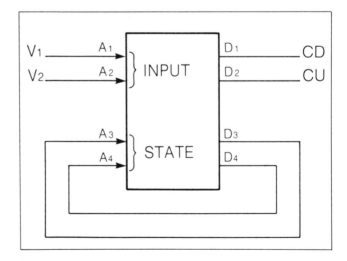

Figure 6. Organization of ROM.

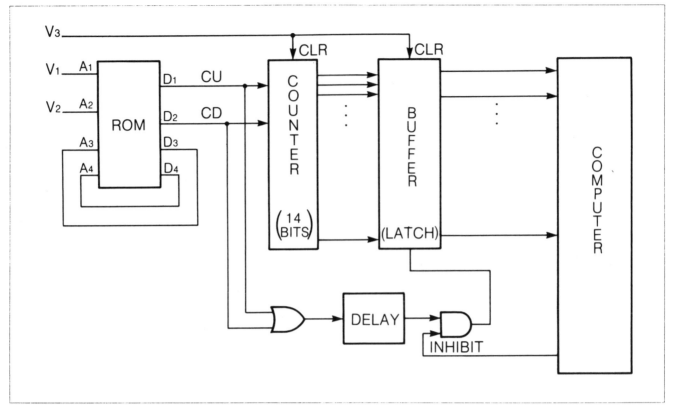

Figure 7. Block diagram of position hardware.

The important observation to be made now concerns the upper two output bits. Initially, the output was 0010, then it changed to 1011, then to 0011. In the process, the upper bit (the count down line) briefly pulsed to a 1 and then back to 0.

We have thus developed a system which decodes the direction and position information from an incremental shaft encoder and requires only one chip. Of course, making use of that data and remembering it requires more chips, a counter and a buffer memory, as shown in Figure 7.

As we mentioned in the introduction, there is a timing/synchronization problem resulting from the possibility that the computer may read the counter output while it is still changing. Users of the TMS 9900 are particularly prone to this problem, since the Communications Register (CRU) input reads bit-serial rather than latching all bits simultaneously. With a parallel input computer, a grey code could be used in the counter to resolve this problem. However, in this design we have used a commonly available binary counter and additional circuitry.

One way to eliminate this timing conflict is to provide an output from the computer which indicates "computer about to do a read," which will stop all counting activity. But then if a count pulse occurs during this time, it will be missed. To avoid this problem, an additional level of memory must be used which will read (in parallel) the contents of the counter each time it is updated, and hold that value for the computer. The "computer about to

read" signal will then inhibit (we will call this signal the INHIBIT from now on) the loading of this memory. Figure 7 shows a block diagram of the complete position detecting system.

The count up and count down pulses operate as described on the counter, their logical OR is taken, delayed (to let the counters settle), and used to strobe the latch. When the computer is about to read, the inhibit signal (normally 1) will be set to 0, preventing the occurrence of the strobe to the memory. After reading, the computer sets INHIBIT back to 1, allowing the latch to be loaded once again.*

Circuit Description

Figure 8 is a circuit diagram of the position circuit as it is implemented. There are some slight differences from the block diagram, which was drawn for clarity, not accuracy. The actual up/down counters used, SN74193's, count on the rising edge of negative-true signals on their count lines; consequently, the CU and CD lines are actually 0 when

---

*One might also consider parallel peripheral circuits which are commonly used with 8-bit microprocessors and will be soon available in the 16-bit field. For example, the Intel 8255 can, in some designs, provide a similar latching function to that provided by the SN4374's in this design. (Latching only on request).

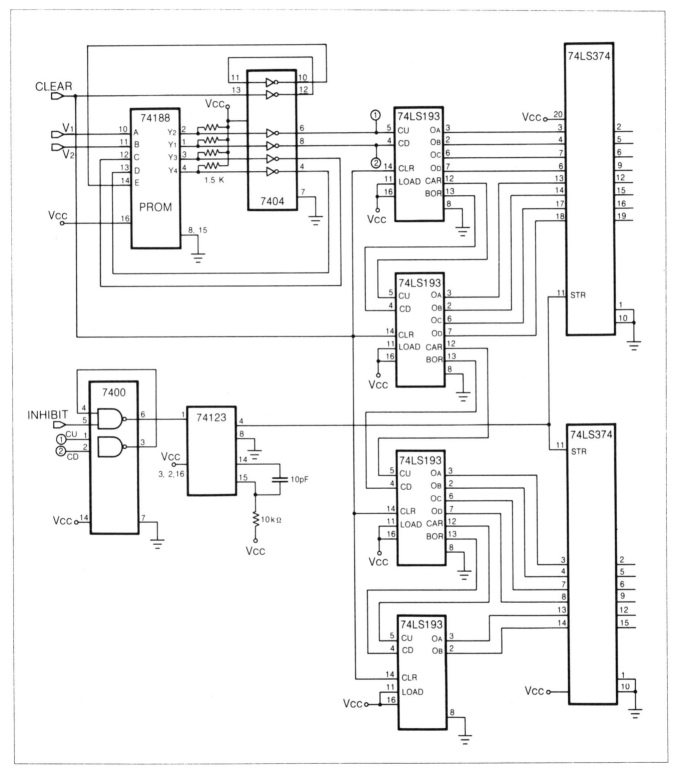

Figure 8. The position measurement circuit.

TABLE 1

	V_1	V_2	A'	B'	CD	CU	A'	B'
	0	0	0	0	0	0	0	0
	0	0	0	1	1	0	0	0
	0	0	1	0	0	1	0	0
Impossible	0	0	1	1	0	0	1	1
	0	1	0	0	0	1	0	1
	0	1	0	1	0	0	0	1
Impossible	0	1	1	0	0	0	1	0
	0	1	1	1	1	0	0	1
	1	0	0	0	1	0	1	0
Impossible	1	0	0	1	0	0	1	0
	1	0	1	0	0	0	1	0
Stable 10 state	1	0	1	1	0	1	1	0
Impossible	1	1	0	0	0	0	0	0
	1	1	0	1	0	1	1	1
Not stable	1	1	1	0	1	0	1	1
Stable 11 state	1	1	1	1	0	0	1	1

asserted. Since we are taking the OR of low-true signals, we can use an AND gate to trigger the strobe. Note that in Table 1, a 1 will cause a *low* output, and consequently the inverter chip is used, which also provides needed additional delay in the feedback loop. In implementation, to meet the low-true requirements of the 74193 one must also require the contents of the D and U columns of Table 1 to be inverted.

The reference pulse from the encoder, which occurs once per revolution, is used to reset the position counter to zero, and is indicated in the circuit by the signal CLEAR. This pulse is delayed as shown and used to address the second half of the ROM, which is filled ones so that the ROM feedback circuit is also reset to a stable starting state for the next counting cycle.

Design Considerations

A good engineer will usually consider making some sort of change to any design he sees. In anticipation of that, we provide a few pointers:

1) A *critical* point in dealing with encoders is that it is possible to receive arbitrarily short pulses from the encoder due to instantaneous direction reversals. This may be induced by mechanical jitter and/or motion around zero speed. Therefore it is essential that no pulses be permitted whose duration is less than the shortest allowable clock period in the electronic counting logic. This may be easily accomplished by using a low-pass filter followed by a Schmitt trigger on the encoder outputs.

2) The CU and CD pulses are very fast, being equal approximately to the propagation delay of the ROM. This pulse *must* be significantly longer than the pulse required for the counters, and don't forget that the counters are synchronous, but cascaded. The pulse length can be increased by adding delay in the feedback of the ROM.

3) The delay from the occurrence of the CU or CD pulse to the rising edge of the strobe must be greater than or equal to the settling time of the counters plus the setup time of the latch. In our case, this is 30 + 16 nanoseconds, and the delay shown is sufficient, but with other parts or logic families, the story may be different.

Hardware for Measuring Velocity

Since velocity is equal to distance divided by time, there are at least two ways to measure it. One could measure how many count pulses occur in a specific time interval, or one could measure the amount of time required by one encoder pulse. There are advantages and disadvantages to both approaches.

In the first case, one must choose a time interval sufficiently long to accumulate a number of pulses, in order to determine velocity accurately. For example, suppose the maximum velocity of the joint is one revolution per second. Then at this rate, our 10,000-count encoder yields one count pulse every 100 μs. For an 8-bit accuracy, we should allow time sufficient to accumulate 255 counts, or 25.5 ms. This gives us a new velocity measurement approximately 40 times a second, an update rate which is marginally fast enough for a small, fast robot, but quite sufficient for larger, more massive machines.

On the other hand, one could measure the length of a single encoder pulse. This yields an update rate which is quite high, but is at the same time rather noise sensitive. Despite this noise sensitivity, this is the approach we have used and will explain in this article. We will also explain methods for dealing with the noise.

The basic idea behind our velocity hardware is the measurement of the width of a single cycle from the encoder. This is accomplished by first dividing the frequency of V_1 by 2 using a simple flip flop,* producing the signal V'_1 and then using a clock whose frequency is much higher than the maximum frequency of V'_1, and counting

---

*We time an entire cycle of V_1 rather than simply the time V_1 is high since the duration of a cycle is much more repeatable than the high or low period.

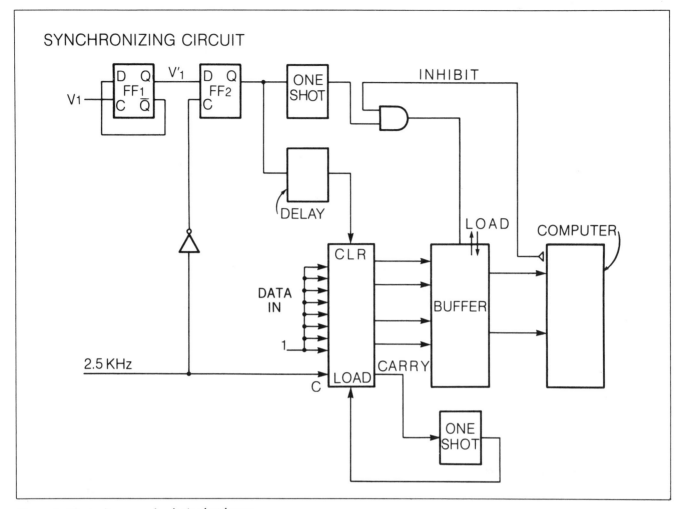

SYNCHRONIZING CIRCUIT

Figure 9. Block diagram of velocity hardware.

the number of clock pulses that occur while V'_1 is low. Again, assuming the same 2500 count/rev. encoder, with maximum velocity of 1 rps, we find the shortest period V'_1 can be low is 400 μs. A 2.5-kHz clock will give one count at this rate.

If we choose to use a 2.5-kHz clock, 255 counts will require 102 ms. Consequently, an 8 bit counter will overflow if it is clocked at 2.5-kHz and the V'_1 pulse is high for more that 102 ms. This corresponds to a velocity of one rotation every four minutes. Thus, with a 2.5-kHz clock and an 8-bit counter, we can measure any velocity from one revolution per second to one revolution every four minutes. We assume faster velocities are impossible, and slower velocities will be counted as stationary. With this system, the accuracy goes down at higher velocities, with a worst case error of $\pm25\%$ at the maximum speed. While this is the worst case error on any individual sample, the 2.5-kHz is uncorrelated with the encoder, and measurement errors will tend to average out, even over a few measurements. Of course, at lower velocities, errors are much smaller.

Figure 9 shows a block diagram of a "quick and dirty"

solution to the velocity measuring electronics. Counting occurs while V'_1 is low. That is, we time the "off" half of the V'_1 pulse.

As long as V'_1 is low, the counter is enabled for counting. Every 400 μs, the counter will be incremented by one. As soon as the rising edge occurs on V'_1 the memory will be strobed (a rising edge triggered latch is used), retaining the contents of the counter at that moment, and then, after the delay, the counter will be cleared. The clear to the counter is an overriding clear, and the counter will stay cleared, even though clock pulses occur, until V'_1 goes low again.

Note that, to avoid the possibility of a load signal to the latch occurring while the counter's outputs are changing, V'_1 must be synchronized with the clock. This is accomplished by the addition of a second flip-flop, whose synchronized output is used instead of the original V'_1. Also, as in the position circuit, the load pulse to the latch is gated by the computer INHIBIT signal.

The only remaining difficulty which we must handle is the possibility that, at low velocities, the V'_1 pulse is so long that the counter overflows. In this case, a carry output from the counter occurs, triggering a one-shot*, which loads the counter with all 1's. At the next cycle of the clock, another carry will again trigger the one-shot and again load all 1's. Consequently the counter will always contain hex "FF" when the arm is stationary.

Determining Direction

The direction of motion needs to be considered at this time, i.e., the sign of the velocity. Our circuit up to this point has considered only the width of the encoder pulse, corresponding only to speed and not to direction. We have handled the question of direction by using one additional flip-flop, and using the CU or CD pulses from the position determining hardware to set or clear (respectively) this bit. There is no race condition to be considered when interfacing this bit to the computer, since erroneous readings could only occur at zero (or very low) velocities, and in that case, it doesn't matter.

Determining Actual Velocity from the Measurements

In designing our velocity hardware, we made a tradeoff between the number of bits in our velocity counter, and the accuracy with which we could determine velocity. Also entering into the calculation were the maximum and minimum measurable velocities. Since we arbitrarily determined to use an 8-bit counter, we introduced significant errors at the high end of the velocity range. These errors could be reduced by using more bits and a higher clock rate, thus measuring the higher velocities more accurately, or by continuing to use eight bits, but with a faster clock, setting a higher minimum measurable velocity.

A third alternative is to anticipate these errors, and to allow them to average out. We have chosen this alternative by using in the computer program a simple low-pass filter. If V = determined velocity and V_M = measured velocity, we compute V by

$$V = 0.5V + 0.5V_M$$

Other choices of coefficients are possible, corresponding to more or less averaging, but we have found this

*The authors normally advise against designing with one-shots, which are particularly sensitive to noise and prone to false triggering. The one-shot approach seemed to fit this application so well that it was used anyway, but readers are cautioned to follow layout rules carefully, and, in noisy environments, to use a sequential logic design approach to the same function.

simple filter to be quite satisfactory in our experiments.

The quantity measured by the hardware of Figure 9 is a measure of time, not velocity. To convert to velocity, one must divide a constant by the measured number. We considered at one time doing this division by a lookup table, utilizing another ROM. However, the $V = 1/T$ calculation has a hyperbolic graph, which at low velocities results in a large number of different time periods, all of which round off or truncate to the same values. Thus, by choosing to use an 8-bit lookup, we would have introduced errors at the low velocity end of the scale, due to round-off. We were already inaccurate at high velocities, and more inaccuracy was clearly undesirable. For this reason, we elected to do the division in the computer.

In fact, we have found it most effective to do the previously described averaging/filtering operation on the measured period data prior to performing the division using 16-bit arithmetic on the computer.

Conclusion

We have described special purpose hardware for determining position and velocity using data from an incremental shaft encoder and for presenting this information to a computer, taking into account the possibility of timing conflicts during reading.

We have included some concepts which are generalizable to a large class of digital hardware problems, including the rather standard but useful two flip-flop synchronizing circuit (Figure 5), and especially the use of a ROM and a delay as a one or two chip implementation of a synchronous machine.

Having provided this information, the best advice we can give is: don't take our advice blindly! That is, it is probably not necessary to use special hardware at all. We have often observed that, in a holdover from the time when computers were very expensive, engineers will sometimes interface to a microprocessor using hardware which is more expensive than the microprocessor itself.

The most effective, easiest to use, and most inexpensive hardware you can use is often the processor itself, or even a second processor. We developed this hardware for an optimal control application which required very tight constraints on software speed. Consequently, the addition of hardware was necessary. In applications with higher rotational speeds, or higher resolution encoders, it may also be impossible to use a microprocessor, due to its speed limitations.

However, many applications do not have such tight constraints, and the processor itself, possibly utilizing interrupts, can be used to count the encoder pulses. If the processor includes a real-time clock, as many do, velocity determination can likewise be done in software. In an earlier report in *Robotics Age* (Vol. 2 No. 1, Spring 1980) we described a system which used a simple processor to implement path control for one joint. In that application, we used hardware similar to that described herein to decode velocity, but used interrupts to keep track of position in software.

In conclusion, one should carefully evaluate the cost-effectiveness of building special hardware versus using the computer itself. In applications which are not pressed for time, the latter is often the better choice.

Acknowledgements

The synchronizing circuit utilizing a single SN7474 which is described in Figure 9 has been around for quite a while and is extremely useful in a wide variety of applications. We suspect it has been re-invented many times.

The concept of using the state transition of the encoder (Figure 4) to provide an accuracy of 1/10,000 from a 2,500 count encoder was developed, as far as the authors know, originally by Clifford Geschke.

The use of the ROM to implement the state diagram of Figure 4 came out of a design projects class taught by one of the authors for seniors in electrical engineering at North Carolina State University. No single member of the group would ever accept credit for the idea.

About the Authors

Dr. Snyder is on the faculty of the Electrical Engineering Department at North Carolina State University, where he is also associate director of the Image Analysis Group. That group is deeply involved in research into computer analysis of images, with a current specialization in analysis of sequences of images as might come from a television camera. Dr. Snyder has just returned from a six month stay in Germany where he did research on computer tracking of objects moving in a television image.

Dipl.-Ing. Schött is an electrical engineer with the DFVLR, the West German Air and Space Agency. He has been involved in a number of projects relating to digital command and control systems, and is currently developing microcomputer-based control hardware and software.

Advances in SWITCHED-MODE Power Conversion

Part I

Prof. Slobodan Ćuk and
Prof. R.D. Middlebrook
Power Electronics Group, 116-81
California Institute of Technology
Pasadena, CA

ROBOTICS AGE is honored to present a significant innovation in solid-state power conversion, described here by the inventors themselves. The new design is certain to have a profound impact on the electronics industry, but, because of its generality and many potential applications in robotics, we wish to familiarize our readers with its design and operation.

Rather than just summarizing the advantages of the new design and suggesting possible applications, we decided to present a thorough discussion of the background, development, and operation of the Ćuk converter, so that even readers unfamiliar with the former state of the art in switching converter design can more fully appreciate the significance of the invention as well as understanding the principles behind it. What emerges is a fascinating account of the process of invention, expressed in the scientists' own words.

Introduction

Switching power supplies and regulators ("switchers") have come into widespread use in the last decade. Because of their much higher efficiency, smaller size and weight and relatively low cost, they are displacing conventional linear power supplies even at low power levels (about 25 Watts). The design of switching converters has been extensively studied, and it is commonly believed that the designs in commercial use today employ the simplest possible switching structures for dc-to-dc level conversion.

However, even though high conversion efficiency has been achieved, present switching converter designs posses several undesirable characteristics. Rapid switching of the input or output currents can cause severe electromagnetic interference (EMI) problems, requiring the addition of appropriate filters that increase both the complexity and cost of the circuit. Also, the implementation of the power transistor switch requires complex drive circuitry in configurations where the emitter voltage "floats" above ground. In switched-mode power amplifier designs, additional problems are imposed by the requirements for dual (bipolar) power supplies, high switching frequencies, and complex feedback circuitry.

A revised look at present converter designs and an analysis of switching circuit topology has led to the discovery of a new design that retains all the desirable properties of conventional designs, with none of their undesirable attributes. Both input and output currents are essentially nonpulsating dc, and, in fact, ripple can be reduced to zero. Furthermore, the new topology may be implemented with fewer parts than comparable solutions, and thus may be said to be an "optimal" design.

Because of its simplicity and generality, the new converter can efficiently raise or lower dc levels, and, with minor modification, bidirectional power flow can be easily achieved, allowing the roles of power source and load to be arbitrarily interchanged without physically switching their connections. A switching power amplifier based on the new converter requires only a single power supply, enabling efficient dc to ac power conversion. The improved performance of the new switching power amplifier design permits the use of a lower switching frequency and simple circuits for both the drive and feedback.

We will begin with a discussion of switching converter design, leading to the development of the new converter. Next, the design of a switched-mode power amplifier using the new converter and a discussion of its performance will be presented.

A Review of Switching Converter Design

In the basic design of the linear power supply, shown in

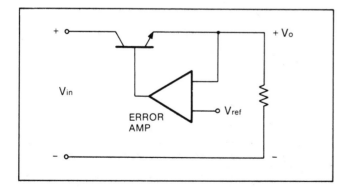

Figure 1. Conventional linear power conversion.

Figure 1, the output voltage, V_o, is regulated to a multiple of the reference voltage, V_{ref}. The difference between the unregulated input voltage and V_o results in power dissipation in the transistor, which, in high current supplies, can result in considerable energy loss and heating. Consequently, the unregulated voltage must be kept as low as possible while still allowing adequate regulation.

Switching power supplies are based on the principle that, by alternately switching the transistor completely off and on, its power dissipation can be held to a minimum. Passive energy storage elements, inductors and capacitors, can then be used to transfer energy from the source to the load, performing the appropriate level conversion in the process.*

Figure 2 shows the most commonly used switching configuration, referred to as the "buck" converter. The ideal switch, S, can be realized by the combination of a bipolar transistor and commutating diode as shown. In operation, the input voltage is connected by S to charge inductor L to the output current necessary to produce the desired voltage, V, across load R. Once V is attained, S disconnects the input and provides an alternate path for the inductor current, which then begins to decay. The output capacitor, C, helps reduce the residual voltage ripple caused by the switching. The cycle repeats at a fixed rate, and the average voltage gain is equal to the duty ratio, D (the fraction of the switching cycle that the transistor is on).

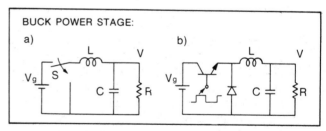

Figure 2. The basic "buck"-type switching converter configuration, showing (a) the topology of the circuit, using an idealized switch, S, and (b) the implementation of the switch using a bipolar transistor and a commutating diode.

* Recall from basic physics that energy is stored by an inductor in the magnetic field produced by the current through its winding. A voltage applied across it either increases or decreases this current to energize or deenergize the field. Conversely, the current through a capacitor either charges or discharges the energy stored in an electric field, respectively raising or lowering the voltage across its terminals.

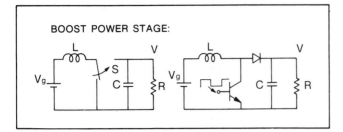

BOOST POWER STAGE:

Figure 3. The common "boost" converter configuration: (a) circuit topology, using an ideal switch, and (b) implementation.

Note that the input current to the regulator alternates between the full output current and zero. This abrupt variation in the input energy flow causes severe EMI, and invariably requires the presence of an input filter to smooth out the substantial current ripple component at the switching frequency. Also, since the emitter floats above ground, isolated drive circuitry (not shown) is required to switch the transistor. A further limitation of this design is that only voltage reduction is possible, since D is less than 1.

For applications requiring a voltage step-up, some of the drawbacks of the buck configuration can be avoided by using the "boost" converter configuration, shown in Figure 3. In this design the inductor is always in series with the input, so that the supply current is continuous. The inductor stores energy when grounded by S, and then the stored energy is released to the output. The output voltage of the boost converter is always greater than the supply, with an ideal gain equal to $1/(1-D)$. Note also that the grounded emitter of the transistor switch simplifies the drive circuit. The drawback of the configuration, however, is that during the inductor charging interval, all the output current must be supplied by discharging the output capacitor, resulting in considerable output ripple.

I. Development of the New Converter Topology

The discovery of the new converter design resulted from the objective of retaining the desirable properties of both types of converters. This goal was realized by examining combinations of the two basic types, with the underlying principles of simplification and optimum interconnection while simultaneously maximizing performance. The details of the analytical technique, described more completely in [1], will be abbreviated here.

Consider the cascade connection of a boost power stage followed by a buck power stage, resulting in the converter shown in Figure 4. This configuration retains the desirable properties of the low input current ripple of the boost converter and the low output current ripple of the buck stage. The voltage gain of the converter is the product of the gains of the two stages. Assuming that S_1 and S_2 are synchronized so that both switch from position 1 to 2 (and back) simultaneously, the resulting ideal gain is

D/D', where $D'=1-D$. Thus, the same converter can be used for both level reduction (D < .5) and increase (D > .5).

The undesirable output ripple of the boost converter is now isolated between the stages, and, in fact, the capacitor serves as the sole energy transfer mechanism between the stages. To see this, note that during the part of the cycle when both switches are in position 2, C_1 is isolated from the output circuit and is charged by the input current through L_1. In the rest of the cycle, C_1 is completely transferred to the output circuit and is discharged by the output current through L_2, reenergizing L_2 while L_1 stores energy from the input. Position 2 of S_2 provides a path to sustain the output current (from energy in L_2) while C_1 is being recharged by L_1.

The issue remains, however, as to whether or not this design represents the optimal configuration of this attractive boost-buck cascade. It is apparent that the two inductors are essential to the continuity of the input and output currents, and that the intermediate capacitor is required for energy transfer. The issue of optimality thus resolves to the following question:

Can the number of switches in this cascade configuration be reduced from two to one and still achieve capacitive energy transfer?

The answer to this question may at first seem surprising. Switches S_1 and S_2 may indeed be combined, resulting in a new optimal converter configuration. The solution is found by considering the topological properties of known converter types and their cascaded combinations. [1, 2] A key to the solution is that inversion of the converter's output voltage is a necessary feature of the new design. Since both of the basic converter types are noninverting, the

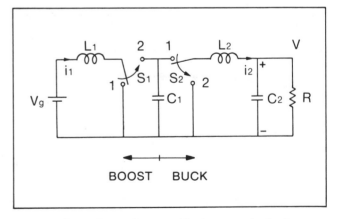

Figure 4. Cascade combination of a boost and a buck power stage.

> "**Even though a true dc-to-dc transformer is physically impossible, the new converter can functionally be considered as such . . .** "

only way to achieve this is by reversing the polarity of the charged energy transfer capacitor when it is switched into the output circuit. Attention to this issue alone leads to the solution shown in Figure 5.

Even though a true dc-to-dc transformer is physically impossible, the new converter can *functionally* be considered as such, since both its input and output voltages *and currents* are very close to true dc quantities, owing to the negligible switching ripple. Moreover, due to the advantages of capacitive energy transfer, the actual conversion efficiency of the new circuit is substantially greater than that of conventional designs. [1] The grounded emitter of the switching transistor also allows the most simple drive circuitry to be used.

In this design, the single switch S alternately grounds the opposite ends of the capacitor, effectively switching it from the input to the output circuits. C_1 is charged by the input current to a positive voltage, as viewed from left to right, with S in position B. With S switched to A (during the charging interval of L_1) the "positive" side of C_1 is now connected to ground, and its "negative" side (since the voltage across it does not change) effectively "pulls down" the voltage level at terminal B. Thus, current flows *from* the grounded load to discharge C_1 through L_2, causing a negative voltage drop across the load, hence a negative output voltage.

The hardware implementation of the new converter, using a transistor/diode combination to implement the ideal switch S as in the other converter configurations, is shown in Figure 6. With the transistor off (open), C_1 is charged by the input current through the forward biased diode. When the transistor turns on, it provides a charging

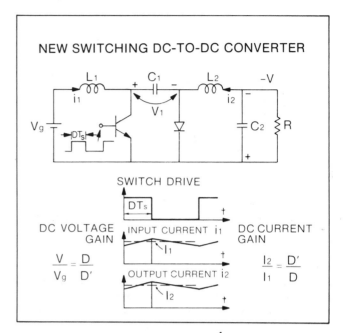

Figure 6. Hardware realization of the Ćuk converter using a transistor and a diode to implement switch S. Ideal dc voltage and current gains are as shown.

path for L_1, but also drops the positive terminal of C_1 to (very near) ground. As in the ideal case, this pulls down the output voltage, reversing the bias on the diode (turning it off) and transferring the stored energy to L_2 in the form of increased current. Note that the dual role of switch S, its inclusion in the input and output circuits simultaneously, requires that the transistor and the diode, when conducting, must carry both the input and the output currents.

Coupled Inductor Extension of the New Converter

It would seem from the above discussion that the simplest possible converter circuit has been obtained, but this is not the case. Consideration of the voltage waveforms across the two inductors L_1 and L_2 over the switching cycle reveals that, for the average dc voltage across each inductor to be zero, (steady state balance condition) the two waveforms must be identical. [3] This means that the two inductors may be coupled by being wound on the same core, without affecting the basic dc conversion property, provided that the resulting transformer has a 1:1 primary to secondary voltage ratio. This is easily achieved by setting $L_1=L_2$, (same number of turns in

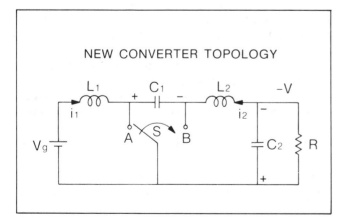

Figure 5. The new switching topology, employing capacitive energy transfer with polarity inversion of the output voltage.

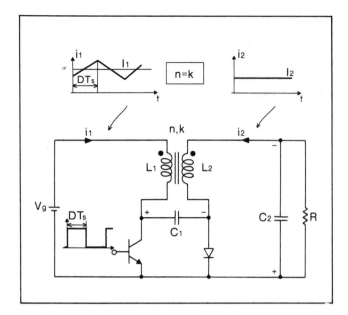

Figure 7. The Coupled Inductor Extension of the Ćuk converter. With appropriate transformer design, the output current ripple can be reduced to zero.

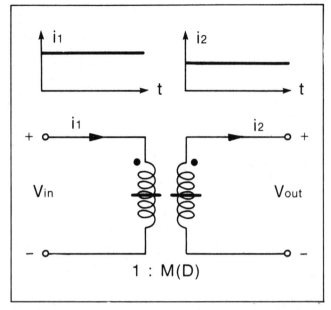

Figure 8. The ultimate objective of switching conversion — a dc-to-dc transformer with constant input and output currents and an electronically controllable effective turns ratio for voltage level change.

each) with the direction of coupling as shown in Figure 7.

The resulting circuit not only eliminates one inductor core, (for further component savings) but has an interesting property that profoundly affects the performance of the new converter. Since the coupled inductors form a transformer, inductive energy transfer between the two windings alters the effective inductance of each. For a 1:1 turns ratio, both inductances are approximately doubled, so that both input and output current ripples are about one half those of the uncoupled converter. Even more significant, however, is that by slightly altering the turns ratio, one of the effective inductance values can be made arbitrarily large, dramatically reducing the current ripple on that side. When the effective turns ratio (primary to secondary) matches the inductive coupling coefficient of the transformer*, the output current ripple can be completely eliminated, resulting in pure dc to the load. [3] In this matching condition, no output capacitor is needed, since the load voltage is also constant.

The coupled-inductor converter thus has the simplest possible structure, consisting of a single transformer, commutation capacitor, and a single switch (realized by two semiconductors), and yet it achieves the maximum performance (both input and output current nonpulsating with one of them even being pure dc with no ripple) in a topology which offers the smallest possible size and weight and highest efficiency.

It is interesting to note that the idea of coupling the

inductors is a new concept which can be implemented with similar benefits to many other switching converter configurations. Nevertheless, the greatest potential of this configuration is obtained through its application to the new converter topology, on which it was originally conceived.

Yet, this still seems to fall short of the ultimate objective, the functional realization of the ideal dc-to-dc transformer shown schematically in Figure 8, which has constant current at *both* input and output. In addition, the dc isolation between input and output shown in Figure 8, often required in many practical applications, is not present in any of the converter extensions discussed so far. This, however, poses no insurmountable problems, since there is an elegant way of introducing dc isolation into the new converter by the addition of a single ac transformer and an extra capacitor, as described in detail in [4]. Furthermore, the same idea of coupling inductors in this dc isolated extension leads also to zero current ripple on one side.

The latest generalization of the new coupled-inductor converter concept has resulted in what may be considered the ultimate solution: a dc-isolated switching dc-to-dc converter with zero current ripple at BOTH input and output terminals, which functionally emulates the ideal dc-to-dc transformer of Figure 8†.

---

* Like the effective turns ratio, the inductive coupling coefficient is a physical property of the transformer construction, with a value ranging from 0 (separate inductors) to very near 1 (windings closely wound on a single ferromagnetic core).

---

† Technical details of this latest development have not yet been publicly released, but are contained in a patent application. The basic Ćuk converter configuration and many of its improvements, as well as new switching power amplifier configurations, are protected by a series of patents. [9, 10, 11, 12]

Modular Concept

A large number of switching configurations are currently available for a multitude of power conversion functions. Heretofore, some circuit configurations have been solely used to perform the dc-to-dc conversion function (converters), others for dc-to-ac conversion (inverters), and yet another for power amplification, with their circuit configurations having nothing or very little in common. With the family of Ćuk converters, however, a vertical integration has been made, and for the first time a logical extension from a single-quadrant to a two-quadrant converter (battery charger/discharger) and finally to a four-quadrant converter (bidirectional power amplifier) has been realized.

Quadrant Classification

Power Processing Systems may be classified according to the nature of their output capabilities. The simplest is a dc-to-dc converter in which output current is delivered in one direction at one output voltage polarity. This means that on a graph of output voltage V versus output current I, only ONE QUAD-RANT is available, as shown for positive V and I in Figure A. This is structurally the simplest switched-mode converter, and can be realized with only one transistor and one diode, as in the Ćuk converter illustrated.

Simply by doubling the number of power devices in the Ćuk converter of Figure A, the output capability of a dc-to-dc converter can be extended so that current can flow in either direction (bidirectional current switch implementaiton), and the system becomes a TWO-QUADRANT converter as shown in Figure B. Thus, INPUT and OUTPUT can be arbitrarily interchanged, and the converter is capable of BIDIRECTIONAL POWER FLOW.

Finally, by the new general concept of converter topological interconnection illustrated in Figure 21, a two-quadrant converter is extended into a FOUR-QUADRANT converter in which the directions of both voltage and current can change independently as shown in Figure C. Hence, a true AC OUTPUT can be obtained together with bidirectional power flow.

Although this vertical integration has been conceived for the family of Ćuk converters, it can be directly applied to other converter types, such as buck, boost, or any other switching configuration.

(A)

+V

ONE QUADRANT

+I

(B)

+V

TWO QUADRANT

-I   +I

(C)

+V

FOUR QUADRANT

-I   +I

-V

Basic Ćuk Converter

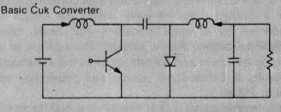

Battery Charger/Discharger

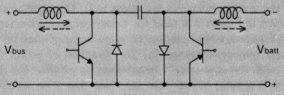

V_bus   V_batt

Bidirectional Power Amplifier

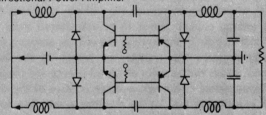

Applications

The Ćuk converters have a broad range of applications in all aspects of power processing. Robotics, in particular, will benefit from the excellent performance of Ćuk converters in motor control. Its high efficiency and low noise make it an obvious choice as a dc motor servo drive, with regenerative braking, in either the two- or four-quadrant configurations. In fact, the high performance of the Ćuk bipolar amplifier make it possible to design servo systems employing ac induction motors, capable of optimum motor drive as well as regenerative braking.

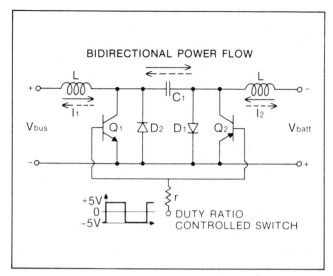

Figure 9. A symmetrical implementation of the Ćuk converter, capable of power transfer in either direction.

Bidirectional Power Flow in the New Converter

Note that the configuration of the new converter shown in Figure 5 is completely symmetrical with respect to the designation of the input and output terminals. Because the ideal switch S allows current flow in either direction, either terminal of the converter can behave as a current source or as a current sink. However, when the switch is implemented by a single transistor and diode, only unidirectional current flow is allowed. The addition of a single pnp transistor and a diode removes this constraint and results in bidirectional current and power flow as shown in Figure 9.

The converter circuit is thus symmetrical, and the input and output terminals can be arbitrarily designated (as long as the voltage polarities are respected). The configuration shown in Figure 9 is ideal for battery charger/discharger applications, since both functions are accomplished by this single converter circuit. [5] The direction of power flow through the converter is determined by whether the duty ratio is greater or less than the value that gives a voltage gain equal to the ratio of the bus to battery voltages. In general, attempting to reduce the voltage to a load that stores energy causes power to be transferred from the load,* just as attempting to raise the voltage across a load normally requires the transfer of power to the load. However, when two sources of stored energy are coupled by a bidirectional converter, precise control of the duty ratio is essential to restrict the resulting power flow to an acceptable limit. The bidirectional current switch implementation is equally applicable to the coupled-inductor extension (Figure 7) of the new converter.

Editor's note: A very useful application of this is to use a dc drive motor as a generator for regenerative braking, recovering some of the energy stored in a robot's momentum or provided by an external source (as when rolling down a slope).

Since both of the transistors in Figure 9 are referenced to ground, the complementary switching of the pair can be accomplished by a single drive source as shown. Moreover, because the two base junctions are tied together, the circuit also automatically prevents the simultaneous turn-on of both transistors, (and thus prevents shorting out capacitor C) in spite of the presence of transistor switch storage time.

Another implementation of the bidirectional current switch is made possible by recent technological advances in Metal Oxide Semiconductor Field Effect Transistors (MOSFETs). Formerly limited mostly to small signal applications, newer MOSFET devices are capable of switching to a relatively low ON resistances, with a substantially higher current rating. It is not widely known that the power MOSFETs are capable of bidirectional current flow. Owing to the device's internal construction, there is effectively an inherent diode connected betweeen drain and source which provides the alternate (opposing) current path. Hence, each transistor/diode pair can be replaced by a single power MOSFET device, resulting in the bidirectional converter shown in Figure 10. If both p-channel and n-channel MOSFETs are used, the source terminals of both devices may be grounded, allowing a simple driving scheme similar to that used for the grounded-emitter transistors in Figure 9.

Besides reducing the component count, power MOSFETs have a number of advantages over bipolar transistors which make them especially attractive for switching converter applications. [6] Whereas bipolars are current-controlled devices, FET's are voltage controlled. As a consequence, they require much lower drive currents and are capable of very fast switching speeds. Also, unlike

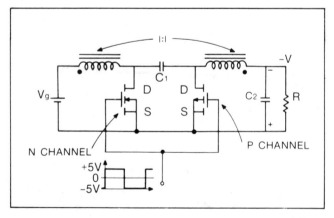

Figure 10. Bidirectional switching converter using MOSFET power switches.

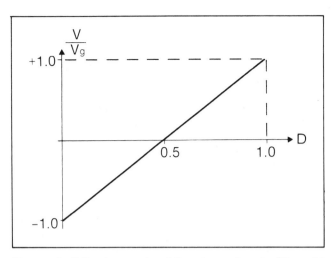

Figure 12. DC voltage gain of the power stage in Figure 11.

bipolars, several MOSFETs can be easily paralleled to distribute switching current. Their switching speed (5ns typical) is one or two orders of magnitude faster than comparable bipolars, (200ns typical) significantly reducing power loss during switching. Although older power MOSFETs had a high saturation voltage compared to bipolars, (3V@ 10A vs. 2V or less for a bipolar) the newest devices have as little as 0.055 ohms R_{SD}, making their use in converter designs extremely attractive, especially for applications requiring a high switching frequency.

II. Application of the New Converter Circuit to Switched-Mode Power Amplifier Design

Although a substantial effort has been made in recent years toward the development of complex switching power supplies, substantially less attention has been devoted to their natural outgrowth — switching power amplifiers. It is quite natural, then, that the principles of the operation of switching amplifiers are not widely known. We will begin, therefore, with a review of present designs, which are all based on a modification of the "buck" power stage of Figure 2. This review reveals substantial performance deficiencies originating from the buck converter itself.

The new circuit topology of Figure 5 eliminates most of the problems associated with earlier switching converter designs, providing greatly improved performance with fewer components. This, together with the extension of the new design to allow bidirectional current flow, suggests that the design of switching amplifiers should be reexamined in light of this new development. Indeed, we will show that a power amplifier design based on the new converter topology provides a much superior solution.

Switching Amplifier Principles of Operation

The major difference between the power stages of a switching amplifier and a switching power supply is that the former must be capable of producing an output of either polarity. In all present designs, this is accomplished by modifying the basic buck power stage to use two power supplies $+V_g$ and $-V_g$ as input, with the switch S switching between positive and negative supplies as shown in Figure 11.

The resulting voltage gain, shown in Figure 12, is a linear function of the duty ratio, D. For D greater than 0.5 the output voltage is positive, while for D less than 0.5 it is negative. Note that since the load voltage may be negative as well as positive, the implementation of switch S shown in Figure 2 is inadequate. Since the output voltage is determined by the inductor current, the hardware implementation of the switch must permit bidirectional current flow, as shown by the arrows in Figure 11. This is readily accomplished by the two-transistor, two-diode circuit shown in Figure 13. This bidirectional implementation is similar to that required for the symmetrical converter in Figure 9, but with the significant difference that neither transistor is referenced to ground, necessitating the use of isolated drive circuitry to accomplish the complementary switching action.

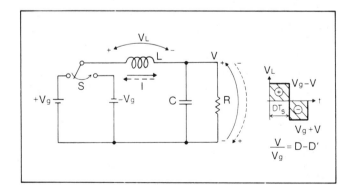

Figure 11. Modified buck power stage with output voltage of either polarity.

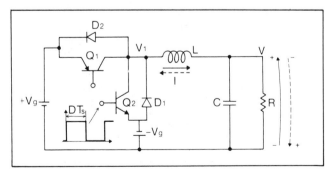

Figure 13. Practical implementation of the converter in Figure 11.

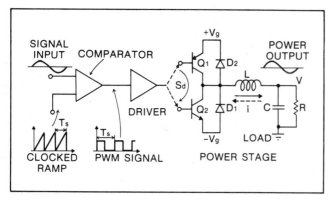

Figure 14. Open-loop buck type switching power amplifier.

To apply this dual-polarity power stage in a switching power amplifier, it is necessary to control the duty cycle so that the output voltage varies in proportion to the input signal. The linear voltage gain of the buck power stage facilitates this function, and is incorporated into the open-loop amplifier configuration shown in Figure 14. The design is the same as that for an open-loop dc-to-dc converter operated at a constant switching frequency $f_s=1/T_s$, with the only difference that a time varying (sinusoidal, for instance) input signal is used at the comparator input instead of a dc reference voltage. When the input signal is positive, a pulse with D greater than 0.5 is generated, producing positive output, while for negative input, D is less than 0.5 and negative output is produced. In fact, comparison of the low frequency input signal and the high frequency sawtooth (clocked ramp), generates a Pulse Width Modulated (PWM) signal, whose low frequency spectrum is, in effect, recovered by low-pass filtering with the inductor. Hence, a close replica of the input signal is generated at the output, but at a high power level.

The comparison of this switching amplifier approach with conventional linear designs with respect to the two foremost constraints in power amplifier design, efficiency

and distortion, now becomes apparent. In terms of efficiency, this approach boasts the usual advantages of switching power supplies over linear — significantly lower power dissipation. Namely, its theoretical 100% efficiency is usually only slightly degraded (often it is over 90%) by losses due to nonzero transistor saturation voltage and switching time and parasitic resistances of storage elements in the power path.

Distortion, however, becomes a function of the switching frequency, rather than being dependent on the linearity of the transistor gain curve. Specifically, for low distortion the switching frequency has to be an order of magnitude or so higher than the signal frequency to avoid overlapping sidebands in the PWM signal and to minimize the effects of output ripple. On the other hand, with increased switching frequency the switching time of the transistor may represent a significant portion of the duty cycle, introducing further distortion and degrading efficiency as well. Other distortion sources arise from nonlinearities in the clocked ramp (which also increases with switching frequency) and variations in the power supply voltages.

To reduce the effects of these sources of distortion requires the use of negative feedback, as shown in the block diagram in Figure 15. Feedback allows a reduced switching frequency and improves linearity at the expense of increased circuit complexity, but the amplifier in Figure 15 still has drawbacks that originate directly from the use of the buck power stage. As in the buck power supply design, the input currents pulsate, at the switching frequency, between zero and the output current, causing severe EMI. The design requires two power supplies of opposite polarity, and complex drive circuitry is required for the transistors, both to translate drive signals to the ungrounded emitters and to prevent simultaneous turn on of the transistors and shorting of the power supplies. Finally, a relatively high switching frequency (300 kHz or so) is still necessary to reduce switching ripple.

As in the case of power supply design, the introduction of the new switching converter topology solves all of these problems, but the objective of achieving maximum performance (wide bandwidth, low noise and distortion, small (or zero) switching ripple, and low switching frequency), for the minimum number and size of parts, is by no means an easy one.

A New Push-Pull Switching Power Amplifier

We now pose the problem of inventing a power stage, based on the new converter of Figure 5, which will produce an output voltage of either polarity, depending upon a

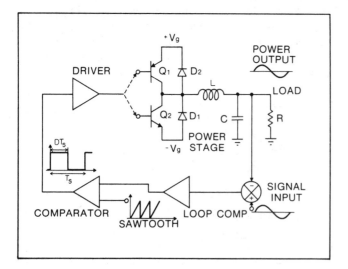

Figure 15. Closed-loop buck type switching power amplifier.

switch duty ratio D. The design should preserve all the good properties of the new converter previously described and possibly add some more. Since the use of dual power supplies, as in the the buck converter power amplifier, adds to the expense and complexity of the design, the additional feature of a single power supply is desired. Consideration of this requirement led to the discovery of the push-pull type topology shown in Figure 16, in which two bidirectional switching converters are operated in parallel from a single supply.

Let us assume that the two converters are operated switching *out of phase*, that is, with complementary drive ratios. Namely, when switch S_1 is in position A_1 for interval DT_s, switch S_2 is in position B_2 for the same interval. Assuming normal operating conditions, the ideal voltage gain of the top stage is D/D', and that of the bottom is D'/D. Thus, the two output voltages are equal only for a duty ratio of 0.5, while one or the other becomes greater for other values of D. Evaluating the difference of the two output voltages, $V=V_1-V_2$, gives a differential voltage gain of:

$$\frac{V}{V_g} = \frac{D-D'}{DD'}$$

which is plotted as a function of duty ratio D in Figure 17 (heavy line). The individual converter gains are shown as dotted lines.

As seen in the figure, the differential gain is just the one needed for a switching power amplifier, since it has the same required polarity change property as that of the modified buck power stage (Figure 11). The only trouble, however, is that the load is not across the two converter outputs. Thus, an interesting question arises:

Is it possible to connect a load between the two outputs without violating any basic circuit laws or disturbing the proper operation of the converters?

The answer to this question is affirmative, and is a key to the success of the new push-pull switching power amplifier design. With the two loads in the converter of Figure 16 replaced by a differential ("floating") load R, the new push-pull power stage of Figure 18 is obtained.

In Figure 16, each converter operates independently, with unidirectional current and power flow as shown. However, this is not so in the new push-pull power stage. Owing to the differential load, current originating from one converter must be "sunk" by the other, resulting in equal and opposite current flow, as shown by the solid arrows in Figure 18. With the opposite polarity of the output current, (dotted lines) the roles are reversed. Thus, the switches S_1

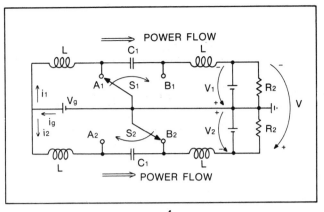

Figure 16. *Two bidirectional Ćuk converters operating in parallel from a single power supply.*

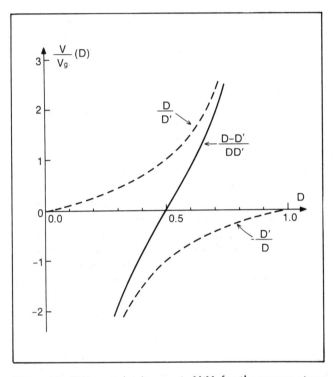

Figure 17. *Differential voltage gain V/V_g for the power stage in Figure 16.*

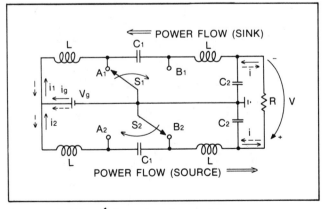

Figure 18. *The Ćuk push-pull switching power stage.*

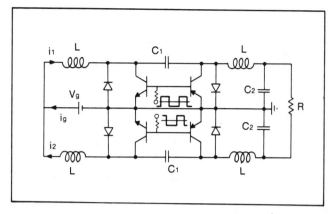

Figure 19. Hardware implementation of the Ćuk power stage.

and S_2 must permit this bidirectional flow, depending upon the duty ratio D. In other words, a part of the energy delivered by one converter is consumed by the load, and the remainder returned by the other to the source. This bidirectional flow is easily accomplished, as discussed previously, as shown in Figure 19.

It may now become evident that the new amplifier power stage may be called a *true push-pull* power stage. Namely, while the lower converter *pushes* the current (and energy) through the load, the upper converter *pulls* it from the load, and vice versa. This is quite unlike the conventional push-pull class B linear amplifiers for which a *push-pull* configuration would be a more appropriate designation.

In addition to the advantages of the bidirectional new converter already described, another very desirable feature derives from this true push-pull configuration itself. Note that since one converter operates as a power sink, the current drawn from the source by the other is reduced by the return current, (i_2 in Figure 18) which is increasing when S_2 is in position B_2 as shown. (The capacitor is discharging return power into the power supply inductor of converter 2.) Since current i_1 is increasing at the same time, the current ripple through the power supply is reduced. *In fact, if the two inductors on the supply side are equal, the current drawn from the power supply is pure dc with no ripple at all.*

The most advantageous configuration is obtained, however, when the coupled-inductor extension of the power stage in Figure 19 is used. The resulting design is shown in Figure 20, which also represents a complete block diagram of a closed-loop switching power amplifier using the new push-pull configuration. When the two transformers (coupled-inductors) are designed to satisfy the matching condition, as described earlier, the output current ripple, and consequently the need for output capacitors, is completely eliminated. Removing the capacitors results in extremely favorable phase-frequency response and permits closing the feedback loop directly, even without any compensation network, and yet with a high degree of stability. Also, there is no longer any need for an excessively high switching frequency to reduce output ripple, thus resulting in further improvement.

It should be emphasized that the new switching amplifier of Figure 19 is actually based on a new concept of significantly broader scope. Namely, this novel technique of connecting a single dc source to a true push-pull configuration with a differentially connected load may be implemented using any other converter type, as illustrated in Figure 21, provided that the converters are capable of bidirectional current flow and are driven with comple-

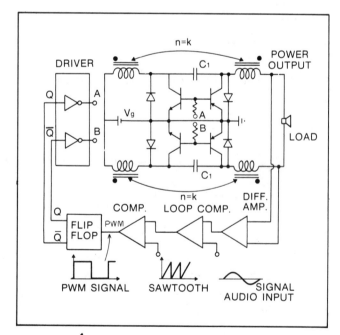

Figure 20. Ćuk push-pull switching power amplifier.

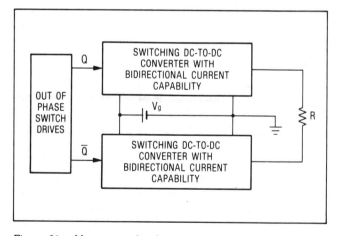

Figure 21. New generalized concept of constructing a switching amplifier power stage from any dc-to-dc converter type.

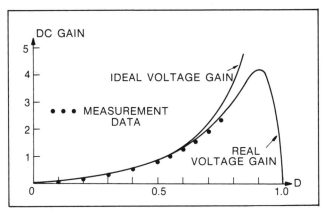

Figure 22. Theoretical and actual dc voltage gain characteristics of a typical test circuit. Experimental measurements confirm the predicted gain curve.

mentary (out of phase) switching drives. [10] In particular, when this configuration is implemented using two buck-type converters, a single-source power amplifier is obtained even with the buck power stage.

Performance of the New Converter and Amplifier

The new switching amplifier design retains all of the optimal characteristics of the new converter topology, while simultaneously offering improvements due to the push-pull configuration itself. However, one important property of the new design, the linearity of the voltage gain, has not been discussed. Note that although the gain curves of the individual converters in the new design are highly nonlinear, as shown by the dotted lines in Figure 17, the differential voltage is very linear near the duty ratio D=0.5, just where it is needed the most — in the center of its useful dynamic range. Nonetheless, the sources of distortion, as well as the effects of non-ideal circuit elements, must now be addressed to more accurately characterize the performance of the new power stage.

The new converter design has been extensively analyzed and experimentally verified. Limitations to the theoretical dc voltage gain arise due primarily to the effects of the parasitic resistance in the windings of the input and output inductors. Consideration of these terms in the analytical circuit model leads to a predicted voltage gain curve of:

$$\frac{V}{V_g} = \frac{D}{D'} \left[\frac{1}{1 + a_1 \left(\frac{D}{D'}\right)^2 + a_2} \right]$$

where a_1 and a_2 are the ratios of the series resistances of the input and the output inductors to the load resistance, respectively, and R is the equivalent resistance of the load. A plot of this curve for a test circuit is shown in Figure 22, and conforms well to the measured experimental results.

The current gain is unaffected and remains equal to D'/D as before. Thus, the power conversion efficiency for a given duty ratio is equal to the bracketed term in the above expression. As the duty ratio approaches one, the power dissipated in the input inductor grows significant and reduces the operating efficiency of the circuit. The power loss due to the output resistance is independent of the operating point, and becomes the most significant detrimental factor when D is less than 0.5 (voltage reduction).

This model does not consider the effects of other resistive losses that occur because of nonzero voltage drops across the semiconductors (when operating) and the finite switching time they require to become fully

conducting. These losses may become significant in applications involving relatively low input or output voltages or high switching frequencies. Nevertheless, all these effects can easily be included analytically by use of the modelling methods described in [7].

When the effects of the parasitic resistances on the performance of the push-pull power stage are included, the differential gain curve of Figure 23 results. As mentioned earlier, the curve for the ideal case (both a_1 and a_2 zero) is fairly linear about the center of the operating region D=0.5. For example, if the variation of the duty ratios in the PWM signal is limited to ± 0.1 (around D=0.5), the resulting total harmonic distortion (THD) in the output will be limited to only 1% [8]. However, a variation of ± 0.2 produces 4% distortion, and increasingly higher THD values result from broader excursions.

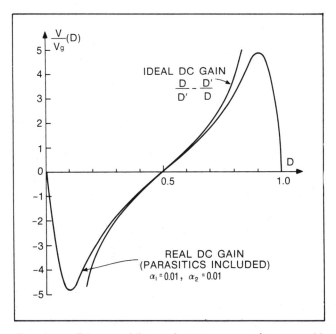

Figure 23. Effective differential gain curve when parasitic resistances are considered.

Furthermore, as can be seen from the figure, the parasitic resistances tend to further linearize the gain around D=0.5. In fact, the values of these resistances can be adjusted to minimize the distortion over the entire operating region. By adding a small resistance in series with the input inductor so that the value of a_1 matches the optimality criterion $a_1 = 0.0718 \times (1 + a_2)$, THD can be reduced to well below 0.1% over the range D=0.5 ± 0.2. [8] Thus, almost perfect linearity in the dc gain curve is achieved even without the use of feedback. Although adding an input resistor results in slightly degraded efficiency, test circuits still perform in the 90% range if the variation in the duty ratio is suitably limited. This need pose no unreasonable restriction, however, if the transformer-isolated version of the converter is used [4].

Equally significant is the effect of the series resistance on the frequency response of the new converter circuits. Although each converter stage has three energy storage elements, which could theoretically introduce complications due to multiple resonances, the damping provided by the input resistance effectively produces a highly favorable "single-pole" frequency response. [8] This completely eliminates feedback stabilization problems and allows closing the loop without a compensation network, greatly simplifying design and resulting in even further savings in circuit size, weight, and cost.

Conclusions

We have shown how a fundamentally new design in switching regulator circuit topology has resulted in dramatically improved performance. The new converter not only offers higher efficiency, low or zero output ripple, and greatly reduced EMI, but at the same time achieves the general conversion function — it is capable of either increasing or decreasing the output voltage depending upon the duty ratio of the switching transistor. The new topology uses capacitive energy transfer between the input and output stages, rather than the inductive energy transfer of other converters, resulting in nonpulsating input and output current. Its implementation requires fewer parts and simpler drive circuitry, with attendant savings in circuit size, weight, and cost.

The excellent frequency response characteristics of the new design allow highly stable feedback regulation to be achieved with simple circuitry. As a result, the basic converter may be used in a new switching power amplifier design that offers performance comparable to linear amplifiers but at much higher efficiency and lower cost —

offering, for the first time, efficient dc-to-ac power conversion.

Thus, the new optimum topology converter is superior to any of the currently known switching converters, outperforming them in every respect.

Prototype version of a Ćuk switched mode audio power amplifier. The circuit delivers 40 watts rms to an 8 ohm speaker load, with a flat frequency response over the 20-20KHz range, and over 90% efficiency. A recently built model has been reduced to less than one quarter this size.

Special thanks go to Robert Erickson and William Behen, members of the Power Electronics Group at the California Institute of Technology, for many devoted hours spent in building the first prototype of the new switching power amplifier and the later improvements.

References

[1] Slobodan Ćuk and R. D. Middlebrook, "A New Optimum Topology Switching Dc-to-Dc Converter," IEEE Power Electronics Specialists Conference, 1977 Record, pp. 160-179 (IEEE Publication 77CH 1213-8 AES).

[2] Slobodan Ćuk, "Modelling, Analysis and Design of Switching Converters," PhD thesis, California Institute of Technology, November 1976. Also, NASA Report CR-135174.

[3] Slobodan Ćuk and R. D. Middlebrook, "Coupled-Inductor and Other Extensions of a New Optimum Topology Switching Dc-to-Dc Converter," IEEE Industry Applications Society Annual Meeting, 1977 Record, pp.1110-1126 (IEEE Publication 77CH1246-8-IA).

[4] R. D. Middlebrook and Slobodan Ćuk, "Isolation and Multiple Output Extensions of a New Optimum Topology Switching Dc-to-Dc Converter," IEEE Power Electronics Specialists Conference, 1978 Record, pp. 256-264 (IEEE Publication 78CH1337-5 AES).

[5] R. D. Middlebrook, Slobodan Ćuk, and W. Behen, "A New Battery Charger/Discharger Converter," IEEE Power Electronics Specialists Conference, 1978 Record, pp. 251-255 (IEEE Publication 78CH1337-5 AES).

[6] L. Shaeffer, "VMOS—A Breakthrough in Power MOSFET Technology," Siliconix Application Note AN76-3, Santa Clara, CA (1976).

[7] R. D. Middlebrook and Slobodan Ćuk, "Modelling and Analysis Methods for Dc-to-Dc Switching Converters," invited review paper, IEEE International Semiconductor Power Converter Conference, 1977 Record, pp. 90-111 (IEEE Publication 77CH1183-31A).

[8] Slobodan Ćuk and Robert W. Erickson, "A Conceptually New High-Frequency Switched-Mode Amplifier Technique Eliminates Current Ripple," Proc. Fifth National Solid-State Power Conversion Conference (Powercon 5), pp. G3.1-G3.22, May 1978.

Patents

[9] Slobodan Ćuk and R. D. Middlebrook, "Dc-to-Dc Switching Converter," U. S Patent Application S. N. 837,532 filed September 28, 1977. Also filed in foreign countries.

[10] Slobodan Ćuk, "Push-Pull Switching Power Amplifier," U.S. Patent Application S.N. 902,725 filed in May 3, 1978. Also filed in foreign countries.

[11] Slobodan Ćuk, "Dc-to-Dc Switching Converter with Zero Input and Output Current Ripple and Integrated Magnetics Circuits," U.S. Patent Application S. N. 026,541 filed in March 30, 1979.

[12] Slobodan Ćuk and R. D. Middlebrook, "Dc-to-Dc Converter Having Reduced Ripple Without Need for Adjustments," U.S. Patent Application S.N. 050,179 filed June 20, 1979.

About the Authors

Dr. S. Ćuk (pronounced "Chook") obtained his BSEE degree from the University of Belgrade, Yugoslavia, MSEE degree at the University of Santa Clara, and a PhD degree at the California Institute of Technology, where he is at present Assistant Professor of Electrical Engineering, teaching courses in Power Electronics and Energy Conversion.

Dr. R. D. Middlebrook obtained his BA and MA degrees at Cambridge University, England, and MSEE and PhD degrees at Stanford University, California. Dr. Middlebrook is a Professor of Electrical Engineering at Caltech, where he is teaching courses in Electronic Circuit Design.

Profs. Middlebrook and Ćuk are leading a highly enthusiastic group of students and researchers in their Power Electronics Group in a strong effort toward finding new, innovative ways for efficient and controlled conversion of electrical energy, covering the whole spectrum from dc-to-dc conversion to dc-to-ac inversion and power amplification.

Advances in SWITCHED-MODE Power Conversion

Part II

Prof. Slobodan Ćuk and
Prof. R.D. Middlebrook
Power Electronics Group, 116-81
California Institute of Technology
Pasadena, CA

In our Winter 1979 issue we presented a new concept in the design of switched-mode power conversion circuitry, which, because of its extreme simplicity, flexibility, and efficiency, will without doubt replace most of the conventional electrical power processing methods currently in use. The demand for compactness and efficiency is especially significant in robot applications, where economy in size, weight, and energy is critical.

In Part II, Profs. Ćuk and Middlebrook, the inventors of the new converter topology, continue their fascinating exposition, presenting extensions to the basic design to provide DC isolation, multiple-output power sources, and a physical realization of the sought-for hypothetical dc-to-dc transformer, a device which converts from pure dc (no voltage of current ripple) at one terminal to pure dc (at a different voltage) at the other terminal. The application of the circuit in a highly efficient amplifier for the servo control of a dc motor or other load will also be presented.

Introduction

In Part I of this series, we provided a review of the basic types of switched-mode power converters and showed how the effort to solve the characteristic problems of these earlier designs led to the development of a fundamentally new converter configuration. Figure 1 shows a summary of the basic converter types and a physical realization of the new converter in its simplest form. The new converter topology embodies all of the desirable features of previous types while retaining none of their liabilities. Whereas other converters rely upon the inductive coupling of energy between the input and output, the new converter uses a capacitor to transfer stored energy between input and output inductors as shown.

Due to the presence of inductors on both the input and output, the current on either terminal remains continuous, avoiding the electrical noise problems associated with switching either the input or output current and resulting in increased conversion efficiency. The new converter has the desirable property that its output voltage can be either higher or lower than that of the input supply, as determined by the *duty ratio* of the switching transistor (the fraction, D, of the switching period (T_s) that the transistor is turned on).

Also presented in Part I were several important extensions to the basic converter design. These include adding the capability for bidirectional power flow between input and output, coupling the input and output inductors via a single transformer core, and a high-performance switched-mode power amplifier configuration using parallel converters driving a differentially connected load.

An important result of the coupled-inductor extension of the new converter, apart from the further reduction in the number of components, is that, by proper adjustment of the magnetic coupling between the input and the output inductors, the residual current and ripple at one of the terminals can be reduced *exactly* to zero, resulting in pure dc. Naturally, this motivated the search for a converter configuration which would achieve the desired characteristic of having zero current ripple at *both* input and output *simultaneously*, thus resulting in a physical realization of an ideal dc-to-dc "transformer." (See Figure 8 in Part I.)

In this article, we will show how, by pursuing the desirable property of dc isolation between the input and the output circuits, the sought-for solution for the ideal converter was found, inherently derived from the new converter topology. Since the new topology provides higher performance than all other converter types but may be implemented using fewer components than comparable solutions, it may be said to be an *optimal* design. To illustrate this point, we will present the design of a working circuit that shows the ease of employing the new topology in practical applications.

DC Isolation in the New Switching Converter

All the basic switching dc-to-dc converter types reviewed in Part I, along with the newly introduced converter topology, are distinguished by the elegant method by which conversion is accomplished through the operation of a single switch (double-pole, single-throw) in an otherwise passive energy storage network, as shown in the converters in Figure 1.

Despite this simple structure, which results in the high conversion efficiency characteristic of "switchers," the basic converter configurations all lack one of the properties often required of power supplies: dc isolation between the input and output ports. There is therefore a strong incentive

Figure 1. The family of four basic switching converter topologies (a-d) and a practical implementation of the Ćuk converter using a transitor-diode combination (e).

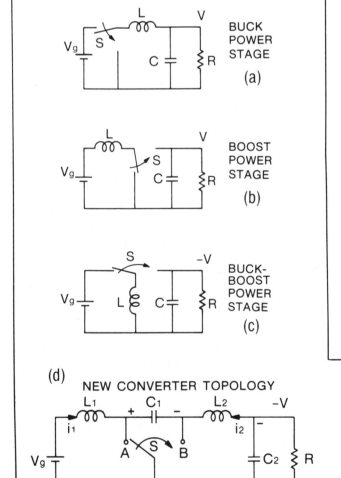

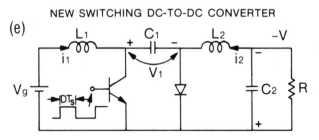

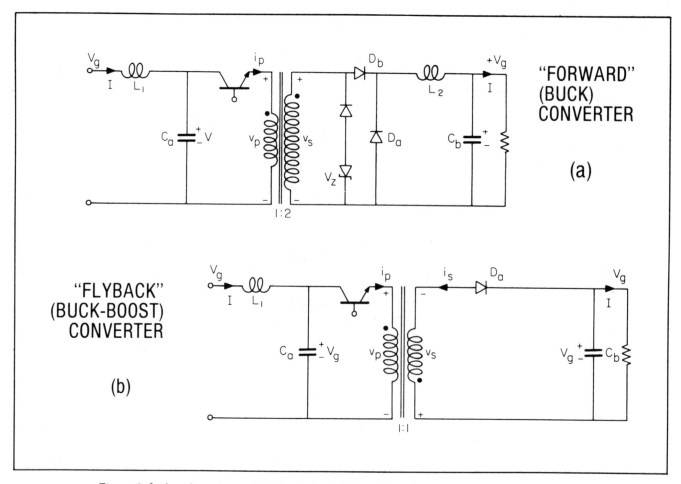

Figure 2. Isolated versions of the basic buck (a) and buck-boost (b) switching converters.

to find a way to introduce an isolation transformer into the new converter design. One could, of course, immediately proceed in a conventional manner and try a "push-pull" configuration using an isolation transformer with a center-tapped primary. However, the consequent doubling of the number of components, along with other practical problems inherent in such a design, favor the search for a less complex single-switch implementation using a "single-ended" isolation transformer.

For example, dc isolation can be introduced into the basic "buck" converter as shown in Figure 2a, which is usually known as the "forward converter." Similarly, simple modification of the "buck-boost" converter results in a "flyback" converter (Figure 2b) by replacing the original single inductor with a "single-ended" isolation transformer. The resulting configuration also allows the output polarity inversion of the buck-boost type to be reversed as shown, where the transformer terminals of like polarity are indicated by the dots. One is therefore motivated to try a similar approach to modify the new converter topology. The obvious place to insert the isolation transformer is somewhere in the inner loop containing the coupling capacitor, transistor, and diode, in which the actual energy transfer is accomplished. There are three key steps leading to a simple, elegant solution to this problem.

The first step is to separate the coupling capacitance C into two series capacitors C_a and C_b, thus making the original symmetrical switching structure divisible into two halves, as shown in Figure 3a, without affecting the operation of the converter. The second step is to recognize that the connection point between these two capacitors, due to its isolation, has an indeterminate dc (average) voltage. This floating potential can then be set at zero by placing an inductance between this point and ground (Fig. 3b). If the new inductance is sufficiently large, it diverts a negligible current from that passing through the series capacitors, so that the converter's operation is still unaffected. The final step is merely the separation of the extra inductance into two equal transformer windings, thus providing the desired dc isolation, resulting in the basic isolated version of the new converter shown in Figure 3c. [1]

One of the main features of the introduction of the isolating transformer is that it has brought the least disturbance to the original converter operation. In fact, with a 1:1 transformer, the voltages and currents in the input and output circuits are the same as in the original non-isolated version. The only difference is that the original switched current loop now has evolved into two loops with equal currents circulating in the same direction.

It is instructive to consider the current paths, voltage

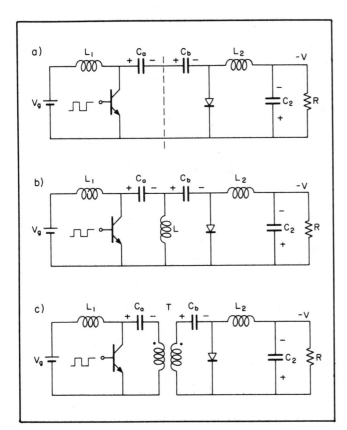

Figure 3. Three key steps leading to the dc-isolated version of the Ćuk converter.

distributions, and energy movements during the two portions of the switching cycle. Figure 4a shows the conditions during the interval $D'T_s$ when the switch transistor is off (open circuit). The input current, having previously stored energy in input inductor L_1, must now pass through capacitor C_a and the transformer primary as shown. This transfers energy from L_1 (in the form of current) to C_a (in the form of voltage, or charge). An equal reflected current in the transformer secondary charges C_b through the now conducting diode. During this interval the output inductor L_2 is releasing its stored energy to the load as shown, and the diode thus carries the sum of both the input and output currents.

In Figure 4b, the transistor is turned on (closed circuit) during interval DT_s, and the input current is allowed to store energy in L_1. Since C_a (as well as C_b) was charged to a positive voltage (viewed from left to right) during $D'T_s$, it now discharges through the transistor and the transformer primary, transferring the stored energy to the output circuit. Note, however, that when the transistor grounds

the positive side of C_a, the voltage drop across the capacitor must (instantaneously) remain the same, so that its right side is pulled down to a negative voltage. This sudden drop is passed by the transformer to the positive (left) side of C_b, and hence to the anode of the diode, cutting it off. Since the output current must now pass through C_b and the transformer secondary, it must match the current through the primary, so that both C_a and C_b discharge their energy into L_2 and the load. One sees that, during this phase of the cycle, it is the transistor that carries the sum of both the input and output currents.

Some of the additional advantages of this new dc-isolated converter become apparent upon reexamining the steps which led to the inclusion of the isolation transformer. Since both windings of the transformer are dc-blocked by C_a and C_b, there can be no dc in either winding. In fact, an automatic volt-second balance is achieved in the steady state, so that there is no problem with "creep" of the transformer core operating point as can occur in push-pull isolation arrangements.

Isolation has thus been achieved in the simplest possible manner by the addition of only the necessary transformer (which is single-ended) and the separation of the original coupling capacitance into

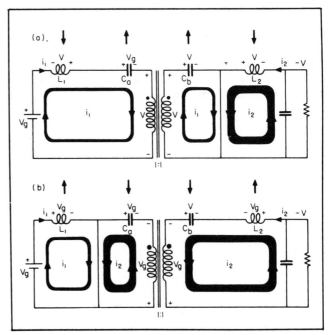

Figure 4. Current and energy distribution in the dc-isolated Ćuk converter: a) interval $D'T_s$ when the transistor switch is open; b) interval DT_s when the switch is closed.

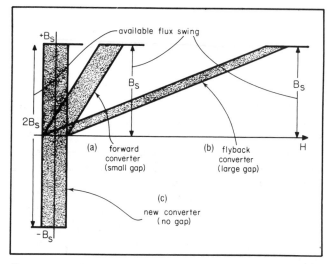

Figure 5. Comparison of the isolation transformer utilization in the Cuk converter and the forward and flyback converters. Twice the flux swing is available in the Cuk converter.

two. Thus, it is apparent that an optimum-topology dc-isolated converter has been obtained which retains all the advantages of the original topology upon which it is based.

Other advantages become apparent upon comparison of the new design with the other popular isolated converters shown in Figure 2. Specifically, for the forward converter the core of the isolation transformer must be gapped, since the magnetizing current is available in only one direction (with an average dc value). The size of the core must be chosen so that the total flux excursion is no greater than the saturation flux B_s of the core material, as shown in Figure 5a. [l]

In contrast, in the new converter (Figure 5c), magnetizing current in the transformer is available in *both* directions, and so a core that is fully utilized (in terms of flux swing) is

only half-utilized in the new converter. Therefore, an ungapped core of half the cross section could be used, with the result that the total flux excursion would be $2B_s$ and the core losses would be correspondingly halved. Similarly, the copper losses would be greatly lowered due to the reduced winding lengths. From the general point of view, these benefits all stem from the fact that in the new converter power is transmitted through the transformer during *both* intervals of the switching cycle, whereas the same average power has to be transmitted during only *one* interval in the forward converter.

Comparison of the transformer properties between the new converter and the flyback converter shows that the disparity is even more extreme, because in the flyback the core gap must be larger than in the forward converter, as illustrated in Figure 4b. This is because the transformer is really an inductor, and the energy transferred through the converter is stored in the magnetic field (principally in the air gap) during one part of the cycle and released during the other. Consequently, the magnetizing current, which is again available in only one direction, constitutes the total primary or secondary current instead of just a small fraction of it.

The Multiple-Output Extension

Once an isolation transformer has been introduced into the new converter, several other extensions become obvious. There is no reason why the transformer should be limited to a 1:1 turns ratio, and therefore a turns ratio factor N_s/N_p is available for either step-up or step-down in addition to gain control obtained by varying the duty ratio D. Thus,

$$\frac{V}{V_g} = \frac{N_s}{N_p} \cdot \frac{D}{1-D}$$

Also, as in the case of the isolated buck-boost converter in Figure 2b, an output voltage of the same polarity as the input voltage may be easily obtained by reversing the polarity of the isolation transformer secondary and changing the direction of the diode and polarity of the coupling capacitor on the secondary side accordingly, as shown in Figure 6. It is interesting to note that this polarity-preserving configuration leads to an implementation of the bidirectional current (two-quadrant) version of the dc-isolated converter using *all npn bipolar transistors,* requiring only positive switch drive signals. (Compare this with the complementary pair, *npn* and *pnp* transistor, version of the bidirectional converter presented in Part I.)

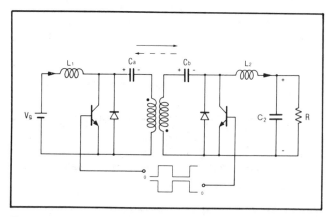

Figure 6. DC-isolated non-inverting bidirectional current Ćuk switching converter, using all npn bipolar transistors. Compare this with Figure 9 in Part I.

The same holds for the power MOSFET implementation in which only n-channel devices are needed. This is of significant practical importance, since npn transistors and n-channel MOSFETs come in all current and voltage ranges, while pnp and p-channel devices are very limited, usually to lower current, voltage, and power ratings.

Multiple outputs of different voltages and polarities are easily obtained from multiple secondary windings, or from a tapped secondary winding as shown in Figure 7. All of the benefits of the basic new converter topology are retained in the multiple-output versions; in particular, all the output currents and the input current are nonpulsating. When

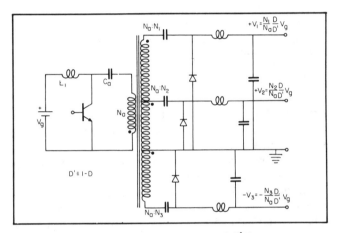

Figure 7. Extension of the dc-isolated Ćuk converter to multiple outputs with arbitrary ratios and polarities.

compared to the conventional approach of designing an isolated, multiple output power supply using a tapped 60Hz transformer followed by several linear regulators, the high performance and cost effectiveness of the new method is obvious. The lower the operating frequency, the more massive a transformer must be. Instead of a bulky 60Hz transformer, a much smaller transformer designed to operate at the switching frequency (20kHz, for example) is used. Instead of the wasteful power dissipation of linear regulation, requiring heatsinks, fans, etc., the new converter typically operates at over 90% efficiency. An additional side benefit is that the latter design is also insensitive to the input line frequency (50Hz, 60Hz, 400Hz, etc.), since the input voltage is supplied by rectifying the ac. Thus, by introducing dc-isolation into the new converter, we have substantially increased its range of application.

The Coupled-Inductor DC-Isolated Converter

The isolation transformer was imbedded in the basic

converter without altering its mode of operation, thus all the modifications and extensions applicable to the basic converter are valid in the isolated case as well. For example, the input and output inductors can be coupled on the same core as shown in Figure 8. As discussed in Part I, zero current ripple may be obtained on either the input or the output through appropriate design of the coupled inductor.

The configuration in Figure 8 may at first seem significantly different from the isolated converter of Figure 3c, apart from the coupling of the inductors. This is because the transfer capacitors C_a and C_b have been relocated to the other side of their respective transformer windings. This change neither alters the converter topology nor affects its conceptual operation. From the practical viewpoint, however, this alteration has important noise reduction advantages. Since the case of an electrolytic capacitor is usually common to its negative terminal, this design effectively grounds the cases of the capacitors, significantly reducing the radiation produced by the switching currents.

While magnetic coupling of the input and output inductors is very desirable for the reduction of ripple, size, and cost, it does introduce an undesirable characteristic, namely the *transient reversal of output polarity* on start-up. Note that in the steady state shown in Figure 8, the average dc currents flow *into* the dots on the coupled inductors, contrary to the behavior of an ordinary ac transformer. During the turn-on transient, however, the ac coupling of the inductors causes the initial current pulse from the input to be reflected in the secondary *in the opposite direction* from the steady-state current, as shown in Figure 9. Although the pulse is of short duration (50 μs. typical), it could result in damage to sensitive loads. The simple addition of an output clamp diode as shown limits the

Figure 8. Coupled-inductor dc-isolated Ćuk converter.

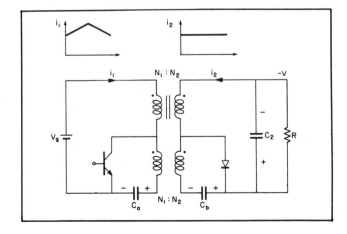

reverse transient to about a volt, or less if a Schottky diode is used.

Just as in the single output case, the same opportunity for coupling exists in the multiple-output isolated converter: any or all of the inductors can be coupled on the same core, as shown in Figure 10. The resulting converter has only two magnetic lumps, one for dc isolation and the other for input and output filtering. Again, by judicious selection of the turns ratios and coupling coefficients, the current ripple may be eliminated on the input or any of the outputs. [1]

These advances notwithstanding, the possibility of zero current ripple on only one side leaves one with a feeling of incompleteness and a desire to accomplish this ideal goal of zero ripple on both input and output ports. That this can indeed be accomplished, with an even further reduction in the size and cost of the circuit, will once again dramatically illustrate the optimality of the new topology.

The Ideal Zero-Ripple Switching DC-to-DC Converter

In pursuit of the goal just posed, there may be many approaches to follow: for example, two coupled-inductor converters could be cascaded, with one set for zero input ripple and the other for zero output ripple. It is quite obvious that this would be undesirable, for not only would the input power be processed twice, with nearly twice the power losses, but the parts count would be doubled and the original advantages of simplicity and reliability would be lost.

Hence the following objective is posed:

Figure 9. Polarity reversal at start-up in a coupled-inductor Čuk converter. The output diode provides a safe path for the transient.

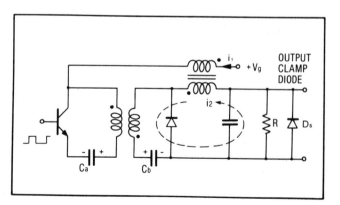

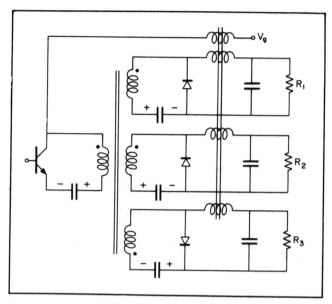

Figure 10. Transformer-isolated Čuk converter with all inductors coupled on a single core.

Synthesize a switching dc-to-dc converter which has the least number of components and yet leads to the ideal dc voltage and current waveforms at both input and output simultaneously.

This problem may at first seem formidable and impossible to achieve, but, based on the background material outlined so far, the coupled-inductor concept may be generalized to satisfy our goal. Recall that the original motivation for coupling the inductors was the 1:1 proportionality of the inductor voltage waveforms of the basic converter (Part I, p.9). The key step becomes that of finding a switching topology in which such proportional inductor voltage waveforms are abundant, hence allowing multiple application of the coupled-inductor concept. However, we have already seen such a topology in Figure 3b, where it was

Figure 11. Switching converter with three proportional (1:1) inductor voltage waveforms.

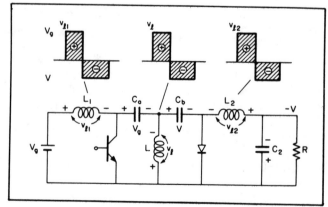

a key intermediate step toward attaining dc isolation. To emphasize the availability of the proportional switching waveforms, this configuration is shown in Figure 11 along with the corresponding inductor voltage waveforms.

It now becomes apparent that the inner inductor L may be coupled with either the input inductor L_1 for zero input current ripple or with L_2 for zero output ripple. However, with a minor modification, *both* current ripples may be simultaneously eliminated. By providing two inductors L_a and L_b in parallel instead of the single L and coupling them as shown in Figure 12a, both ripple currents may be transferred inside the converter, resulting in pure dc at each port.

The two inductances L_a and L_b are not coupled magnetically, but are only connected in parallel electrically, hence the configuration has two magnetic lumps as shown in Figure 12b, in which gapped "U-cores" are used as one possible implementation. However, these two lumps could be even further reduced by coupling the windings L_a and L_b *magnetically* as well. This is equivalent to magnetically coupling the single inner winding of Figure 11 to both input and output inductors simultaneously, as shown in Figure 13. Note that, as in Figure 12, the input and output inductors L_1 and L_2 are *not* directly coupled, as indicated by the two separate core lines for the input and output sides. However, *unlike* the configuration in Figure 12, the adjustments for zero ripple are no longer independent, due to the magnetic coupling through the common core (L). Changing one air gap alters the ripple characteristics of both sides. Nevertheless, the adjustments are highly convergent and the ideal zero ripple case is readily obtained. [2]

None of the switching topologies shown so far in this section has the important dc-isolation property. However, this feature is now easily obtained by replacing the single winding of the inner leg of the magnetic circuit (L) with separate, isolated windings for the input and output, as shown in Figure 14, with no additional magnetic material required. This step is completely analogous to the one described earlier which led to the original dc-isolated version of the converter (Figure 3b-c). However, in the single magnetic circuit of Figure 14, additional benefits are obtained:

The added isolation transformer now has a double role—it not only provides the isolation, but also, through magnetic coupling with the input and output inductors, results in ideal dc current at both input and output simultaneously.

From the magnetic circuit realization in Figure 14b, it may seem that a rather special magnetic core configuration is

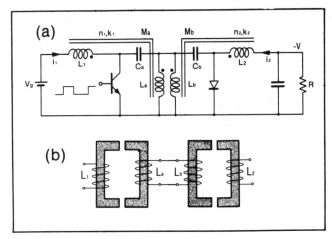

Figure 12. Converter with zero current ripple at both input and output (a) and the realization of the magnetic circuits (b).

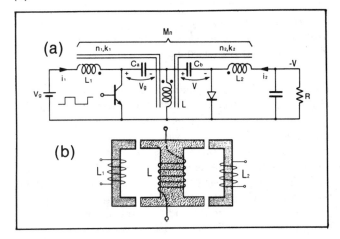

Figure 13. Zero-ripple switching converter with shared inner inductor (a) and one possible core implementation (b).

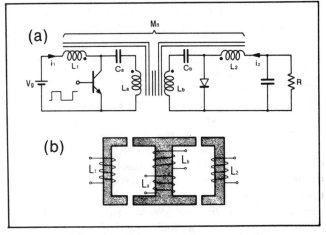

Figure 14. DC-isolated version of the zero-ripple converter obtained by adding a single winding to the magnetic circuit of Figure 13b.

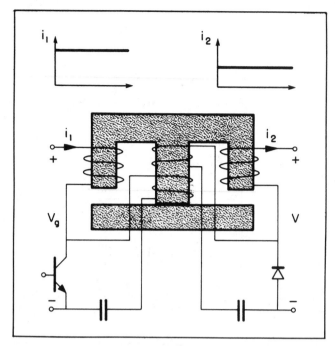

Figure 15. The simplest realization of an ideal dc-to-dc transformer: a four-winding magnetic circuit, a transistor and diode, and two capacitors.

needed, one comprising more than a single core structure. However, the structure illustrated was used only for conceptual reasons to clearly explain each step of the development. A number of widely used standard magnetic cores, such as common E-I cores, for example, can be used for the practical implementation of the switching converter in Figure 14a and its single magnetic circuit. In Figure 15 the E-I core implementation of the integrated inductor-transformer differs from standard cores only in that each of its outer legs contains an air gap, as opposed to the usual case in which only the inner leg is gapped. The fact that three bobbins are used (one winding on each leg) is not unusual and is, in fact, common in three-phase transformer designs using E-I cores.

This resemblance in core structure is not a mere coincidence. Merging the three single-phase transformers into a single three-phase transformer results in corresponding savings in core material due to the merging of the magnetic fluxes and the sharing of the common flux paths. Similarly, the core configuration of Figure 15, by merging the two separate inductors L_1 and L_2 and the isolation transformer into this single magnetic structure, results in substantial savings in size, weight, and efficiency.

Thus, a truly optimum dc converter is obtained which has both outstanding features: true dc currents at either port, as well as dc isolation, in the simplest possible topology consisting of a single magnetic circuit, a transistor and diode, and two capacitors. Under the control of a clock signal with variable duty cycle, D, the converter becomes the physical realization of a dc-to-dc "transformer" with an electronically variable "turns-ratio" (Figure 16).

A natural extension of this zero-ripple converter configuration (Figure 15) to multiple outputs is shown in Figure 17a. There, a single magnetic circuit M_m with six windings appropriately coupled is capable of providing zero current ripple not only at the input but simultaneously at both outputs as well. The actual physical implementation of its magnetic circuit is a simple three-dimensional extension of the one-output core configuration, as illustrated in Figure 17b, providing an additional flux loop common to the central core but separate from the other legs. The three windings of the isolation transformer are placed on the central leg, while the input and two output windings are placed on the three outer legs. Interestingly enough, such a configuration has recently been proposed for a different application, the "Variable Leakage Transformer," in which the third leg is used by a control winding. [3] The extension of this zero-ripple configuration to provide additional outputs (three or more) is performed similarly.

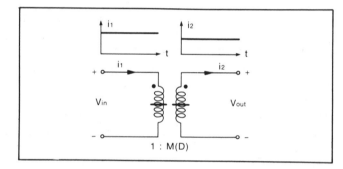

Figure 16. Equivalent circuit symbol devoting the ideal dc-to-dc transformer with constant input and output currents, used as a model of the converter in Figure 15.

The Concept of Integrated Magnetic Circuits

Apart from the usefulness and practical advantages of the various zero-ripple converter configurations previously outlined, their significance extends even further, since their development has led to a new and general integrated magnetic circuit concept.

The coupled-inductor extension of the new converter was the crucial first step in that direction. There, from the conceptual viewpoint, two magnetic components (inductors), which until then had been used only separately to perform their function in circuits, were for the first time *integrated into a single magnetic circuit* (single core) with two windings. While it looked like a classical ac transformer from the structural viewpoint, it indeed performed the function of two separate inductors, when polarity marks

The Evolution of Zero Ripple DC-to-DC Switching Converters

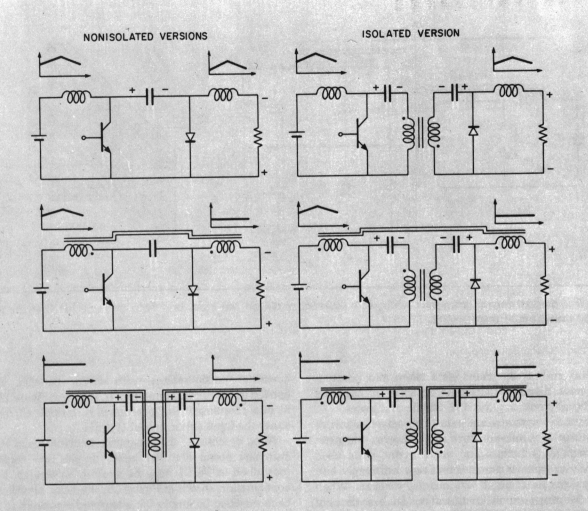

NONISOLATED VERSIONS   ISOLATED VERSION

The search for new efficient, low noise switching converter topologies was originally motivated by the desire to reduce or possibly eliminate some of the major problems which limit "switchers," such as the large pulse currents at either the input or the output port or even both. For example, in a conventional buck-boost (flyback) converter, both input- and output-port currents are large pulses, leading to severe conducted and radiated electromagnetic interference (EMI). The basic Cuk converter (nonisolated version) was a substantial step in the direction of reducing this problem by creating nonpulsating input and output currents. However, the unique topology of the Cuk switching converter and in particular the proportional voltage waveforms on their two inductors has led to a powerful new concept: the coupling of inductors. Besides the obvious reduction in complexity, yet another significant advancement has been acomplished: current ripple at either input or output is not only reduced but completely eliminated. The natural outgrowth of this new technique is a switching configuration which employs a single magnetic circuit with three windings and achieves zero current ripple at both ports simultaneously.

For many practical applications, dc isolation between source and load is essential. The next key step in the development is the incorporation of an isolation transformer in an optimum single-ended manner, which leaves the fundamental features of the basic Cuk converter intact. Hence, coupling of the input and output inductors leads again to zero current ripple at either end. Finally, the crowning achievement of the converter development is its final evolution into a topology which truly emulates the desired electronically controllable ideal dc-to-dc transformer. This ultimate switching dc-to-dc converter configuration possesses both outstanding features: true dc currents at both ports and dc isolation in the simplest possible topology, consisting of a single magnetic circuit with four windings, two capacitors and a single switch implemented by the usual transistor-diode combination.

In summary, a long journey in the development of optimum switching converter topologies has been successfully completed: from the conventional buck-boost with large pulsating currents at both input and output ports, through new converters with bolt currents nonpulsating, to an ideal dc-to-dc switching converter, all featuring nonisolated as well as isolated versions. The basic Cuk converter, as well as many of its extensions and applications, are protected by a series of patents. [7-10]

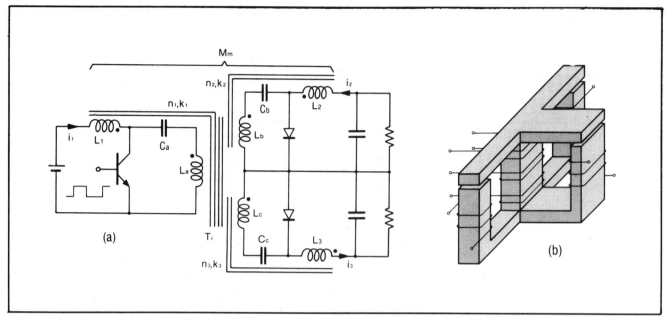

Figure 17. Multiple-output converter with zero current ripple on the input and both outputs (a) and the single magnetic circuit implementation (b).

and actual current directions were taken into account. Furthermore, it even outperformed the separate inductors by significantly reducing the size, weight, and losses.

The natural outgrowth, and in fact the generalization, of this concept is achieved when such diverse magnetic components as inductors and transformers, again used exclusively as separate elements until now, are merged into an integral magnetic circuit with multiple windings. When put into the proper circuit configurations (such as those of Figs. 13-15,17), they not only perform the original functions, but even tremendously improve the performance of the converter (ripple currents reduced to zero), with additional savings in size and weight and increased efficiency.

An analogy with the development from discrete electric circuits to modern integrated electronics can also be drawn. Just as electronic components and their interconnections were integrated on a common silicon medium, with electron current as the binding carrier, we now have diverse magnetic components integrated into a single magnetic circuit, with the core material as the medium and magnetic flux as the binding carrier. Whereas previously we could describe the current distribution in a complex electronic circuit, we now, in the case of switching converters, have a complex magnetic circuit, with a corresponding flux distribution, which is superimposed on and related to its electrical complement. For this process of reducing the separate magnetic components into a single magnetic circuit with accompanying savings in size and weight, and owing to its similarity with the integration of electrical circuits, a generic name *integrated magnetics* is proposed.

This new concept is applicable to a broad spectrum of electrical circuits involving magnetic components. To illustrate this, we will demonstrate its use in different converter configurations, with similar benefits. When applied to the SEPIC converter [4], shown in Figure 18a, or its dual counterpart (b), it results in zero ripple current on either the input (a) or output (b) side.

Also, of course, the magnetic elements used in the switching power amplifier based on the new converter (described in Part I) may be similarly integrated. In the configuration shown in Figure 19, the basic amplifier has been modified by employing integrated magnetics to have dc isolation (for use with ac line voltage input, for example) and zero ripple on both the input and the load. The resulting circuit has only two magnetic circuits, each with four windings.

These examples illustrate how, in a complex switching configuration consisting of a number of storage elements (inductors, transformers, and capacitors) and switches, interconnected to perform some useful function, such as dc-to-dc conversion, dc-to-ac inversion, the otherwise separate magnetic components can be merged into an integral magnetic circuit with multiple windings. The prerequisite for such a simplification is the existence of synchronized and proportional voltage waveforms on the inductors and transformers, as in the case of dc-isolated version of the new converter previously described.

Conclusions

A long and most rewarding exploration of switching dc-to-dc converter topologies has come to its fruition in establishing a practical switching configuration which truly emulates the ideal dc-to-dc transformer function. Let us conclude by summarizing some of the major milestones encountered along the way.

The traditional approach taken by many researchers and practicing engineers is indeed very simple: let us first get the required dc-to-dc conversion function by using some simple switching mechanism (such as buck or buck-boost) and then cure its problems (such as pulsating input or output currents and severe EMI) later, such as by adding complex filter and storage elements to "clean up" the signals.

Our approach, however, has been just the opposite: from the very beginning we considered that storage and energy transfer elements, together with the switches, are an intergral part of the problem. Moreover, we have not only attached extreme importance to converter topologies but now have extended the previous topology of *electrical* connections into its complementary topology of *magnetic* connections through the newly introduced *integrated magnetics* approach.

At every point in the development, we introduced additional complexity by adding circuit elements, electrical or magnetic connections for only two reasons: because it was the optimal method of obtaining a desired feature or function—or as an intermediate step towards a further simplification. Even when a seemingly simple configuration was achieved in the basic converter, the desire to still further simplify and improve performance led to the new coupled-inductor concept, which has now been generalized

Figure 18. The integrated magnetic circuit concept, when applied to two converter configurations, the SEPIC converter (a) and its dual counterpart (b), results in zero current ripple at the input (a) or output (b).

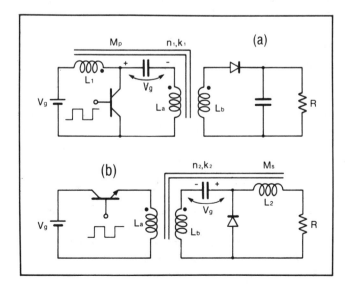

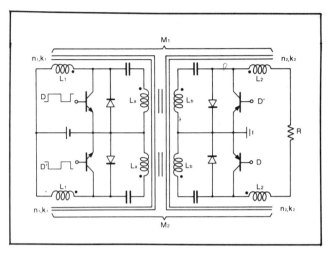

Figure 19. Integrated magnetics applied to the Ćuk switching power amplifier. Compare with Part I, Figure 19.

into the integrated magnetics concept.

As a result of this approach, dc-to-dc conversion topologies have now been extended to include bidirectional power flow (two-quadrant converters) and dc-to-ac inversion and power amplification (four-quadrant converters) as presented in Part I, and now to the dc-isolated and ideal zero-ripple converters presented here. Thus, the approach has been generalized to include the whole spectrum of Power Electronics applications.

Appendix: Circuit Specifications for a Switched-Mode Power Amplifier

In Part I of this article (pp. 14-18) the application of the Ćuk power stage to a new push-pull power amplifier configuration was presented, along with a detailed discussion of its performance. The Ćuk amplifier features broad frequency response (flat over 20Hz-20kHz), low distortion (less than .1% THD), and high efficiency (appx. 90%), using inexpensive, commonly available components. Its capability for bidirectional power flow, as with the other bidirectional configurations described in Part I, make it highly suitable for use in driving loads from which stored energy can be recovered, such as for performing regenerative braking of dc motors.

Interestingly enough, the same amplifier may be used for the servo control of ac induction (Tesla) motors. When an ac signal source at a constant slip frequency relative to the motor's speed is used as input, optimum torque control, as well as the desirable regenerative braking, may be obtained.

We present here the design of a working Ćuk amplifier

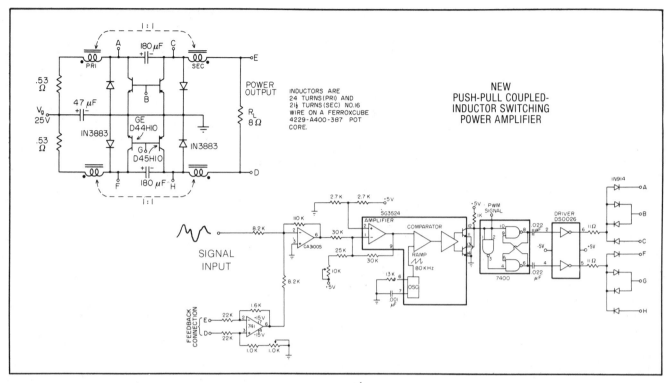

Figure 20. A push-pull switching power amplifier based on the Ćuk coupled-inductor power stage.

suitable for use either as an audio amplifier or a servo driver for small dc motors (up to 40 Watts). The circuit, shown in Figure 20, is based on the coupled-inductor, non-isolated version of the Ćuk converter. [5]. Note that the load is connected differentially across the two power stages, instead of relative to ground. (A photograph of the assembled amplifier was shown at the end of Part I.) Each of the coupled inductors uses a single ferrite "pot" core assembly, wound as specified on the single bobbin, secondary first. Observe the polarity markings on the diagram so that current flowing "into the dot" on each winding will travel in the same direction around the core.

The voltage across the load (E-D) is differentially sensed by the 741 op amp and compared with the input signal using a RCA CA3100S op amp. R_1 is adjusted to obtain the best differential output from the 741. The excellent frequency and phase response of the Ćuk power stage eliminates the need for any frequency compensation of the feedback. A switching regulator integrated circuit SG3524, by Silicon General, is used in the feedback loop to generate a pulse-width-modulated (PWM) signal with an 80kHz switching frequency. R_2 provides DC balance and is adjusted to produce zero output voltage in the absence of an input signal. Logic gates (7400 NAND) connected as a flip-flop produce two out-of-phase input signals for the DS0026 drivers, which produce the ±5V drive signals for the switching transistors. Diodes (1N914) were used in a

modification of the "Baker clamp" to improve the transistor switching times, while still retaining the feature of *automatically* preventing the overlap of the transistor on times.

In many dc motor control applications, it is desirable for the servo input voltage to determine the *current* to be supplied to the motor instead of the voltage. Current sensing for use in feedback control can be accomplished by inserting a small (0.1 ohm) sense resistor in series with the load and applying the voltage across it to the feedback inputs E and D, with the gain of the differential amplifier suitably increased.

The 0.53 ohm resistors on the input side of the switching power stage serve to optimize the harmonic distortion characteriestics of the amplifier at a slight expense in operating efficiency. (See Part I.) For applications in which distortion is not a critical factor, these resistors may be omitted.

The use of coupled inductors in the power stage, although eliminating the switching ripple current on the output, imposes a limitation on the maximum available output current, as *both* the input and output currents produce magnetizing flux with the shared core that must be kept below the core's saturation point. Using a non-coupled configuration, Cuk converters capable of up to one kilowatt (over 1HP) output power have been constructed. [6]

References

[1] R. D. Middlebrook and S. Ćuk, "Isolation and Multiple Output Extensions of a New Optimum Topology Switching DC-to-DC Converter," IEEE Power Electronics Specialists Conf. 1978 Record. (IEEE Pub. 78CH1337-5 AES).

[2] S. Ćuk, "A New Zero-Ripple Switching DC-to-DC Converter and Integrated Magnetics," Proc. IEEE Power Electronics Specialists Conf, (May 1980).

[3] H. Hirayama, "Simplifying Switched Mode Converter Design with a New Variable Leakage Transformer Topology," Proc 7th Natl. Solid-State Power Conversion Conf. (Powercon 7, March 1980).

[4] R. Massey and E. Snyder, "High-Voltage Single-Ended DC-DC Converter," IEEE Power Electronics Specialists Conf. (1977).

[5] S. Ćuk and R. Erickson, "A Conceptually New High-Frequency Switched-Mode Power Amplifier Technique Eliminates Current Ripple," Proc 5th Natl. Solid-State Power Conversion Conf. (Powercon 5, May 1978).

[6] L. Rensink, A. Brown, S.-P. Hsu, and S. Ćuk, "Design of a Kilowatt Off-line Switcher Using a Cuk Converter," Proc. 6th Natl. Solid-State Power Conversion Conf. (Powercon 6).

Patents

[7] Slobodan Ćuk and R. D. Middlebrook, "DC-to-DC Switching Converter," U.S. Patent 4,184,197, January 15, 1980. Foreign patents pending.

[8] Slobodan Ćuk, "Push-Pull Switching Power Amplifier," U.S. Patent 4,186,437, January 29, 1980. Foreign patents pending.

[9] Slobodan Ćuk, "DC-to-DC Switching Converter with Zero Input and Output Current Ripple and Integrated Magnetics Circuits," U.S. Patent Application, March 30, 1979.

[10] Slobodan Ćuk and R. D. Middlebrook, "DC-to-DC Converter Having Reduced Ripple Without Need for Adjustments," U.S. Patent Application, June 15, 1979.

MICRO-COMPUTER BASED PATH CONTROL

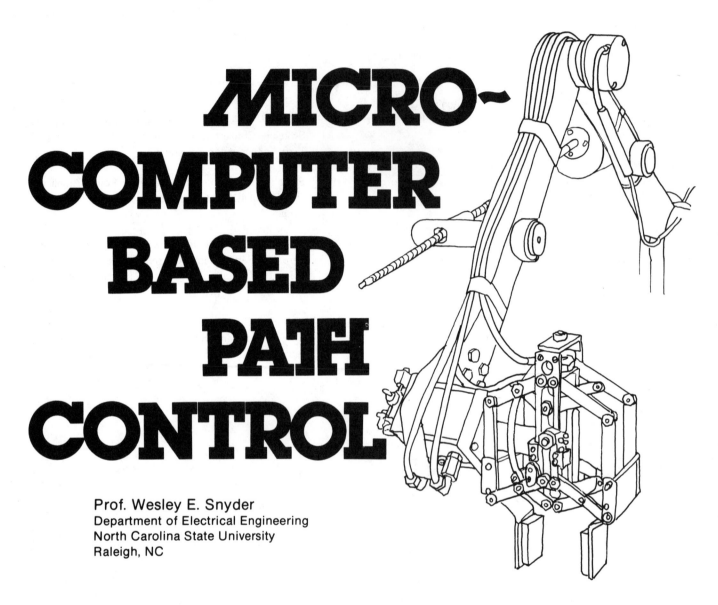

Prof. Wesley E. Snyder
Department of Electrical Engineering
North Carolina State University
Raleigh, NC

Introduction

The most frequent demonstration of industrial robots consists of a performance in which the robot is ordered to "Go to a point, stop, go to another point, stop, close the hand, go to another point, stop," etc. The robot is permitted to travel between stopping points along any path, usually unconstrained.

Programming a robot by specifying its task in terms of a series of points at which it must arrive and stop is not always possible. If the robot was using a paint sprayer, a sander or an arc welder, or carrying a tray of drinks, such start-stop motion would be totally unsatisfactory, if not disastrous.

For many applications, it is necessary for the robot's hand to move smoothly in space along a set of points which define its path (or "trajectory"). While point-to-point control can be implemented simply in hardware through relatively simple circuits, path control requires more complex algorithms which are generally most conveniently implemented in software. As we will show in this paper, an expensive computer is not necessarily required to implement this path control software. For many path control applications, a relatively cheap microprocessor is adequate. We will describe a system utilizing a slow eight-bit microprocessor, a Motorola M6800/D2 system, with a trivial amount of external hardware, which performs fairly well.

There are two principal ways to specify a manipulator

Reprinted from Robotics Age, *Volume 2, Number 1, Spring 1980* path: as a set of hand positions and orientations, or as a set of joint angles (see Figure 1). Which representation is most appropriate depends on the task.

In the most general of robot tasks, the robot must track and acquire some object moving in the robot's working area or apply a controlled force. To coordinate with the external world in this way requires the ability to control the hand in some fixed workspace reference frame. To move in this way, given the desired location in space, the computer must determine the proper torque to apply to each joint of the arm. Typically, this transformation from "Cartesian space to joint space" is a matrix operation in which the elements of the matrices are themselves trigonometric functions of the arm geometry. [1]

Fortunately, these extensive computations are not always necessary. Many applications require only that the robot follow a memorized path, such as the contours of an object to be painted. The operator may program the robot by manually guiding it through the desired path. A computer controlling each joint will remember the angles through which the joint moves (relative to some absolute zero reference point). These angles define the joint's position at every (discrete) point in time. Since this path may be memorized in terms of joint angles, real time computation of these coordinate transformations is consequently not required. In these cases, the path can be specified as a set of paths, one for each joint. The path for a joint is then simply a list of desired positions and corresponding times at which the joint should be in that position.

We will consider in this paper only the control of a single joint moving along a defined path. This implies using either several microprocessors, with one specifically assigned to each joint, or a time-division multiplexing system on a single processor. Both systems are in use today in different commercial systems.

A brief overview of the general control problem and servo design will precede a discussion of path control techniques which use these servos. We will discuss traditional (as distinguished from optimal) servo design, which is based on the concept of proportional error (PE) control to determine both the direction and magnitude of the torque to be applied to a joint. With traditional controllers come the traditional problems—steady state error and overshoot—and we will mention techniques for dealing with both of these.

We will then discuss the specific problem of path control and how one might design a suitable servo method for a path control system. We will describe two techniques, one based on control of position and one based on control of velocity.

The concepts described herein assume that the robot is operated by DC motors. We chose this context since it makes the ideas easier to explain. When one switches to another type of actuator, a hydraulic system, for example, the concepts are still valid, but some of the details must be changed.

Principles of Robot Manipulator Control

The basic principle of control is very straightforward: move the system in the direction that minimizes some error function. An example error function might be $E=\theta_d-\theta$, where θ_d is the desired angular position and θ is the actual position. (We will in future sections refer to θ_d as the "set point." When $E=0$, the joint is at the desired position. If E is negative, then the joint has moved too far and must reverse its motion. Thus, always moving in the direction which makes E appoach zero will provide a type of control.

Besides the drive direction, we should also be concerned with its magnitude. That is, not only must we ask, "In which way should I move the motor?," but also, "How much power (torque) should I apply to the motor?" Again, the error signal $E=\theta_d-\theta$ provides an answer. Let us apply a drive signal (a control) which is proportional to E. This rule

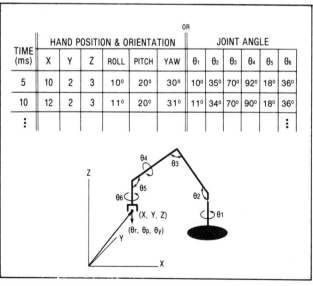

TIME (ms)	HAND POSITION & ORIENTATION						OR JOINT ANGLE					
	X	Y	Z	ROLL	PITCH	YAW	θ_1	θ_2	θ_3	θ_4	θ_5	θ_6
5	10	2	3	10°	20°	30°	10°	35°	70°	92°	18°	36°
10	12	2	3	11°	20°	31°	11°	34°	70°	90°	18°	36°
⋮												⋮

Figure 1. Two alternate ways of representing the path of a robot manipulator. This example assumes the robot arm has only angular joints. For a machine with linear joints, simply replace the appropriate θ_i by an "r_i."

defines a feedback control system as shown in Figure 2. Such a system is called a "proportional error (PE) control" system.

There are some problems with the proportional error control system which we have described. The first of these is the steady state error problem.

Consider the equation we have proposed for this servo: $T=K(\theta_d-\theta)$. That is, power applied to the motor (torque) is equal to some constant (K) multiplied by the error. Now, suppose we must hold a load against gravity. To do so requires a torque, so we cannot hold against gravity without an error, since no error would imply no torque. This is known as the steady state error problem.

A second problem with proportional error control is overshoot. That is, a manipulator operating under control of a proportional servo has only friction to slow it down. To see this, suppose the arm is close to its destination, then $T=K(\theta_d-\theta)$ is quite small, but not negative. If there is much friction, the arm may stop short of θ_d due to lack of drive, but if friction is small and inertia is large (relatively), then the arm may move on past θ_d. Now the error signal is negative, torque is backwards, and the arm will be driven back to θ_d. In the meanwhile, however, it has "overshot" its goal. If the task to be performed requires critical positioning, as in moving television picture tubes, for example, the occurrence of overshoot can be disastrous.

Let us now consider some techniques for dealing with steady state error and overshoot.

The Steady State Error Problem

One approach to dealing with steady state error is to output a torque $T=L+K(\theta_d-\theta)$ where L is a constant sufficient to hold the load when $\theta_d-\theta$ is zero. Use of this

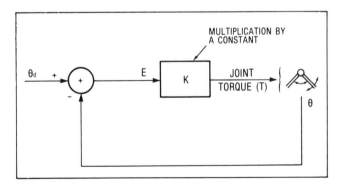

Figure 2. A Proportional Error (PE) control system, in which the control torque (T) is in proportion to the difference between the desired joint angle (θ_d) and its actual angle (θ).

approach requires that the load be known precisely. In the case of robots, this knowledge is difficult to achieve since the load on a particular joint is usually a function of the positions and motions of the other joints.

An alternative is to make the drive signal equal to the integral of the error with respect to time. That is, allow the output of the servo (the motor torque) to accumulate with time, and make the rate of accumulation proportional to the error signal. Such a controller essentially *finds* the constant "L" defined above by the experimental technique of increasing L slowly until the load can be held stationary with no error.

A robot operating with "proportional integrating" (PI) controllers on its vertical axes can be observed to droop when a load is suddenly applied, and then rise back to the desired position.

Of course, the improved performance of an integrating controller does not come for free. We have been discussing how the integrator helps the steady state error problem. That is, integration is of assistance when the arm is stationary or moving slowly. When an integrating controller is used to achieve fast motion, however, it tends to increase the overshoot. In fact, under certain loading conditions, such controllers can be unstable and oscillate about the desired point. Thus, while reducing one problem, steady state error, we have made another problem, overshoot, even worse. Such are the joys of engineering (or government-controlled economies, for that matter)!

The Overshoot Problem

One approach to this problem is to write differential equations which describe a PE or PI controller and its load (inertia, gravity, etc.). The solutions to these equations would then provide a mathematical insight into the conditions under which overshoot occurs. (Overshoot occurs when the solution is "underdamped.") However, in this paper we will pursue a more intuitive argument.

Overshoot occurs because the controller has an insufficient mechanism for "applying the brakes." A PE or PI controller in fact has nothing to stop the arm other than friction. If there is any positive error at all, $\theta_d-\theta$, is positive, and the motor will have positive (although small) drive applied right up to the point where the error goes to zero. If friction is small and inertia large relative to the friction, a joint driven by such a controller will overshoot.

To provide a degree of braking, we can use the following concept:

1) If error is large (we are a long way from the goal)

and the velocity is small, apply a large drive.
2) If error is small (we are close to the goal) and the velocity is high, apply a negative drive.

The simplest way to achieve this is to make the drive torque $T=K_1(\theta_d-\theta)-K_2\dot{\theta}$ where $\dot{\theta}$ is the angular velocity, the derivative (rate of change) of position with respect to time.

The ability of this controller to handle overshoot then depends on the *gains* of the controller, K_1 and K_2, and the inertia and friction of the load. One cannot always guarantee zero overshoot unless something is known about maximal values for inertia. Choice of optimal K_1 and K_2 is then possible. However, these constants are most often determined experimentally. Increasing K_2 is equivalent (for purposes of control) to increasing the friction of the system.

A controller with derivative feedback can then be combined with the concept of integration to yield a "proportional integral derivative" or PID controller. There are several ways in which one could configure such a controller. One such configuration is shown in Figure 3.

With a PID controller, one trades off the possibility of overshoot against the speed of the joint motion. Increasing K_2 tends to slow the arm down since it increases the negative contribution to torque due to velocity.

Because of their relatively simple implementation and "robustness" (ability to adapt to changing loads), PID controllers are probably the most commonly used controllers today, even though their performance is not optimal.

So far, we have discussed the servo implementation as if the total computation of the joint torque, T, was to be done in software. More and more robot manufacturers are going in this direction, although a variety of mixes of hardware and software are available. For example, one could close the servo loop in analog hardware such as operational amplifiers and use the microprocessor to adjust the gains for improved performance and to change the set point at the appropriate time. Alternatively, one could perform some of the servo computation in dedicated digital hardware.

Designing a Path Control System

Point-to-point control requires simply that the robot move to a specified point and stop, and then move on to the next point. To avoid the jerkiness provided by this type of control, we must develop a method to enable the robot to slide smoothly from one stopping point to the next along a specified path. Such path control can be accomplished in several ways by making use of the servo techniques discussed in the previous sections and imposing over those servos a control structure which dynamically modifies the set point.

In this section, we will discuss two techniques for achieving path control of a single joint. The first of these we call the "dog race" technique, in which the control computer simply moves the set point as the servo moves the joint towards the set point.* The second technique involves controlling the path by controlling the velocity along the segments of the path, thus insuring smooth transitions. We will provide a Motorola M6800 program listing of this method.

The Dog Race Technique

One way to achieve a continuous path is the "carrot and horse" or "dog race" technique. A computer tracks the moving robot and moves the goal before it is reached. Thus the robot, like the racing greyhound in pursuit of the mechanical rabbit, follows the path of the moving point.

However, the time required for the robot joint to move from one pont to another is affected by its interactions with other joints, its own inertia and friction and the inertia

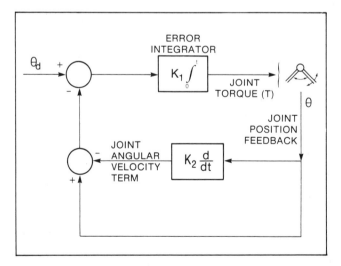

Figure 3. A Proportional Integral Derivative (PID) control system.

---

*Note: it is likely that the servo is simply software in the same computer as the path control algorithm. We distinguish between them only for pedagogical reasons.

of any load it might be moving. For example, if the robot is carrying a light load or if friction is less than was anticipated, the robot will tend to overshoot the goal (catch the rabbit) and stop. This can result in jerky start-stop movement. On the other hand, the friction may be extremely large, causing the robot to move more slowly. If a complicated path with forward and backward movement is designated, problems can develop since the robot may fall so far behind that it no longer moves along the desired path. (The greyhound may get so far behind that he finds the shortest distance to the rabbit is across the center field rather than along the rabbit's path on the track.).

There are numerous techniques for dealing with these problems. The most frequently used methods are:

1) Using a sampling rate (spacing between set points) which is appropriate for the robot in question. As a general rule, the time between set points should not exceed the mechanical time constant (response time) of the arm.

2) Making estimates of the load and adjusting the control gains (K_1 and K_2) appropriately. For example, one might compute loading at both ends of a trajectory

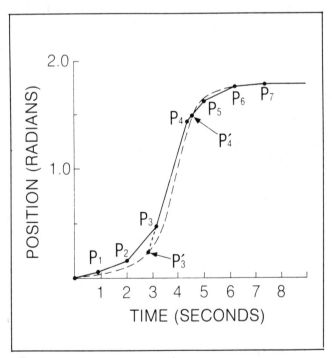

Figure 4. A typical trajectory for a manipulator joint, shown as a sequence of angular positions vs time. A possible joint response is shown by the broken line. The use of the intersection points P_3', P_4' will be explained later.

and estimate loading along the trajectory by extrapolation.

Quite good performance can be achieved by such systems when properly tuned.

A Technique based on Velocity Control

With the velocity control technique the time required for the robot to complete the path is divided into straight line segments. The robot is programmed to move at an assigned velocity which may increase or decrease at each time segment.

For example, Figure 4 shows a set of points (P_1, P_2, P_3...) at which the joint being controlled should be at particular times (T_1, T_2, T_3,...respectively). We have connected these set points with straight lines.

Each line segment has a constant slope. Since the slope is equal to the angular velocity, we can control the path by specifying the desired angular velocity at the appropriate time and then making use of a servo which controls velocity.

This method allows us to specify fewer points along the path than must be specified in the dog race technique and guarantees a reasonably smooth path.

There are some problems involved with correcting tracking errors in this method, and we shall deal with those in a subsequent section.

Since this method requires a servo which controls not position but velocity, some adjustments need to be made to the control algorithm given earlier. Thus, we will first consider how we might control velocity.

We will take the same PID controller developed in section 2 and apply it to control of velocity rather than position. In so doing, some computational simplifications result.

If we are controlling velocity rather than position, nothing about the structure of the PID controller must change, only the names of the variables and the quantities being measured.

The error signal e(t) is now equal to the difference between desired velocity and actual velocity:

$$e(t) = w_d - w(t)$$

where w(t) is the velocity at time t. The PID control signal is then:

$$k_i \int_0^t e(t')dt' + k_i k_d \int_0^t w(t')dt'$$

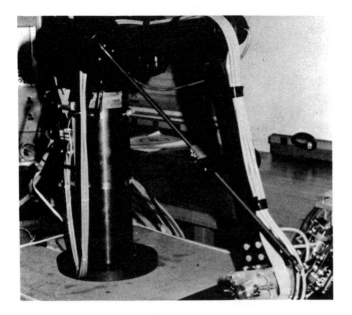

The microcomputer-controlled manipulator used in the NCSU experiments. Note the use of a lead-screw drive on the elbow joint. Apart from the arm itself, the system requires very little external circuitry, since most of the control functions are performed in software.

Note, however, that the term

$$k_i k_d \int_0^t \dot{w}(t')dt'$$

is obtained by first taking the derivative of the velocity feedback to find its rate of change $\dot{w}(t)$ (i.e., acceleration) and then integrating it. Since these two operations cancel one another, the same result can be obtained by using the velocity feedback in the control loop directly. Also, the process of differentiation tends to enhance noise and errors and should be avoided whenever possible. This can be accomplished by avoiding both the differentiation and integration, utilizing the control structure shown in Figure 5, where the new term $K_d' = k_i k_d$. This system implements a control method identical to that provided by the PID controller shown in Figure 3, but avoids the differentiation operator. Consequently it is called a "pseudo-derivative feedback" [2] controller.

Another simplification to the controller can be made by observing that the integral of error $e_I(t)$ is

$$e_I(t) = k_i \int_0^t e(t')dt' = k_i \int_0^t (w_d - w(t'))dt'$$

$$= k_i \int_0^t w_d(t')dt' - k_i \int_0^t w(t')dt'$$

but, since the integral of velocity gives the position, the term

$$\int_0^t w_d(t')dt'$$

is just the desired position (where the arm should be at any instant of time), and

$$\int_0^t w(t')dt'$$

is the arm's actual position, which can be measured directly by feedback. Thus, we can also avoid integration entirely by calculating the joint's positional error and multiplying it by k_i. The desired position can be obtained by solving a straight line equation of the path segment being servoed.

Consequently, we have implemented a "proportional, integrating, differentiating" controller without the necessity of either integrating or differentiating.

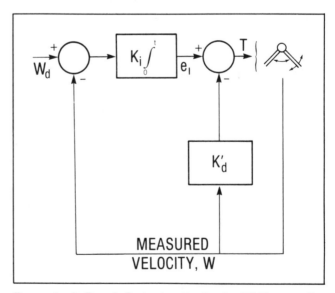

Figure 5. A Pseudo-Derivative Feedback (PDF) controller. Differentiation is avoided by using the feedback term directly, after the error integrator.

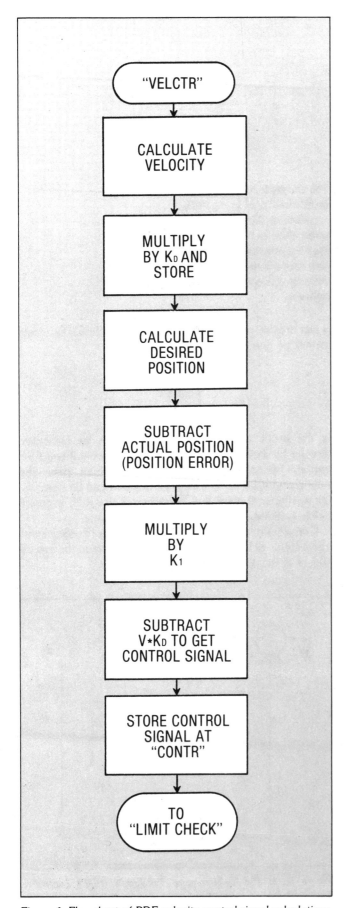

Figure 6. Flowchart of PDF velocity control signal calculations.

MOTOROLA 6800 PROGRAM

```
*       CALCULATION FOR PDF CONTROL SIGNAL
VELCTR  JSR     VELCY       READ VELOCITY
        JSR     DISPLAY     DISPLAY IT
*       MULTIPLICATION OF VELOCITY BY KD1
*       SINCE THE VELOCITY IS MEASURED AS
*       1/16th OF THE ACTUAL VELOCITY, KD1=1/4
*       IS OBTAINED BY MULTIPLYING THE
*       VELOCITY WITH 4.
*       THE RESULT IS A 16 BIT BINARY NUMBER
CNTCAL  CLR     A           CONVERT THE
        LDA B   VELTY       MEASURED VELOCITY TO
        BPL     K2          A 16 BIT NUMBER.
        COM A
K2      LDX     #$2         MULTIPLICATION BY 4
K3      ASL B               DONE BY 2 ROTATE
        ROL A               LEFTS
        DEX
        BNE     K3
        STA A   CONTR       SAVE KD1•VEL AT
        STA B   CONTR+1     CONTR
        LDX     TPNTR
*       CALCULATION OF
*       INTEGRAL(VDESIRED-VACTUAL)•K1
*       WHERE K1 IS A CONSTANT
*       SINCE ABOVE IS EQUAL TO
*       K1•(DESIRED POSITION-ACTUAL POSITION)
*       WE CAN AVOID INTEGRATION BY CALCULATING
*       P. DESIRED AND SUBTRACTING P. ACTUAL
*       FROM IT.
*       CALCULATION OF K1•INTEGRAL(VDES-VACT)
        SEI
        LDA A   CLOCK       GET X (TIME)
        LDA B   CLOCK +1
        CLI
        SUB B   1,X         PUT X-X1
        SBC A   0,X         IN THE MULTIPLICAND
        STA A   MULTND
        STA B   MULTND+1
        LDA A   4,X         PUT DESIRED VELCTY
        JSR     THETA       GET VEL•(X-X1)
        LDX     TPNTR
        LDA A   2,X
        LDA B   3,X         GET Y1
*       Y=Y1+VEL•(X-X1)
*       ASSUMING VEL•(X-X1) IS LESS THAN 32K
*       IN MAGNITUDE, THE RESULT WILL BE
*       IN LEAST SIGNIFICANT BYTES.
        ADD B   RESULT+2    GET DESIRED
        ADC A   RESULT+1    POSITION
*       CALCULATE INTEGRAL ERROR AS YDES-YACT
        SEI
        SUB B   COUNT+1
        SBC A   COUNT
        CLI
*       CALCULATION FOR VELOCITY ERROR.
        LDX     #2
INTDVD  ASL B
        ROL A
        DEX
        BNE     INTDVD
*       ASSUMING THAT THE POSITIONAL ERROR IS
*       ALWAYS LESS THAN 8K IN MAGNITUDE,
*       THE RESULT WILL BE IN LS BYTES.
        SUB B   CONTR+1
        SBC A   CONTR
        STA A   CONTR
        STA B   CONTR+1
        •
        •
        •
        (PERFORM LIMIT CHECKING)
```

Figure 7. PDF control method implemented in M6800 assembler code. Implementations for other microcomputers would use similar logic.

The flow chart for PDF velocity control is given in Figure 6, and a Motorola 6800 program is given in Figure 7.

Choosing the Change Point

In the last section, we proposed to implement path control by describing the path as a series of straight lines (regions of constant velocity).

This rather naive approach suffers from the fact that velocities cannot change instantaneously. Consequently, due to inertia, there is a time lag before the system achieves the desired velocity. Thus, there is always a positional error.

Figure 4 shows a desired joint path (solid line) for a typical motion, and the expected response (broken line).

One method to partially compensate for this lag is to command a new velocity earlier than specified. For example, rather than command velocity 4 at time P_3 in Figure 4, one might command it at time P_3', where the actual trajectory meets the extrapolated next segment.

We can determine when we cross the extrapolated next segment by the following procedure:

The equation of a line passing through a given point (x_1, y_1) and having a known slope "c" is

$$y = y_1 + c(x-x_1)$$

or

$$y-y_1-c(x-x_1) = 0$$

To determine whether a given point is above or below the given line, we substitute the x and y values for the point in the left hand side (LHS) of the equation above and test the sign of the result.

If LHS>0, the point is above the line,
If LHS=0, the point is on the line,
If LHS<0, the point is below the line.

To determine the point where the actual trajectory cuts the next segment, we first have to decide the region from which we are approaching the line. Once we know the region of our approach, to find the change point, we just look for the trajectory to move into the opposite region. The region of approach is determined by comparing the present desired velocity with the next desired velocity. This essentially tells us whether we are about to accelerate or decelerate. The two cases are:

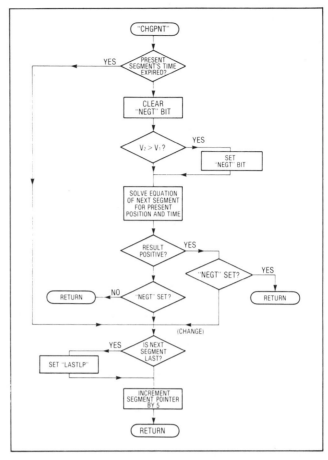

Figure 8. Flowchart of the program to find the change point between path segments.

1. If the next desired velocity is greater than the present desired velocity, we are approaching from a positive region, and should look for a negative value of LHS to change the desired velocity.
2. If the next desired velocity is smaller than the present desired velocity, we are approaching from a negative region and should look for a positive value of LHS to change the desired velocity.

There are some problems with the criteria given above for changing segments. In segment P_3–P_4, the joint acquires a very high velocity and if we wait until point P_4' to change velocities, we will be guaranteed to overshoot. A simple solution to this problem which yields satisfactory results is given by the following heuristic: Always change the desired velocity at the end of a segment, even if the actual velocity has not reached the next segment yet. On the other hand, if the actual trajectory reaches the next segment earlier than the "end of segment" time, the change is made as described earlier. Essentially, the heuristic says to always change at the earlier time, either when you cross the next segment or when the schedule specifies.

A flow chart of the algorithm for determining the change point is given in Figure 8.

Implementation and Performance

Our hypothesis was that one can do path control very inexpensively and simply. To test this hypothesis, we implemented this path control algorithm on a Motorola M6800/D2 system. This is an eight-bit microprocessor with no multiply or divide instructions, running at a reduced clock speed (650kHz). The entire program resided in less than 1k of ROM and required less than 512 bytes of RAM. The microprocessor board had no user communication features other than a hexadecimal keyboard/display.

Positional information was derived from a 2500 count/rev. incremental optical shaft encoder, which was connected to the computer to provide an interrupt once every ½ cycle of the encoder (approximately every 0.07 degrees). The computer incremented (or decremented) a memory location every interrupt and consequently kept track of position totally in software. Time information was also kept totally in software, derived from an oscillator chip which provided an interrupt every millisecond.

Velocity information was derived by some special purpose hardware which measured the width of the encoder pulses. This consisted of essentially a counter, a delay, and a latch. The total amount of off-board hardware used was eight TTL chips.

The total cost of the control system hardware was under $200, less than the cost of a single encoder.

The performance of the system is shown in Figures 9, 10, and 11. The desired trajectory is shown by the dotted lines with set points indicated. The actual path is given by the solid line.

Conclusions

In this paper, we have dealt with a small aspect of robot control, control of the path of a single joint.

Controlling the path of the robot as a whole can be decomposed into control of the paths of the individual joints. Thus, if we can control the paths of the individual joints as functions of position and time, we can control the path of the hand in space.

We have discussed two techniques for performing path control at the joint level. One technique based on velocity control has been developed in considerable detail. This technique has the advantages of extremely simple implementation even with a very limited hardware configuration, and reasonably good performance.

We have tried to show that path control is achievable with simple hardware and minimal software. We hasten to point out however, that this technique, at least in the form described here, is far from the solution to the world's problems. In particular, it requires the path to be known in

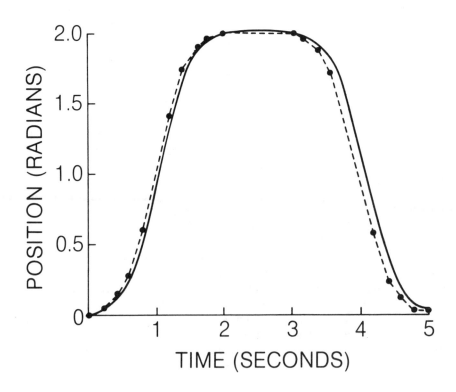

Figure 9. Response of the system to a bell-shaped trajectory.

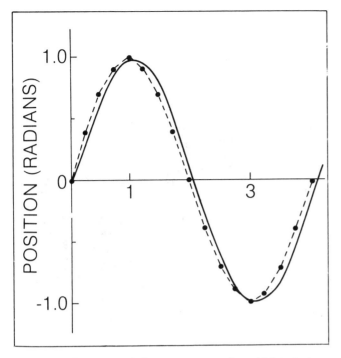

Figure 10. Response of the system to a sinusoidal trajectory.

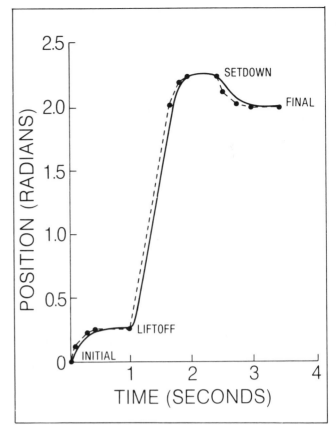

Figure 11. Response of the system to a typical joint motion.

advance. This is possible in many applications where the program is essentially "canned," but not in others, such as conveyor tracking, where the path is continuously modified.

With the additional computational power achievable today, many robot manufacturers are moving in the direction of performing their servo calculations in Cartesian space, thus providing a capability for controlled interaction with moving objects, and introducing a whole new set of interesting concepts. [3,4]

Acknowledgements

The experimental portion of this work was sponsored in part by the National Science Foundation (grant ENG-77-06771). The computer programs were written by Maroof Mian while he was a student at NCSU, and more detail can be obtained from his thesis. [5] The concept of piecewise linear velocity control originally evolved from a suggestion by Richard Paul. Many useful suggestions about the control theory involved were provided by my colleague at NCSU, Bill Gruver.

About the Author

Dr. Snyder is an assistant professor of Electrical Engineering at North Carolina State University, where he is an associate director of the image analysis group.

At present he is on leave with the West German space agency (Deutsche Forschungs-und Versuchsanstalt fuer Luft-und Raumfahrt). His specialties are computer applications to automation and in particular to analysis of images.

Prof. Snyder has taught courses in applications of microprocessors at IBM, Raleigh, and such faraway places as Warsaw, Poland. His current assignment in Germany involves research into use of microcomputers in the real time analysis of television images.

He is a member of the IEEE, the ACM, a senior member of the SME, and a charter member of the Robot Institute of America.

References:

[1] Berthold K. P. Horn, "Kinematics, Statics, and Dynamics of Two-Dimensional Manipulators," in *Artificial Intelligence: An MIT Perspective,* Winston and Brown, eds., MIT Press, Cambridge, MA (1979).

[2] R. M. Phelan, 1977. *Automatic Control Systems.* Cornell University Press, Ithaca, New York 14850.

[3] R. Paul, "The Mathematics of Computer Controlled Manipulators," Joint Automatic Control Conference, San Francisco, 1977.

[4] K. Takase and T. Okada, "Robotics Research in Japan," *ROBOTICS AGE,* Vol. 1 No.2 (Winter 1979).

[5] Maroof Mian, "Trajectory Control of a Robot Joint with a Microcomputer," MS Thesis, EE Dept., NCSY (1978).

CONTINUOUS PATH CONTROL WITH STEPPING MOTORS

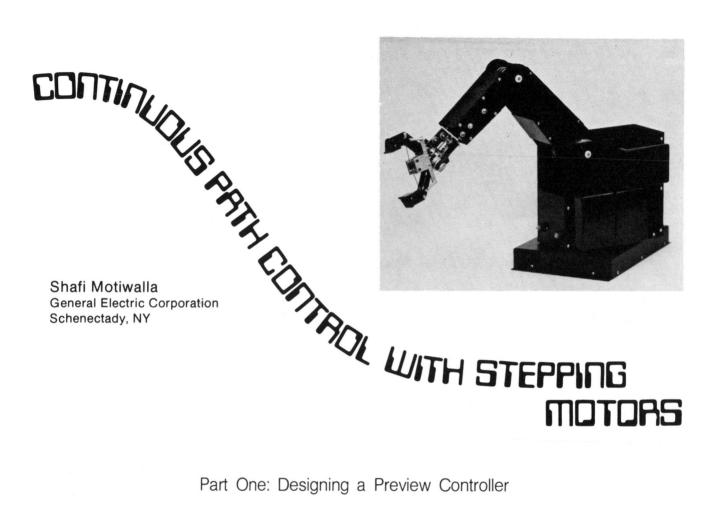

Shafi Motiwalla
General Electric Corporation
Schenectady, NY

Part One: Designing a Preview Controller

The capabilities of today's industrial robots vary over a broad range, from simple *pick and place* machines to sophisticated multi-processor controlled systems. Yet all industrial robots consist of two major components, the manipulator and the controller. The moving parts in the manipulator, which make up the arm, wrist and end-effector (gripper, etc.), must all be driven by some kind of actuator. At present, pneumatic and hydraulic actuators are the most commonly used. Hydraulic drives have a large lift capacity and offer good control in a compact package, but are expensive, have limited speed, and may pollute their workspace with hydraulic fluid. Pneumatics are relatively inexpensive, but the compressibility of the air makes them suitable only for movement between fixed mechanical stops. For the newer robots designed for assembly and other precision work, electric drive is becoming increasingly popular. The use of servo motors or stepping motors can provide fast, accurate control of the manipulator's position and velocity throughout its range of motion.

Apart from the physical limitations of the manipulator, the design of the robot's controller determines the capabilities of the system and its range of applications. The most common controllers can position the hand very accurately, but may offer little control of the robot's path and speed between target points. This latter feature, called *continuous path* control, is essential for many manufacturing operations that require the robot to follow a line or curve in space, as in the case of arc welding.

In this article I will describe a method of attaining continuous path control of a robot driven by stepping motors. This work was done in the Mechanical Engineering Department at the University of California at Berkeley. After first discussing the operation and advantages of using stepping motors, I will present a control scheme that uses "look ahead" to anticipate changes in path direction. This approach, using "preview" or "feed-forward" control, demonstrates the performance possible with stepping motors in an area thought to be restricted to expensive servo motor systems.

Stepping Motors as Actuators

Because it is possible to design very simple control systems for stepping motors, they are being used in a wide variety of applications. The idea of using stepping motors in control systems was proposed in the 1930's, where they were employed by the British Navy as remote positioning devices. [1, 2] Today the diverse applications of steppers include machine tools, process control, computer peripheral equipment, and recently, robots. In general terms, a stepping motor is an electromagnetic incremental actuator which converts digital pulse inputs to discrete motion steps. (*See sidebar: "Stepping Motor Design."*) In the familiar rotary stepper, the shaft rotates in equal increments in response to an appropriate input pulse train. There are a variety of stepping motor designs, the most popular of which is the permanent magnet rotary type. For a discussion of the capabilities and limitations of each type, see [3].

A stepping motor has several operating modes, depending upon the stepping rate and load conditions under which it operates. If the stepping motor is pulsed at a sufficiently slow rate, it comes to rest at the end of each step, moving in discrete steps as shown in Figure 1. In applications where the distance to be traveled is small and the response time is not critical, this mode of operation provides the simplest solution.

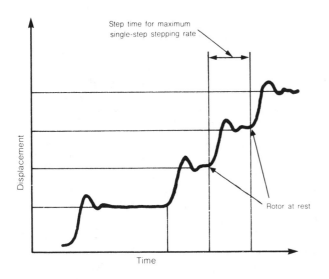

Figure 1. Displacement vs. time for single-step operation. For each step, the inertia of the motor and its load result in the step response shown, which has a momentary oscillation at the end of the step.

As the input pulse rate to the motor is increased, the motion changes from discrete steps to a continuous motion referred to as *slewing*. In the transition between these operating modes, the motor does not come to rest between steps, thus subsequent switching pulses may come when the motor has either a positive or negative velocity. Unless the pulse rate is chosen carefully, this can cause the motor to behave in a somewhat unpredictable and erratic manner. As shown in Figure 2, the velocity of the stepping motor in its steady-state slewing rate generally goes through oscillations about an average value.

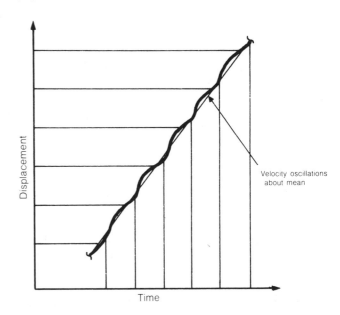

Figure 2. Displacement vs. time for steady-state slewing.

Figure 3 shows the torque vs. step rate curves for a typical stepper in each operating mode. Below the dotted curve for the discrete stepping mode, the load on the motor is low enough for it to step at the given rate without "missing" any pulses and with the motor returning to rest after *each* step. Operating in the slewing mode, however, the motor can produce greater torque at a given step rate, but changes to the step rate must be made carefully so that the peak torque is not exceeded. Above the peak torque, the motor may fail to step in response to an input pulse. One of the key advantages of using a stepping motor is that, by observing its torque/step rate limitations, you can use the motor "open loop" with full confidence that you can tell where its shaft is (relative to some initial position) merely by keeping track of the number of steps sent to the motor.

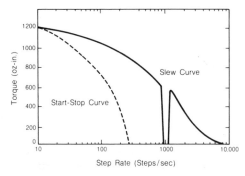

Figure 3. *Speed-torque curves of a permanent magnet stepping motor. The notch in the slew mode curve is due to the mechanical resonance of the motor, which in this case occurs at a 1KHz step rate.*

Control of Stepping Motors

The block diagram of Figure 4 displays the basic elements of a stepping motor control system. These consist of sequencing logic, power drive circuitry, power supply, current or voltage limiting circuitry and the motor. Position/velocity feedback may be used if necessary. The sequencing logic accepts the input pulses with the direction command and supplies low level signals to each of the power drivers. The power driver circuits take this low level signal and control the sequence of high currents from the power supply. A current and voltage limiting circuit is provided to buffer the power delivered to the motor phases to provide the discrete motion. A working stepper motor control circuit is described in "A Homebuilt Computer-Controlled Lathe" which appears in this book.

Figure 5 illustrates the open-loop use of a stepping motor in a numerically controlled machine tool. Stepping motors are natural for this type of application since if properly controlled the positional error does not accumulate and no feedback is necessary. An example of feedback control using a stepping motor is shown in Figure 6. Although steppers are typically used open loop, feedback may be necessary for some applications, for example, when the load is unpredictable (which could cause missed steps), or when there is an inexact relationship between the movement of the motor shaft and the position of the load (due to gear backlash, for instance).

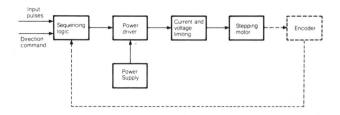

Figure 4. *Basic elements of a typical stepping motor control system. The use of feedback can usually be avoided.*

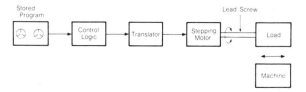

Figure 5. *Open loop control of a machine tool.*

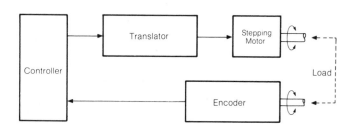

Figure 6. *Closed loop control.*

Although control systems using conventional a.c. or d.c. motors and prime movers can be designed to control incremental motion satisfactorily, a system with stepping motors offers the following advantages: (a) A stepping motor is inherently a discrete motion device. You can more easily interface it to other digital components; (b) The position error in a properly controlled stepping motor is non-cumulative, in that you can achieve accurate position and speed control with a stepping motor in an open-loop system. Transducers such as tachometers or position encoders can usually be eliminated; (3) Stepping motors can be induced to produce high holding torque during quiescent periods. This can eliminate the need for brakes or clutches; (d) If you stay below the single-step load curve, it is possible to start and stop stepping motors instantly; (e) Since a stepper uses digital power pulses, expensive linear power amplifiers are eliminated; (f) The design procedure for a stepping motor controller is normally simpler than for a servo motor controller. Recently, manufacturers have been putting increased effort on the development of improved stepping motors of smaller size and higher power, speed, accuracy and resolution. This has diversified the application of stepping motors to include those areas formerly limited to d.c. servo motors.

The Continuous Path Control Problem

Continuous path control techniques can be divided into three basic categories, based on how much information about the path is used in the motor control calculations. These are illustrated in Figure 7. The first is the conventional or *servo control* approach. This method uses no information about where the path goes in the future. The controller may have a stored representation of the path it is to follow, but for determining the drive signals to the robot's motors, all calculations are based on the past and present path tracking error. This is the control design used in most of today's industrial robots and process control systems. (An introduction to basic servo control has appeared in *Robotics Age*. [4])

The second approach is called *preview control*, also known as "feed-forward" control, since it uses some knowledge about how the path changes immediately ahead of the robot's current location—in addition to the past and present tracking error used by the servo controller. A simple example of preview control is the case of driving a car. You don't wait until you're starting to drive off the road before starting to turn the wheel. Instead, by noting the curvature and grade of the road immediately ahead, you prepare for the change and initiate an appropriate action before you start to go astray. The controller I'll be describing here works somewhat like this, but, of course, in a much more restricted domain.

The last category of path control is the "path planning" or "trajectory calculation" approach. Here the controller has available a complete description of the path the manipulator should follow from one point to another. Using a mathematical-physical "model" of the arm and its load, it precomputes an acceleration profile for every joint, predicting the nominal motor signals that should cause the arm to follow the desired path. This approach has been used in some advanced research robots to achieve highly accurate coordinated movements at a high speed. [5]

A Preview Controller for Stepping Motors

In traditional control theory, variables are usually regarded as continuous analog quantities, and feedback loop designs also treat time as continuous. Analog control systems can be close approximations to these designs. When designing control systems that use digital logic, including computers, we must quantize time and allow for the time it takes to perform the control calculations. If the control loop is "running" at a constant cycle time,

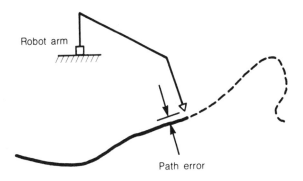

Robot arm

Path error

(a) Basic Servo Control

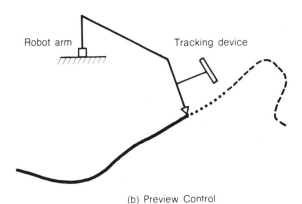

Robot arm

Tracking device

(b) Preview Control

Robot arm

(c) Trajectory Calculation

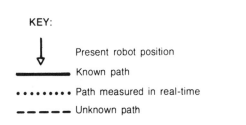

KEY:

Present robot position

——— Known path

.... Path measured in real-time

— — — Unknown path

Figure 7. Three categories of continuous path control. In (a), the controller uses present and past error to determine motor drive signals. In (b), future path information is added to improve tracking. In (c), nominal motor drive signals are precomputed for the entire path, and modified in real time if tracking errors occur.

determined, let's say, by the computer's real-time clock, we can measure time by counting loop cycles. In the following discussion, we will assume that all time-varying quantities are measured at the start of some particular interval, designated by an integer (j through m).

The key ingredients used in our calculations for the control of a single robot joint are the position where the joint *should* be at a given time, $x_d(k)$, and where the joint *actually* is or was at some particular time, $x(k)$. For control of a stepping motor, our goal is to compute the appropriate step rate command, V_c, that will correct any tracking error and keep the robot on the desired path.

Readers familiar with stepping motors might wonder why, since the computer is in control of the step rate sent to the motor, there should be any tracking error at all. Why not just compute the number of steps necessary to go to the desired position and send that many pulses to the motor during the next interval?

Unfortunately, it isn't that simple. We may have to allow for missed steps by providing feedback for closed loop control. Alternatively, we must place restrictions on the change to the step rate to stay within the torque limitations of the motor. In this case the *actual* rate we send to the motor may be higher or lower than the desired rate, with the result that the error will not be corrected in one cycle.

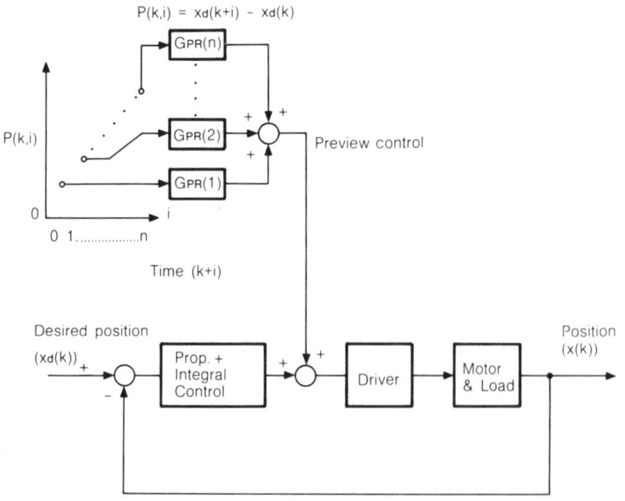

In [4], Snyder described the basic ideas of "proportional" and "integral" correction terms for correcting tracking errors. For our case of discrete-time control of a stepping motor, a standard "PI" control computation would have the form:

$$V_c(k) = G_P\,E(k) + G_I \sum_{j=0}^{k} E(j)$$

where $E(i)$ is the tracking error at the time of the "i"th cycle: $x_d(i)-x(i)$. The proportional and integral gain coefficients, G_P and G_I, respectively, determine the response of the system to errors. They must be "tuned" for the best performance, and their values are influenced by the choice of the loop cycle time, which should be kept constant. The integral term incorporates the influence of past tracking error by summing up all the error amounts since the beginning of the path.

If we have some description of the path ahead, perhaps from vision or tactile sensing, we can compute values x_d for a few intervals beyond the current one and use these for preview calculations. The idea of preview path control is not new—it has been applied to servo motor control for some time. [6,7] We can introduce preview correction into our simple PI stepping motor controller by including preview "error" terms,

$$P(k,i) = x_d(k+i) - x_d(k),$$

into the step rate calculation.

One way of including these terms is to add in several of them, each weighted by its own preview gain coefficient, $G_{pr}(i)$. This tactic is illustrated in Figure 8. The equation for this controller would be:

$$V_c(k) = G_P E(k) + G_I \sum_{j=0}^{k} E(j) + \sum_{i=1}^{n} G_{pr}(i)\,P(k,i)$$

where n is the number of preview samples available. To see the effect of these preview terms, suppose the path for this joint moves in the $+x$ direction on the next few cycles. In this case, the preview terms $P(k,i)$ would be positive. With positive preview gain coefficients, this path change would result in an increased step rate command *before* the point where the path changes. In effect, the controller "anticipates" the change and starts to speed up before a tracking error occurs. Of course, the reverse occurs if the path changes in the $-x$ direction. Instead of a simple PI controller, we now have a "*P*roportional + *I*ntegrating + *P*review" (PIP) controller.

Figure 8. One approach to introducing preview correction terms is to give each preview "error" P(k,i) its own weight, G_{pr}(i).

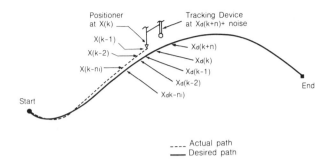

_____ Actual path
_____ Desired path

Figure 10. Path information used by the modified simple PIP control scheme.

Applying the PIP Control Scheme

The hardware we used to test the PIP controller design was a six axis welding robot in our lab at Berkeley. (This machine will be described in more detail in Part II of this article.) Two axes of the robot are realized by an x-y table that moves the workpiece in a plane under the welding tip. The table is positioned by lead screws that are turned by stepping motors. To simplify matters, we limited our experiments to controlling these two motors only. To avoid having to worry about missed steps, we used optical encoders for position feedback on each lead screw. Figure 9 shows an architectural diagram of the experimental setup.

At the time of our tests, no vision or tactile sensing was available to use as the source of preview knowledge for the controller. Instead, we used a representation of the welding path stored in memory, much in the way traditional "blind" industrial robots are programmed. We could simulate the effects of using sensory evidence by superimposing a randomly generated error on the preview data derived from the stored path. This let us test the PIP path control system's response to noisy position measurements typical of those made by real tracking devices.

The problem of applying a feedback control scheme is largely one of choosing the gain coeffients that give the best performance, which, in this case, means getting the robot to accurately track the commanded path. Of course, minimizing the computations necessary to accomplish this in real time can also be a problem. We tried several different PIP control schemes, corresponding to different choices of the preview gain terms and their relationship to the other gains. For more details on the procedure, see [8].

The control equation that gave the best results, minimizing computation time without sacrificing good path control, was a "modified simple PIP" control scheme that used only one preview data point and ignored integral error effects older than a few cycles. The formula for this controller is:

$$V_c(k) = G_P E(k) + G_I \sum_{j=k-n_I}^{k} E(j) + G_{pr} P(k,n)$$

where n_I is the number of past samples used for integral correction and n is the number of steps ahead of the current position that the preview sample is taken from. Figure 10 shows the path information used in this scheme.

The first term is the standard proportional correction, the second is a modified integral action term. Integral control action has the effect of minimizing steady state error at the expense of increasing the transient settling time. When the integral term was computed from the start of the path (i.e., from $j=0$), the settling time was excessive. Using only the last n_I terms gave a good compromise between settling time and steady state error when n_I was about 10. The last term in the control calculation is the simple preview correction term. Unlike the earlier expression that summed the effects of several points, this rule uses the nth preview term and ignores the others in between.

The remaining factors to be determined are the three control gain coefficients and the number of steps ahead, n, that the preview term will look. Appropriate gains for servo systems can be derived analytically or by trial and error experimentation. Often a combination of the two methods is used. To simplify matters, we assumed that $G_P = G_{pr}$, and selected a trial value for these terms and the integral gain G_I. I'll talk more about the "tuning" procedure for adjusting these gains in Part II. For now, let's assume that we have already done this, and look at the performance of the controller for different values of n.

Performance of the Simple PIP Control Scheme

We made several test runs using the simple PIP control rule to drive the steppers on the x-y table. The same pre-

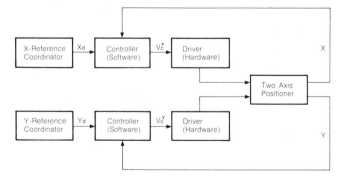

Figure 9. Architecture of the experimental setup used to test the PIP tracking controller.

programmed path was used in each case, consisting of a 10″ straight line followed by a 90° turn onto a full circle of 5″ radius, so that the straight line formed the diameter of the circle. The robot was supposed to follow this path at a constant velocity. Obviously, this would require an instantaneous velocity change at the 90° turn, so some tracking error was expected. Figure 11 shows three test runs, with the solid line indicating the desired path and the dotted line the path actually followed. In each case, the desired velocity was 2 in/sec. The operating frequency of the control loop was 10 Hz.

In case 1, we used a single step look-ahead ($n=1$ in the equation). Here the performance was similar to a conventional feedback controller, with the sharp turn resulting in a classic "step-response" damped oscillation. In case 2, the preview sample was taken three steps ahead ($n=3$). The actual path showed marked improvement, with only a little overshoot on the turn. When the preview was taken five steps ahead, the controller "saw" the turn five steps ahead, and actually rounded off the corner a little.

In welding, the velocity of the welding tip must be carefully controlled to get the best results. Figure 12 shows plots of the position and velocity errors over the course of the path for the three tests of Figure 11. Position error is

defined as the distance between the desired x-y position and the actual one. Velocity error is the magnitude of the vector difference between actual and desired table velocities, with the sign determined by comparing the actual (scalar) speed with the commanded speed. The first two seconds for all three cases were identical. This is due to the limited accelleration of the steppers to get the table up to the desired speed. The superiority of the PIP controller can be seen when the corner is encountered at 5 sec. into the path. The errors were excessive for Case 1 which, used little preview, but were drastically reduced in Case 3.

We also investigated the effects that changing the desired tracking speed had on the performance of the controller and of introducing noise into the preview measurements. Figure 13 shows the results of a number of test runs made with different values on n and at speeds of 1 and 2 ips. From these we saw that at low velocities, a little preview can reduce velocity error at the expense of increased position error. At higher velocities, preview improved performance significantly unless the controller looked too far ahead. A look-ahead of four steps seemed to be best for our system. Introducing a random noise of .1″ average error into the preview measurements increased the tracking error by much less than .1″.

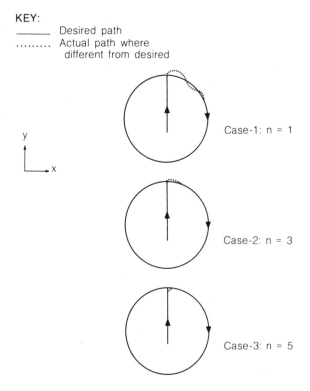

Figure 11. The results of three test runs using different values of n, the number of steps ahead that the preview sample is made.

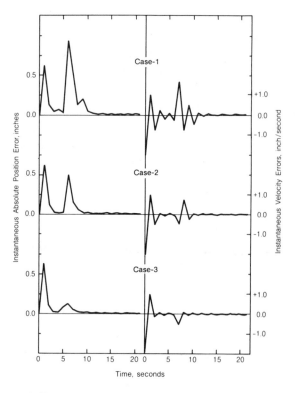

Figure 12. Position and velocity errors for the three test runs of Figure 11.

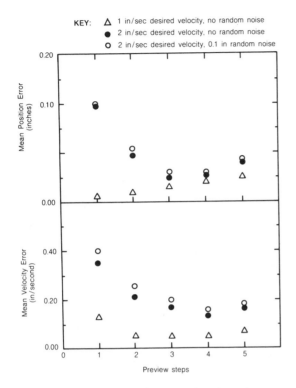

KEY: △ 1 in/sec desired velocity, no random noise
      ● 2 in/sec desired velocity, no random noise
      ○ 2 in/sec desired velocity, 0.1 in random noise

Figure 13. The effects of tracking speed and preview measurement noise on controller performance.

Conclusions

Because they are so easy to interface to computers, stepping motors are getting a good deal of attention for use as actuators for robots. When some sort of position/velocity feedback is available, a stepper can be used like a servo motor—without the analog power amplification that servo motors usually require.

The real advantage possible with steppers is the elimination of the requirement for getting direct feedback from the controlled mechanism. If the mechanical loading and inertia are well defined, proper attention to the limitations of the stepping motor can keep it from ever missing a step (provided that collisions are avoided), and position can be measured just by counting output steps. (The motor's physical characteristics need not be quantitatively known, just well-behaved, so that satisfactory constants can be obtained experimentally.)

Regardless of the architecture used, the PIP control scheme described here gives a stepper-driven robot the ability to track a continuous path accurately. A key advantage of the preview scheme is that it can make use of measurements not only from a pre-programmed path, but from path sensors such as vision, touch, or other seam-tracking devices.

In this article I hardly mentioned the actual hardware and software design of our robot and its control system. I will give a more complete description of this in Part II. Work on improving the performance of the PIP controller and the stepping motor control circuitry is still in progress at Berkeley's Mechanical Engineering Department.

STEPPING MOTOR DESIGN

Because of their construction, stepping motors feature easy interfacing and simplified control with digital circuitry. The key to stepping motor performance lies in the design of a rotor and stator combination that has regularly spaced equilibrium positions created by alternating magnetic poles. In its most common form, the rotor is constructed of a ceramic permanent magnet that has a fixed pattern of alternating north and south poles and a stator made from two sets of toothed soft-iron cups energized by separate windings. The teeth on the cups are folded so that when energized they form alternating poles that interact with the rotor fields, producing either movement or a holding torque.

The pole patterns created by the two windings are 90° out of phase. Therefore, by selectively switching the windings the motor will either step forward or in reverse. A simple logic circuit is all that's needed to convert pulses from a copmputer output port to appropriate stepper winding control signals. That circuit, as well as the power circuits to drive the windings of a typical stepper, are described in "A Homebuilt Computer-Controlled Lathe" which appears in this book.

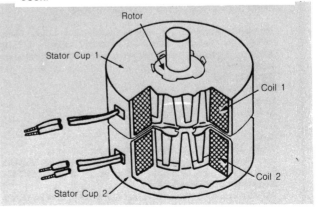

References

[1]   Porter, J., "Stepping Motors Move In," *Production Engineering*, vol. 34, February, 1963, pp 74-78.

[2]   Kieburtz, B. R., "The Step Motor—The Newest Advance in Control Systems," IEEE Trans. on Automatic Control, January, 1964.

[3]   Kuo, B. C., *Theory and Application of Step Motors*, West Publishing Co., 1974 Ed.

[4]   Snyder, W., "Microcomputer-Based Path Control," *this book*.

[5]   See discussion in the article "Robotics Research in Japan," by K. Takase, *Robotics Age*, Winter 1979-80.

[6]   Tomizuka, M., Whitney, D. E., "Optimal Discrete Finite Preview Problems (Why and How is Future Information Important?)," Trans. of ASME, *Journal of Dynamic Systems, Measurement and Control*, Vol. 97, No. 4, December, 1975.

[7]   Tomizuka, M., Dornfeld, D., Purcell, M., "Application of Microcomputers to Automatic Weld Quality Control," Trans. of ASME, *Journal of Dynamic Systems, Measurement and Control*, Vol. 102, No. 2, 1980.

[8]   Dornfeld, D., Tomizuka, M., Motiwalla, S., Tseng, R., "Preview Control for Welding Torch Tracking," *Proc. 1981 Joint Automatic Control Conference*, University of Virginia, June 1981.

[9]   Engelberger, E. F., *Robotics in Practice*, Amacom publication, 1980 edition.

CONTINUOUS PATH CONTROL WITH STEPPING MOTORS

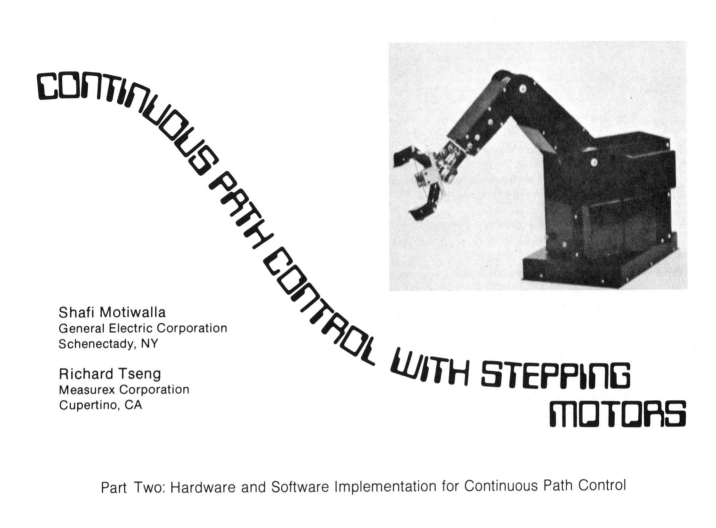

Shafi Motiwalla
General Electric Corporation
Schenectady, NY

Richard Tseng
Measurex Corporation
Cupertino, CA

Part Two: Hardware and Software Implementation for Continuous Path Control

In Part I, we showed how a robot arm driven by stepping motors can accurately track a continuous path in space, using the motors in their slewing mode. The slewing mode gives you the fastest stepping rates you can get from the motor, whereas in the "single-step" mode, the motor and its load must come to rest after each step. By limiting the motor's acceleration you can even drive the motor "open-loop" without missing steps. With the *Preview Path Control* scheme, the robot can use "look-ahead" evidence from either a path-tracking sensor or from a preprogrammed path stored in memory. This approach can give you performance comparable to that of feedback-controlled servo motor robots, but for far less effort and expense.

In this article, we will discuss some of the details involved in building a working system based on the Preview Control method. In particular, we will describe some of the circuitry and software techniques we used in a two-axis welding positioner controlled by a microcomputer. Rather than trying to document our entire system we

will concentrate only on some of the key design choices, including the sequencing logic, the mechanical design, the interface to the computer and a joystick controller, and the method of limiting acceleration.

The Sequencing Logic

The windings of a stepping motor must be pulsed in a particular sequence to cause the motor to move in a given direction. (See the discussion of stepping motor design in Part I.) Generating these pulses is the function of the sequencing logic, which can be implemented either in hardware or software. Our goal was to control the entire robot with a single micro. And given the burden of the path control computations, we decided to perform the pulse sequencing in hardware.

There are two basic types of pulse sequences that will cause the motor to step. In *wave drive*, only one phase of the stator winding is powered at a time (Figure la). In the

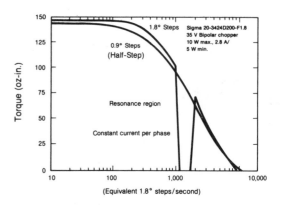

Figure 2. Torque output comparison between two-phase and half-step drive. With two-phase drive, torque drops when the step rate approaches the mechanical resonance frequency of the motor. Since half-stepping produces a more uniform drive sequence, the problem is avoided.

other sequence, called *two-phase drive*, two stator phases are energized simultaneously (Figure 1b). With two windings providing power, the latter step sequence will produce a greater torque output from the motor. Thus, it is the most commonly used of the two sequencing methods. Either of these sequences will cause an *N* degree/step motor to take a full step, although there is an angular offset of *N*/2 degrees between the actual positon of the steps on the motor's shaft.

This half step difference between the outputs produced by the two sequences is the key to a third method of sequencing that combines the two basic types. A drive that alternates between wave drive and two-phase drive will cause the motor to take steps of half the usual increment. Reference [2] gives working circuits for all three types of sequencing logic. Half-step drive doubles the position resolution of the motor and produces smoother slow-speed motions. Also, as shown in Figure 2, it has the useful feature of reducing the effects of mechanical resonance in the motor.

Two successive steps produced by a half-step driver have differing characteristics that can result in missed steps when the load is near the limit. This motivated us to build a sequencer that would let us select either two-phase or half-step drive, so that we could experiment with both types.

Power Drive Circuitry

The function of the power drive circuit for the stepping motor is twofold: it should provide a high voltage to produce a fast rise-time on the current pulse to the winding, and then it should drop the voltage to sustain the current at the correct level for the duration of the pulse. Drive requirements for steppers vary from a few volts at 50ma or so to over 100 volts at 20 amps. The shape of the current pulse affects the torque curve of the motor—a slow rise time will limit the torque available at higher speeds. The windings behave as an inductance in series with a restistance, with a time constant typically on the order of 10ms. Clearly, the higher the initial voltage, the faster the rise-time, yielding better high-speed performance (up to practical limits). It is not unusual for the starting voltage of the pulse to be 30 times the steady-state voltage.

There are several methods of power conditioning to produce the current pulse in the stator winding. The simplest and least expensive is to limit the steady-state current with a resistor, as shown in Figure 3a. Although this is the simplest and least expensive method, it fails by a wide margin to yield the maximum torque at high speeds, since the resistor increases the current rise-time. It also wastes considerable energy as heat in the resistors.

High-efficiency drives offer dramatic improvements in stepping motor performance—but at the expense of complexity and cost. A good example of such a drive is "chopper" current limiting, shown in Figure 3b. Full power supply voltage is applied at the start of a step, until the current builds up. Then, the control circuit starts switching the voltage applied across the winding off and on. The inductance of the winding filters the voltage pulses to an average DC level, producing the desired steady-state current.

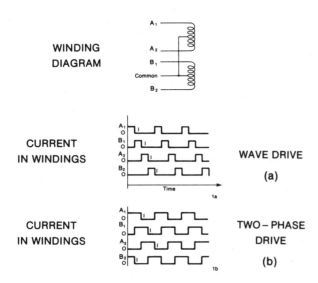

Figure 1. The basic sequences for pulsing the stator windings of a stepping motor. When the plot for a particular winding is high, the winding is energized.

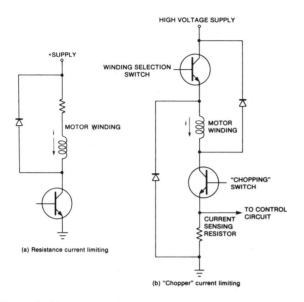

Figure 3. Two types of stepping motor power drive circuits. Resistance current limiting (a) is simplest, but wastes energy in the resistor. In a "chopper" driver (b), current is limited by pulse width modulation. (PWM).

Another high-efficiency scheme involves switching between two power supplies: one high enough to give a good rise time and a low voltage one to sustain the current. A fourth method uses a switched current source that will maintain the correct current almost regardless of the stepping rate.

The peak stepping rate we needed in our experiments was 1000Hz (full steps), which would produce a platform speed of 5 in/sec. At this rate, the torque loading of the motor was well within the capabilities of a resistance-limited drive scheme. Since our main goal was to experiment with the software for preview control, we decided to keep things simple and use that method, with a 24V power supply as the source.

Mechanical Design

We built the structure of the welding positioner out of surplus aluminum stock. It stands 6' high with a base of 8' × 5'. The total welding area is about 17 sq. feet. So far, we've only automated two of the machine's six axes. Each of the two axes runs on linear bearings mounted on case-hardened steel rails. We used surplus stepping motors with 200 steps/revolution and 200 oz.-in. torque. The motors position the welding carriage along each axis by turning a lead screw, which has ball-bearing couplings on

the carriage (a "ball-screw" design). A full step of the motor will move the carriage 5 mils (.005"), and half-stepping gives 2.5 mil resolution. Figure 4 shows a diagram of the welding positioner.

Even though it is possible to drive the steppers in slewing mode without position fedback, we decided that a closed-loop system would allow for far greater flexibility in our experiments. Having position feedback makes it easier to evaluate the performance of the controller, since the computer can detect missed steps by comparing the carriage location with the actual value. In the case of industrial processes such as arc welding, feedback may be essential for dealing with unpredictable variations in loading and with collisions. We elected to use a linear incremental optical encoder on each axis, running the entire length of travel. Although a linear encoder is more expensive than a rotary shaft encoder, it is not subject to the errors caused by "backlash" in the ball screw drive. The encoder provides standard quadrature pulse outputs that are translated into up/down position count signals by a logic circuit. (See "Using Optical Shaft Encoders" in this book.)

At the end of each axis we mounted limit switches that were wired into the stepping logic as failsafes. To provide for unusual failures, we also put mechanical dead stops at the extreme ends of each axis. To protect the steel rails, position encoders, and wiring from weld spatter, we attached spring-tensioned polyvinyl covers over each axis.

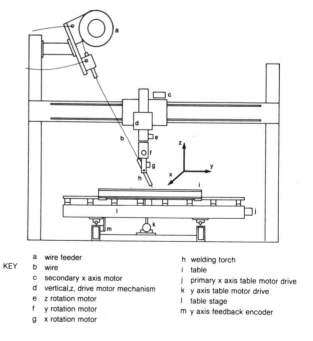

KEY			
	a	wire feeder	
	b	wire	
	c	secondary x axis motor	
	d	vertical,z, drive motor mechanism	
	e	z rotation motor	
	f	y rotation motor	
	g	x rotation motor	

h	welding torch
i	table
j	primary x axis table motor drive
k	y axis table motor drive
l	table stage
m	y axis feedback encoder

Figure 4. The six-axis arc welding positioner.

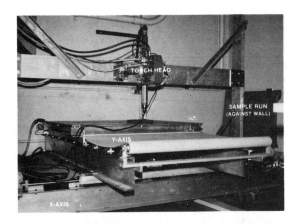

Figure 5. Photo of the welder. Vinyl covers are stretched over the carriage and kept taut by window-shade springs.

Although large molten drops can burn through, these covers offer adequate shielding for normal operation. Figure 5 shows a photograph of the welder.

Digital to Pulse-Train Conversion

In keeping with our decision not to burden the CPU with low-level logic functions that could be done in hardware, we decided upon using a logic circuit to convert from a digital pulse rate command to the pulse train that would drive the sequencing logic. We also wanted to have a manual joystick controller that would be able to drive the *xy* table at varying speeds.

There are two convenient ways of generating a pulse train of variable frequency. One is entirely digital, using a standard TTL rate multiplier chip. The 74LS97 will scale down a reference clock in proportion to the 6 bit binary number on its input pins. You can cascade several of them to form an *n* bit rate multiplier. The *TTL Data Book* by Texas Instruments shows how. To use this approach in our controller, we could have a built a clock with a frequency equal to our maximum stepping rate and then run the latched output lines from a parallel interface directly to the multiplier. Its output could drive the sequencing logic directly.

We used a different method, however—one that made adding the joystick controller quite simple. Instead of scaling the pulse rate digitally, we used a digital-to-analog (DAC) converter to produce a voltage proportional to our 8 bit pulse rate command and then used that voltage as the input to a voltage controlled oscillator (VCO). Although this approach lacks the inherent accuracy of a purely digital circuit, the error in the output pulse frequency was

at most a few percent—entirely adequate for our experiments. Figure 6 shows the circuit that converts the analog voltage from the DAC or joystick to the stepping pulses. A DAC input of 0FF(hex) produces a 2000Hz pulse train. The addition of a simple switching circuit (Figure 7) let us select between manual or computer control of the platform.

Electrical noise can be a severe problem for circuitry anywhere near an arc welder. In an earlier arc welding project, lines from the optical encoder picked up radio frequency interference (RFI) from the welder that showed up as extra pulses. This threw off the controller's position

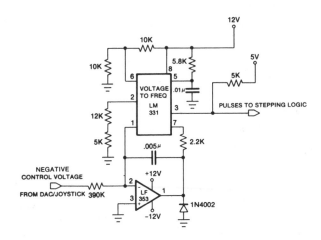

Figure 6. Circuit for generating a pulse train with frequency proportional to an analog voltage.

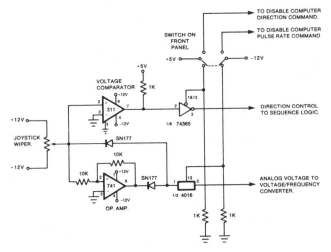

Figure 7. Circuit for switching between joystick controller and computer. The 741 op-amp inverts the joystick voltage and the two diodes insure that only the negated absolute value of the joystick voltage is applied to the 4016 CMOS analog switch.

feedback measurements and caused poor performance. Output lines from the interface can be affected too. We used optical isolators and shielding on all I/O lines on the interface. Figure 8 shows the Hewlett-Packard HP2630 opto-isolator circuit, which has a throughput near 1MHz.

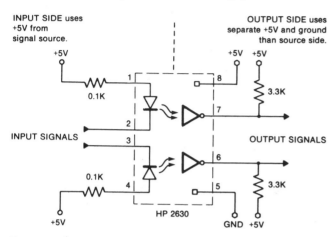

Figure 8. Optical isolation circuit used on interface lines.

The Control Software

The computer we used to control the system was an LSI-11 system by Digital Equipment Corp. With only FORTRAN, BASIC, and assembly language available as the source language for the control software, we picked FORTRAN. Interpreted BASIC would have been too slow to perform the preview control calculations in real time, and assembly language would have taken too much coding and debugging. DEC FORTRAN has "Ipeek" and "Ipoke" commands for direct access to the I/O ports, as well as provisions for handling interrupts for real-time control, so we could use it for the entire system.

Part I gave a description of the Preview Path Control scheme we used to calculate the pulse rate commands to be sent to the stepping motor control logic. As described in Part I, the basic operations in this procedure are as follows. For each axis: (1) Determine the component of position error along the axis. We used a representation of the desired path stored in memory and the position feedback from the encoders. (2.) Compute the "feed-forward" term from the preview information. Here we used the stored path again, but measurements from a seam tracking sensor would be used if available. (3) Update the integral error term. (4) Multiply each term by its "gain" and sum the results to get the "ideal" stepping rate command for the axis. (5) Modify the command if necessary and replace the old rate command (and direction, if necessary) in the output latches for the axis.

These operations are repeated at a fixed rate, determined by the *cycle time* of the control loop. The LSI-11 has a 60Hz clock interrupt, which the program "divides down" by counting interrupts. We normally used a divisor of 6—every 6 interrupts, the program would repeat the control calculations. Interrupts may occur and be counted during the calculations, which must, of course, be completed before the 6th interrupt occurs. This cycle time is the key to setting the gain terms used in the calculations. The "steady-state" gain of the system must be high enough so that the controller could correct some maximum position error within one cycle. Initial choices for the various gains can be derived analytically (see [1]). In practice, however, these terms are adjusted to get the best performance.

Step 5 of the cycle, modifying the ideal rate command, requires knowing the torque and load characteristics of the system. Comparing the new rate with the old gives the acceleration during the next interval. If, given the inertia and friction of the load, this acceleration would require a torque greater than the maximum available from the motor at the current stepping rate, the controller must reduce the new rate appropriately. We experimentally determined all the factors necessary to perform this acceleration limiting—the computation reduces to little more than a few simple conditional assignment statements in the real-time loop. With our closed-loop system, we could afford to make some approximations. An open-loop system would need a fairly detailed model to get the most from the motors and still be sure that they won't miss steps.

With the exception of some floating point calculations to simulate the effects of random noise from a seam-tracking sensor (see part I), we did all the calculations in integer arithmetic for maximum speed. With the random noise, the controller required 14ms for all the computations in the cycle. Without it, it needed only 11.4ms. Thus, the control calculations required only 11.4% of the LSI-11's compute power, based on the 100ms control cycle time we normally used. The rest of the time could be spent processing signals from a seam-tracking sensor or controlling the other axes of the robot when they are added.

Performance of the System

Over 85% of the time we devoted to this project was spent on designing, building, and debugging the hardware. Developing the control algorithm took about 5%, and the rest went towards writing and debugging the software.

Much of the hardware development time was consumed dealing with RF noise problems from the arc welder. The first time we turned on the welder, it destroyed a flip-flop in the motor sequencing logic. The long lines from the limit switches, which were wired into the logic as fail-safes, apparently acted as antennas, picking up RFI. We added opto-isolators to these lines and shielded them, like the lines between the computer interface and the control logic. The welder also caused large transients in the AC power lines. We added AC line filters, shielded the cable from the welding power supply to the torch, and isolated the welder's ground. All this cleared up most of the problems. Adding additional filter capacitors to the sequencing logic helped.

If we were designing the hardware over again, we would definitely use one of the stepping motor control ICs now commercially available. Sigma Instruments of Braintree, Mass. has a chip that will perform most of the sequencing logic functions. A more expensive chip by Cybernetic Micro Systems of Los Altos, California functions as a complete motor control computer. You can program it for half- or full-stepping, set the stepping rate and acceleration numerically, and perform numerous other functions. It will control the drive circuits directly.

As we showed in the run-time plots in Part I, the Preview Control method proved quite satisfactory for continuous-path control of stepping motors. We are still working to improve the system, and we plan to add four more stepper-driven axes to the welding positioner by the end of the year. We are also continuing our research in the area of adaptively controlling the welding parameters (arc power, speed, and weaving), in addition to the welding path, based on signals from arc sensors and from advanced seam-tracking sensors.　　　ⓡ

References

[1]　Shafi Motiwalla, "Design, Development and Implementation of a Preview Control Scheme for On-Line Seam Tracking for Welding Automation," MS Thesis, Dept. of Mechanical Eng., U. California, Berkeley, Dec. 5, 1980.

[2]　B.C. Kuo, *Theory and Application of Step Motors*, West Publishing Co., 1974.

Fast Trig Functions for Robot Control

Carl F. Ruoff
Robotics Research Program
NASA Jet Propulsion Laboratory
California Institute of Technology
Pasadena, CA

How do you tell your robot where to go? If you're controlling a robot manipulator directly—with a joystick, for example—you can watch the result of your commands and use your own visual feedback to get the arm on target. If the robot is recording the target points you teach it as a programmed movement sequence, it must have a way to represent each target so that it can return to it later. If the robot has position feedback sensors in the arm, (encoders, pots, etc.) it can simply read the position of each joint or linkage and store these values. (See Figure 1). Later it can move the arm to the identical configuration by matching the recorded values. As long as neither the target point nor the robot has moved, the arm will be at the target.

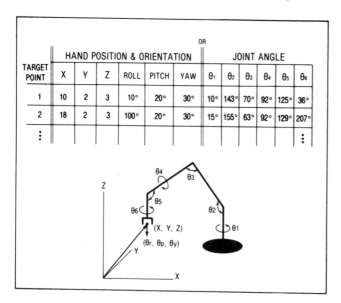

TARGET POINT	HAND POSITION & ORIENTATION						OR JOINT ANGLE					
	X	Y	Z	ROLL	PITCH	YAW	θ_1	θ_2	θ_3	θ_4	θ_5	θ_6
1	10	2	3	10°	20°	30°	10°	143°	70°	92°	125°	36°
2	18	2	3	100°	20°	30°	15°	155°	63°	92°	129°	207°
⋮												⋮

Figure 1. Two alternative ways to represent target points for a robot manipulator. On the left, the targets are defined by their coordinates in an external frame. On the right is their representation in terms of the joint angles of the arm. If the machine has any linear joints, replace the appropriate "θ_i" by an "r_i."

Another way of identifying a target to the robot is to give its location in some external coordinate system. This can be especially useful if the target's location is measured by vision or some other sensory system. It is also an integral feature of many high-level robot programming languages.

For the robot to use such a description, however, it must be able to *transform* a location given in external coordinates into the manipulator joint angles that correspond to that target. (I'll use the terms *joint angles* or *joint displacements* interchangeably to refer to the complete set of joint angles and/or linear displacements that uniquely characterize a given arm configuration.) Similarly, the robot may need to know, given a set of joint angles, where the arm is in that external frame.

The equations defining these relationships are referred to as the *kinematic solution* of a robot arm. A robot's kinematics describe its geometrical relationship with its environment. Since many of these relationships may involve angular rotations, both of the robot's joints and of the axes of the external frame, the transformations naturally require the evaluation of trig functions. A real-time robot control program needs a fast, accurate method of computing trig functions, especially if the transformations are included in the servo loops. In this article, I'll describe an efficient approach to trig computations that uses a combination of table lookup and linear interpolation.

Consider a robot manipulator with the cylindrical geometry shown in Figure 2. The location of the robot's wrist can be determined by measuring z, r, and θ, directly. Additional angles describe the orientation of the hand about that point, as shown in the figure. To find the location of the wrist in the Cartesian frame at the base of the robot, you must use the transformation:

$$x = r \cos \theta, \ y = r \sin \theta, \ z = z$$

Conversely, if you want to find the joint displacements that correspond to a given wrist location in that frame, you must use the *reverse* kinematic solution:

$$r = \sqrt{x^2 + y^2}, \quad \theta = \arctan(y,x), \quad z = z.$$

The arctangent includes both x and y as arguments so that the appropriate quadrant is determined. The square root looks time consuming, but if you find θ first, then you can make use of the fact that $r = x\cos\theta + y\sin\theta$ to eliminate it. (I obtained this relation for r by rotating the x and y axes through the angle θ.) Since we can easily compute both the tangent and cotangent from x and y, we can factor this relation for r to make the calculation easier:

$$r = \cos\theta \, (x + y \tan\theta) = \cos\theta \, (x + \frac{y^2}{x}), \quad x \neq 0.$$

$$r = \sin\theta \, (x \cot\theta + y) = \sin\theta \, (\frac{x^2}{y} + y), \quad y \neq 0.$$

The fastest way to evaluate trig functions is simply to use the input argument (angular measure) as an index which points directly to the correct value. Unfortunately, if the angular resolution is high, this table can grow extremely large. If you measure angles with 16 bits, for example, the table must have 65,536 separate values! An alternative approach used by most mainframe computers is to evaluate polynomials which correspond to truncated power series. This approach is compact because it uses no lookup tables; but obtaining sufficient accuracy requires evaluating several terms of the series, each with two floating point multiples and an add. On small computer, this may be time consuming, and error propagation can become a problem.

The approach I will describe offers a reasonable tradeoff between accuracy, memory requirements, and speed. The method is so effective that it is used by researchers and industry in some of the most sophisticated robot systems.

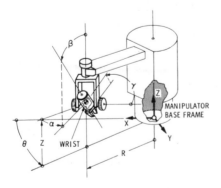

Figure 2. A robot with cylindrical geometry. The robot controls the r, θ, and z axes directly, as well as the hand angles. It must use kinematic transformations to relate measurements of r, θ, and z to the wrist position in the xyz frame at the base of the robot.

Linear Interpolation

The basic idea of linear interpolation is to approximate the function over an interval by passing a straight line segment between the known values of the function at the endpoints of the interval, as shown in Figure 3. If we know $f(x_1)$ and $f(x_2)$, and x lies between x_1 and x_2, then:

$$f(x) \approx f^*(x) = f(x_1) + (x - x_1) \frac{f(x_2) - f(x_1)}{x_2 - x_1}$$

To get a complete description of the function over a larger interval or to increase accuracy, we can approximate the function by a series of such segments, as shown in Figure 4. We can simplify the evaluation of this by storing for each interval the *divided differences*:

$$D(x_i) = \frac{f(x_{i+1}) - f(x_i)}{x_{i+1} - x_i}$$

Using these values in addition to the stored values of the function, we can compute the approximated function as:

$$f^*(x) = f(x_i) + (x - x_i) \, D(x_i),$$

for x in the interval between x_i and x_{i+1}. The calculation has only one add and one multiply, so it can be performed very rapidly.

This approach works nicely for functions which are "well behaved" (finite, continuous, etc.) in the interval of interest. Trigonometric functions, except for tangent near 90° and cotangent near zero, fall into this category. By using trig identities we can handle the exceptions as well.

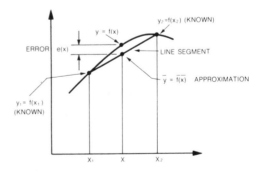

Figure 3. Linear Interpolation. The function f(x) is approximated over the interval from x₁ to x₂ by assuming that the approximate value lies on the line segment connecting (x₁,y₁) to (x₂,y₂). This results in an error e(x) which can be made as small as needed by choosing x₁ and x₂ sufficiently close together.

Reducing the Error

Purists may be concerned that the results of the interpolation are approximate. When you consider that the position feedback has only finite accuracy and that the

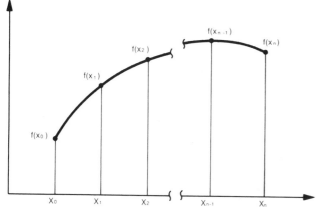

Figure 4. A function can be accurately approximated over a large range by linking a series of interpolation intervals. The first step is finding the interval i from x_i to x_{i+1} that contains x and then applying the linear interpolation procedure.

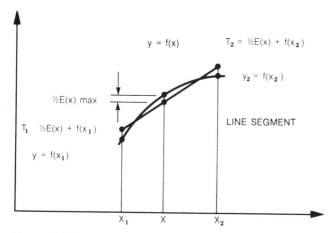

Figure 5. The maximum error in an interval can be cut in half simply by adjusting the table values T_1 and T_2 upward by one-half the maximum error. This distributes the error over the region.

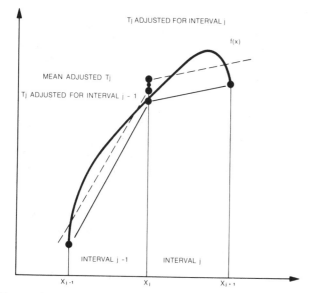

Figure 6. The mismatch caused by adjusting table values can be eliminated by setting the table value to the average of the adjusted values computed for the adjacent intervals.

control computer has only a finite word length, it's easy to show that these worries are unfounded. It is possible to select reasonable sizes for the interpolation intervals that will preserve all the accuracy available. It is pointless to maintain precision that cannot be used.

There are two sources of error in the approximation: that due to the rounding of arithmetic results and that due to the interpolation. The biggest contribution to rounding errors is caused when the divided difference $D(x_i)$ is multiplied by the interpolation distance $x-x_i$. If we assume for the moment that we're using fixed-point arithmetic with rounded n-bit fractions, each table value will have an error bound of $\pm 2^{-n-1}$. The total error due to rounding will be approximately $\pm 2^{-n-1} [2 + (x - x_i)]$.

The interpolation error $f(x) - f^*(x)$ depends on the shape of the function over the particular interval. Assuming that the function is monotonic in the interval (this is true for the functions and intervals we are considering), we can find the point on the (fixed) interval with the maximum interpolation error by differentiating the expression for the error with respect to x. The point with the most error is where the slope of the actual function is equal to the slope of the approximating segment. This is a useful result that we can use to reduce the interpolation error.

It is also important to find the interval with the greatest maximum interpolation error in the interpolation region. Noting that the maximum error in an interval is a function of the interval location, we see that this can be accomplished by fixing the interval length and differentiating the expression for interpolation error with respect to the interval location. Equating the derivative to zero and expanding the resulting expression in a power series around the interval location, we find that an interval has extremal maximum interpolation error when it is located where the *third* derivative of the function is zero. This fact is useful later in choosing the (common) width of the interpolation intervals. In the nterval that contains the point where the third derivative is zero, we can express the maximum interpolation error as a function of the width of the interval and then find the width that results in an acceptable bound for the error. The error in all other intervals will be smaller.

The maximum interpolation error in an interval can be cut in half (at the expense of increasing the error at the endpoints of the interval) simply by raising the line segment by half the maximum error, as shown in Figure 5. This spreads the error over the interval, but it creates a mismatch between table values on adjacent intervals that could result in discontinuities in the approximation. We can handle this by adjusting a table entry by half the *average* of the error maximums of the adjacent intervals (see Figure 6), so that

$$T_j = f(x_j) + \tfrac{1}{4} [e_j(max) + e_{j-1}(max)].$$

This won't work on every interval, however, since limit conditions like $\sin(\theta)=0$ have special significance for many kinematic calculations. On such intervals, just adjust *one* of the endpoints.

Making the Calculations Fast

We've seen how the trig computations can be reduced to a table lookup, a subtraction, a multiplication, and an addition. We can speed up the calculation even more by a clever way of representing angles in the computer, one that proves much better than either degrees or radians.

The first step is to use a unit of angular measure such that the total number of units in a circle is an integral power of two. This is easy if the robot uses optical encoders for position feedback, since the number of pulses or states per revolution is usually a power of two. Similarly, we choose interpolation intervals that partition a quadrant of the circle into a number of intervals also equal to (a lesser) power of two. This yields several simplifications. The quadrant number, the number of the interval within the quadrant, and the interpolation displacement $(x=x_i)$ can all be directly masked out of the angular argument by logical operations. This amounts to using *turns* and *binary fractions* of a turn to represent angles, where one turn equals 360 degrees or 2π radians.

If turns are stored as two's-complement fixed point numbers with the integer representing the number of full turns and the fractions representing a partial turn, the fractional part has the familiar properties of angular displacements—a positive ¾ turn and a negative ¼ turn, for example, have identical representations, as do a full turn and a zero turn.

In this representation, the two most significant bits (MSBs) of the fractional part give the quadrant the angle lies in, and the next several bits identify which interpolation interval the angle lies in within that quadrant. The exact number of bits will depend upon the number of intervals in the table. The remaining least significant bits (LSBs) give the *absolute* displacement of the angular argument within the interpolation interval. (See Figure 7.) All this follows from making the number of interpolation intervals in a quadrant equal to a power of two. Since the trig functions are periodic, you only have to store lookup tables for one quadrant instead of four, and then use trig identities to evaluate the function in the other three quadrants.

Another feature of this approach is that it is easy to turn the absolute displacement of the angular argument (the LSBs of the fractional turn) into the *fractional displacement* of the angle within the interval, $(x-x_i)/(x_{i+1}-x_i)$, simply by a shift operation. All that's necessary is to shift the displacement field so that its MSB moves to the MSB of the fixed-point fraction (that is, the bit corresponding to ½). The resulting fraction equals the fractional displacement. This lets you store in the lookup table just the (adjusted) difference of the function across the interval instead of the divided difference $D(x_i)$ mentioned earlier.

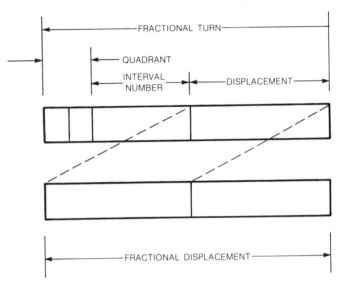

MOST SIGNIFICANT BIT = 2^{-1}

Figure 7. The quadrant, interval number, and the displacement of the argument within the interval can be extracted directly from the "turn" representation of an angle, provided that the number of interpolation intervals in the quadrant equals an integral power of two. The displacement of the argument can be converted to a fractional displacement simply by a shift operation.

Evaluating the Trig Functions

Sine(x)

To evaluate the sine function using a table that covers the first quadrant, use the following identities: (remember that the units are *turns*, so that ¼ corresponds to 90°)
$$\sin(-x) = -\sin x, \text{ and}$$
$$\sin(\tfrac{1}{2} - x) = \sin x$$

To find sine values in quadrants 3 and 4, negate the argument, calculate the sine in quadrant 1 or 2, and negate the result. Use the second relation to find sine values in the second quadrant—subtract the argument from ½ and use the result as the argument to evaluate the sine in quadrant 1. Figure 8 shows these symmetry relations.

The sine function will have the greatest interpolation error when x equals ¼ or ¾ (90° and 270°). We can estimate this error as a function of the interval width, d, by supposing that the interval is centered about $x=¼$, as shown in Figure 9. Remembering that the table values can be adjusted to cut the maximum error in half, we can say that:
$$2e = 1 - \sin(¼ - \tfrac{d}{2}) = 1 - \cos(\tfrac{d}{2}). \quad , \text{ or}$$
$$d = 2\arccos(1 - 2e)$$

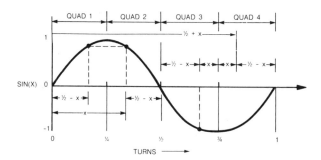

Figure 8. The sine curve. The curve in quadrant 2 is the reflection of that for quadrant 1 about the line $x = \frac{1}{4}$ turn.

If we want the interpolation error to be about the same as the error in representing the number $\pm 2^{-n-1}$, as described earlier, choose an interval width of:

$$d = 2\,arc\,cos\,(1 - 2^{n})$$

To find N, the number of intervals in a quadrant, calculate:

$$N = \frac{1}{4d}$$

which is unlikely to be an integral power of two. Just choose the nearest power of two that gives sufficient accuracy—use the expression for e above to check it. Also, remember that to find the total error in the function evaluation you must add the rounding error as well.

Cosine(x)

Since $cos(x)=sin(\frac{1}{4}+x)$, (arguments in turns, again) you can get cosine values by adding $\frac{1}{4}$ to the argument and calling the sine routine.

Tangent(x) and Cotangent(x)

Just use the defining relations:

$$tan\,x = sin\,x/cos\,x \quad and \quad cot\,x = cos\,x/sin\,x.$$

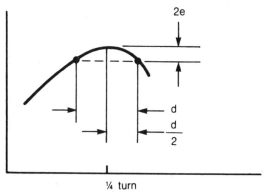

Figure 9. Estimating the maximum interpolation error sine.

Arctangent(y,x)

Our arctangent function needs two arguments (an input vector), so that it can tell which quadrant the input vector is in by examining the signs of x and y:

$$y \geq 0,\ x \geq 0: \text{quadrant 1}$$
$$y \geq 0,\ x < 0: \text{quadrant 2}$$
$$y < 0,\ x < 0: \text{quadrant 3}$$
$$y < 0,\ x \geq 0: \text{quadrant 4.}$$

Save the quadrant number for later, then set the sign of each to plus to place the angle in the first quadrant. Next, find the actual tangent of the angle to use for the function evaluation. Since tangent blows up whenever x is too small, we can first find the cotangent and then use the relation $cot\theta = tan\,(\frac{1}{4} - \theta)$.

A convenient way to program this is to choose the relation with the largest denominator: If $x < y$ compute:

$$Q = y/x = tan\theta.$$

Then find $\theta = arctan(Q)$ by interpolation from the lookup table. If $x = y$, then we know right away that $\theta=\frac{1}{8}$, $(45°)$ and we don't need any lookup or interpolation. If $y > x$ then compute:

$$Q = x/y = cot\theta.$$

Then arctan $tan\,(\frac{1}{4} - \theta)] = arctan\,(cot\,\theta) = arctan\,Q$, so that $\theta = \frac{1}{4} - arctan\,Q$.

Now we need to put θ back into the correct quadrant. Using the quadrant number saved earlier, compute the output angle θ to return as the result of the function:

$$\text{quadrant 1:} = \theta$$
$$\text{quadrant 2:} = \frac{1}{2}-\theta$$
$$\text{quadrant 3:} = \frac{1}{2}+\theta$$
$$\text{quadrant 4:} = -\theta.$$

Since Q is always less than one, the interpolation argument is just a fixed-point fraction (that's why it's convenient to exclude the case $x = y$). The interpolation region corresponds to the angles from zero to one count less than $\frac{1}{8}$ turn.

We can perform an error analysis for the arctangent interpolation similar to what we used to find the interval size for the sine table. Arctangent has the greatest interpolation error for $Q = 1/\sqrt{3} = .57735$, which corresponds to $1/12$ turn or $30°$. By centering the interval d around this point and solving for d as a function of e like before, we get: $d=\frac{8}{3}\sqrt{2|e|\sqrt{3}}$

Just as we did for the sine function, we must divide the interpolation region ($\theta \leq Q < 1$) into N intervals, where N is an integral power of two. Once you've picked an interval width, you can solve the above equation for e in terms of d and use it to verify your choice. Like before, the total error includes this interpolation error, plus the rounding error, and also, for arctangent, the error introduced by the division used to compute the tangent or cotangent.

Using the Routines: An Example

Listing 1 shows a coding of the sine, cosine, and arctangent subroutines written in the MACRO-11 assembly language for the DEC PDP-11 series of computers. This code assumes that the calling PDP program represents angles in the two word (32 bit) format described in the comment statements. However, since the PDP-11/34 we used has floating point hardware, we decided to use real numbers for the sine and cosine results and for the arguments to arctan. This was much more convenient than using fixed-point arithmetic throughout the program. The routines can be called either by an assembly language program or by PASCAL programs compiled by the OMSI pascal compiler. It shouldn't be difficult to adapt this code to any other computer or compiler.

The routines assume that the main program has loaded the function value and difference tables into memory. The subroutine TABLEREAD, at the end of the listing, shows an example of how to read a binary data file containing the tables to load them into their proper locations. For both the sine and arctangent tables, we used 64 interpolation intervals. The sine table, Table 1, lists the adjusted values of sine in the first quadrant in steps of 1/256 of a turn, together with the corresponding differences. The arctangent table, Table 2, shows the adjusted arctangent values and their differences.

Conclusions

The methods I've described here are extremely useful for evaluating trig functions, and can be applied with little difficulty on any type of computer. You can also adapt this method of linear interpolation to other functions beside trig. I've used this method for trig calculations in two robotics projects with very successful results. In one case, using an IS-1000 minicomputer with 16 bit fixed-point arithmetic, we were able to calculate 15 bit sines and cosines in just 35 microseconds without even using difference tables. This let us control *two* robot arms and still have compute power left over for vision functions and extensive geometric calculations. Using these algorithms on your microcomputer will help make it possible for your robot to do fast coordinate transforms and attain higher levels of intelligent behavior. ®

Acknowledgement: I wish to thank John Wedel of JPL for his work in programming these algorithms.

Listing 1. The sine, cosine, and arctangent routines programmed for the PDP-11/34, using fixed point angle representation and floating point for the interpolation tables. Bit 15 is the MSB of a PDP-11 word.

```
        .TITLE PCMATE
        .ENABL LC
;       Interpolation subroutines for trig functions
;       and table reader to read in the tables from file DK:SINTAB.DAT
;
;       latest version 22-OCT-80
;
AC0=%0          ;definition of the floating point registers
AC1=%1
AC2=%2
AC3=%3
AC4=%4
;
        macro to push and pop registers on and off the stack
        .MACRO PUSH   A
        MOV     A,-(SP)
        .ENDM
        .MACRO POP    A
        MOV     (SP)+,A
        .ENDM
        .PSECT TRIGRT
;
;  The two word angle format is a 32 bit twos-complement number.  Therefore
;  the sin and cos routine ignore the first word which is number of complete
;  turns.  The second word is a fraction of a turn as follows:
;    Bits 15-14 indicate quadrant.  This is a value from 0 to 3 for the
;    quadrants 1 through 4.  Bits 13 through 8 are the table index
;    having a range of 0 to 63.  In octal format, 1 part in this field
;    has a value of 400 and the value ranges from 0 to 37400.  Bits 7
;    through 0 are the part used for interpolation between table values.
;  The sine and cosine output and the arctan input are single precision
;  floating point numbers in DEC format.  This routine is callable from
;  either a macro main program or a PASCAL main program and is in format
;  compatible with the calling sequence used by OMSI (Oregon Software, Inc.)
;  PASCAL.
;
;  PASCAL callable as if declared:
;          TYPE ANGLE = ARRAY[1..2] OF INTEGER;
;          function PSIN(arg:ANGLE):real;
;          function PCOS(arg:ANGLE):real;
;          function PARCTN(sin,cos:real):real;
;  Note that parctn returns a thing declared real which is really an angle
;  in the two word angle format.  The main program declarations will need
;  a type union (variant record) to accomodate this.
;  The Macro callable interface uses different names: TSIN, TCOS, TARCTN.
;  There are no FP registers saved, others are.  Call for TSIN and TCOS:
;          Mov ^angle, R0        (^angle means a pointer to the angle)
;          Mov ^result, R1
;          JSR PC,TSIN or TCOS
;  Call for TARCTN:
;          Mov ^sin, R0
;          Mov ^cos, R1
;          Mov ^result, R2
;          JSR PC,TARCTN
;
        .PAGE
;
;  COSINE ROUTINE    (COS(X) = SIN(90+X) with X MOD 360)
        .GLOBL  TCOS,PCOS
        .GLOBL  $B74,$B76
; $B74 and $B76 are routines in the PASCAL runtime system which save the
; registers and other context variables.
        .ENABL LSB
PCOS:   JSR     R0,$B74
        MOV     36(6),R0         ;move the argument to R0
        MOV     #-1,-(SP)        ;indicate that this was a PASCAL call
        BR      1$
TCOS:   PUSH    R0               ;save some registers
        PUSH    R2
        MOV     2(R0),R0         ;get the argument from the pointer
        CLR     -(SP)            ;indicate that this was a MACRO call
1$:     ADD     #40000,R0        ;add 90 DEGREES to argument
        BR      FMCOS            ;and jump to sine routine
        .DSABL LSB
;
;    SINE ROUTINE
        .GLOBL  TSIN,PSIN
PSIN:   JSR     R0,$B74
        MOV     36(6),R0         ;get the argument (word 2 only used)
```

```
        MOV     #-1,-(SP)       ;a PASCAL call
        BR      FMCOS
TSIN:   PUSH    R0              ;save registers
        PUSH    R2
        MOV     2(R0),R0        ;get argument
        CLR     -(SP)           ;a MACRO call
FMCOS:  MOV     R0,R2           ;SAVE ANGLE
        CLRB    R0              ;GET TABLE LOOKUP PART
        ASL     R0
        ASL     R0              ;GET RID OF QUADRANT INFO
        BCC     1$              ;IF QUADRANT IS NOT 2 OR 3
        NEG     R0
        BEQ     2$
        NEGB    R2              ;change sign of interpolation part
        BEQ     1$
5$:     SUB     #2000,R0        ;index table from other end if interp part
                                ;is non zero
1$:     SWAB    R0
        LDF     TBL(R0),AC1     ;get table entry into floating pt acc
        LDF     DTBL(R0),AC2    ;get difference table entry
        MOV     R2,R0
        BIC     #177400,R0
        LDCIF   R0,AC3          ;interp part
        MULF    AC2,AC3         ;mult by difference table
        ADDF    AC3,AC1         ;add in table entry
4$:     TST     R2              ;is this quadrant 3 or 4?
        BPL     3$              ;NO
        NEGF    AC1             ;change sign
3$:     TST     (SP)+           ;was this a PASCAL or a MACRO call
        BPL     6$
        STF     AC1,40(SP)      ;store argument allowing for stuff $B76 has
        JSR     R0,$B76         ;put on the stack
        MOV     (SP),4(SP)      ;restore calling context
        ADD     #4,SP           ;fix the return address and the stack
        RTS     PC              ;the end
6$:     STF     AC1,(R1)        ;MACRO return sequence
        POP     R2
        POP     R0
        RTS     PC              ;the end
2$:     NEGB    R2
        BNE     5$
        LDF     #^040200,AC1    ;this is a DEC floating 1.0
        BR      4$
;
;       ARCTAN OF A/B
;       The table gives the sign of the arguments and tangent for each quadrant
;       QUADRANT DATA: (QUAD:SIN/COS,TAN) Q1:++,+ Q2:+-,- Q3:--,+ Q4:-+,-
;       For Q3, Q4 we return a negative angle.
;
        .GLOBL  TARCTN,PARCTN
PARCTN: JSR     R0,$B74         ;save things according to PASCAL
        LDF     40(6),AC0       ;get arg A
        LDF     34(6),AC1       ;get arg B
        MOV     #-1,-(SP)       ;indicate PASCAL call
        BR      CMTN            ;to common routine
TARCTN: LDF     (R0),AC0        ;load arg immediately to floating acc
        LDF     (R1),AC1        ;as above
        PUSH    R0              ;save registers
        PUSH    R1
        PUSH    R3
        CLR     -(SP)           ;indicate MACRO call
CMTN:   CLR     R3              ;quadrant info stored here
        CLR     R1              ;used to indicate which 45 deg octant
        MOV     #40000,-(SP)    ;hope this really saves time, it saves space.
;       test for which quadrant we are in and for zero arguments.
        TSTF    AC0
        CFCC                    ;put floating condition codes into PSW
        BLT     1$              ;ARG A IS NEGATIVE
        BGT     2$              ;ARG A IS POSITIVE
        TSTF    AC1             ;ARG A is zero
        CFCC
        BPL     10$             ;ARG B IS POSITIVE
73$:    MOV     #100000,R3      ;ANGLE IS 180 OR -180 (gives -180 for result)
10$:    TST     (SP)+           ;GET RID OF #40000
        TST     (SP)+           ;PASCAL/MACRO SWITCH
        BEQ     12$             ;go to MACRO exit
        CLR     44(SP)          ;SET + ANGLE
        TST     R3              ;Q3,4 CAUSE NEGATIVE ANGLE
        BGE     11$
        COM     44(SP)
11$:    MOV     R3,46(SP)       ;STORE RESULT WD 2
        JSR     R0,$B76
        MOV     (SP),10(SP)     ;RETURN ADDRESS
        ADD     #10,SP
        RTS     PC              ;This is one exit from routine
;
12$:    CLR     (R2)
        TST     R3
        BGE     13$
        COM     (R2)
13$:    MOV     R3,2(R2)
        POP     R3
        POP     R1
        POP     R0
        RTS     PC              ;This is other exit from routine
;       if you get beyond here arg A was not zero
1$:     NEGF    AC0             ;ABS OF A
        MOV     #100000,R3      ;THIS IS Q3 OR Q4
2$:     TSTF    AC1             ;now check ARG B
        CFCC
        BLT     3$              ;B IS NEGATIVE
```

```
        BGT     4$              ;B IS POSITIVE
        BIS     (SP),R3         ;ANGLE IS 90 OR 270 (B=0)
        BR      10$             ;set angle and exit
3$:     NEGF    AC1             ;ABS OF B
        TST     R3              ;Q2 OR Q3?
        BMI     41$             ;Q3
        BIS     (SP),R3         ;Q2
        BR      41$
4$:     TST     R3
        BPL     41$
        BIS     (SP),R3
41$:    CMPF    AC0,AC1         ;which octant (use table as TAN or CTN?)
        CFCC
        BGT     5$              ;WE ARE IN OCTANT 2 MOD 90 DEGREES
        BLT     6$
        BIS     #20000,R3       ;(ARG A = ARG B) 45 DEG. (or odd multiple)
        BR      10$             ;set result and exit
5$:     DEC     R1              ;INDICATE OCTANT 2
        DIVF    AC0,AC1
        STF     AC1,AC0         ;GET RESULT INTO AC0 FOR NEXT OPERATION
        BR      7$
6$:     DIVF    AC1,AC0         ;RESULT NOW IN AC0 FOR NEXT OPERATION
7$:     MODF    8$,AC0          ;INITIAL AC0 IS LESS THAN 1
        STCFI   AC1,R0          ;AC0 OUGHT TO BE <64 OR THERE IS AN ERROR
        ASL     R0              ;convert to index for table
        ASL     R0
        LDF     TNTBL(R0),AC2   ;GET TABLE PART
        LDF     DTNTBL(R0),AC1  ;and difference table part
        MULF    AC1,AC0
        ADDF    AC2,AC0         ;sum table and interp value
        STCFI   AC0,R0          ;convert to integer for result angle
        BIT     (SP),R3         ;IN Q2 OR Q4 ?
        BNE     71$             ;NO
        TST     R1              ;(check octant) CTN ?
        BEQ     9$              ;NO - TAN IN Q1 Q3
        BR      72$
71$:    TST     R1              ;CTN ?
        BNE     9$              ;YES- CTN IN Q2 ,Q4
72$:    TST     R0              ;IS THE ANGLE TOO SMALL?
;       the interpolation may lead to a zero angle. In some cases this angle
;       is in a different quadrant than the one indicated by the stored
;       quadrant information which is valid for non-zero angles.
        BNE     91$             ;NO
        BIT     (SP),R3         ;AT 0 OR 180?
        BEQ     91$             ;NO
        BIT     #100000,R3      ;DO WE WANT 180 DEGREES
        BEQ     73$             ;YES
        CLR     R3              ;NO, CLEAR EXTRANEOUS BITS
        BR      10$             ;R0 AND R3 CLEAR
91$:    NEG     R0
        ADD     (SP),R0         ;R0 AND R3 CLEAR
9$:     BIS     R0,R3
        BR      10$
;
8$:     .FLT2   64              ;a floating point 64
;
;       TABLE LOADER PROCEDURE
;       This procedure with no arguments is callable by TABLRE (or TABLREAD)
;       in PASCAL. Dont use it with MACRO - its sets up its own stack which will
;       probably overlap the normal MACRO stack and lose badly
;       It can be used with either v1.1 or v1.2
;       Destroys R0,R1
        .GLOBL  RTAREA          ;external in the PASCAL runtimes
        .GLOBL  TABLRE          ;internal in this routine
        .MCALL  .LOOKUP,.READW,.CLOSE,.PRINT,.EXIT,.WAIT
;
TABLRE: MOV     #5,R1           ;Only looks at first 6 channels
1$:     .WAIT   R1              ;is this RT-11 channel available?
        BCS     2$              ;we got one
        DEC     R1
        BPL     1$
3$:     .PRINT  #MSG1           ;all errors print same message
        .EXIT                   ;and abort the program
;       Dont reserve channel since we close it before exiting routine
2$:     MOV     SP,SPSV         ;save stack pointer
;       the pascal stack is at top of core and the USR will swap this part out
        MOV     #1000,SP
        .LOOKUP #RTAREA,R1,#FBLK
        BCS     3$              ;error in finding file
        .READW  #RTAREA,R1,#DTBL,#256.,#0
        BCS     3$              ;error in reading file
        .READW  #RTAREA,R1,#DTNTBL,#256.,#1
        BCS     3$
;       we should also check number of words read but dont do so
        .CLOSE  R1              ;release channel
        MOV     SPSV,SP         ;restore PASCAL stack pointer
        RTS     PC
        .NLIST  BEX
SPSV:   .WORD   0
FBLK:   .RAD50/DK SINTABDAT/    ;name of file to read for the data
MSG1:   .ASCIZ/? error reading SINTAB.DAT/
        .EVEN
;
; KEEP FOLLOWING SIZES AND ORDER THE SAME. Data is stored here from the file
        .GLOBL  DTBL
DTBL:   .BLKW   128.
TBL:    .BLKW   128.
DTNTBL: .BLKW   128.
TNTBL:  .BLKW   128.
;
        .PSECT
        .END
```

Table 1. Sine values and differences for the first quadrant, with an interval of size of 1/64 of a quadrant.

SINE AND SINE DIFFERENCE TABLE

Index	Sine	Difference	Index	Sine	Difference
1	0.0000000	0.0000959	33	0.7071338	0.0000670
2	0.0245426	0.0000958	34	0.7242748	0.0000653
3	0.0490700	0.0000957	35	0.7409793	0.0000635
4	0.0735678	0.0000955	36	0.7572377	0.0000617
5	0.0980213	0.0000953	37	0.7730399	0.0000599
6	0.1224158	0.0000950	38	0.7883765	0.0000581
7	0.1467365	0.0000947	39	0.8032380	0.0000562
8	0.1709688	0.0000943	40	0.8176159	0.0000542
9	0.1950981	0.0000938	41	0.8315012	0.0000523
10	0.2191099	0.0000933	42	0.8448856	0.0000503
11	0.2429898	0.0000927	43	0.8577612	0.0000483
12	0.2667233	0.0000921	44	0.8701200	0.0000462
13	0.2902960	0.0000914	45	0.8819547	0.0000442
14	0.3136940	0.0000907	46	0.8932582	0.0000421
15	0.3369030	0.0000899	47	0.9040235	0.0000399
16	0.3599090	0.0000890	48	0.9142445	0.0000378
17	0.3826983	0.0000881	49	0.9239145	0.0000356
18	0.4052570	0.0000872	50	0.9330282	0.0000334
19	0.4275716	0.0000862	51	0.9415798	0.0000312
20	0.4496287	0.0000851	52	0.9495642	0.0000290
21	0.4714149	0.0000840	53	0.9569765	0.0000267
22	0.4929172	0.0000828	54	0.9638125	0.0000244
23	0.5141225	0.0000816	55	0.9700679	0.0000222
24	0.5350182	0.0000804	56	0.9757390	0.0000199
25	0.5555916	0.0000791	57	0.9808223	0.0000175
26	0.5758303	0.0000777	58	0.9853149	0.0000152
27	0.5957221	0.0000763	59	0.9892139	0.0000129
28	0.6152552	0.0000749	60	0.9925170	0.0000106
29	0.6344176	0.0000734	61	0.9952223	0.0000082
30	0.6531978	0.0000718	62	0.9973281	0.0000059
31	0.6715846	0.0000702	63	0.9988331	0.0000035
32	0.6895669	0.0000686	64	0.9997365	0.0000010

Table 2. Values of arctangent and its differences over 64 equal intervals in the range 0-1. The first two columns of turns and differences are given as a normalized fixed-point octal fraction, and the second two columns are the actual decimal fraction values.

ARCTAN AND ARCTAN DIFFERENCE TABLE

INDEX	NORMALIZED (OCTAL) TURNS	DIFF	ACTUAL (DECIMAL) TURNS	DIFF
1	0	242	0.00000	0.00249
2	242	242	0.00249	0.00249
3	505	242	0.00497	0.00248
4	750	242	0.00746	0.00248
5	1213	242	0.00993	0.00247
6	1455	241	0.01241	0.00247
7	1717	241	0.01488	0.00246
8	2160	240	0.01734	0.00245
9	2421	240	0.01979	0.00244
10	2661	237	0.02224	0.00243
11	3120	236	0.02467	0.00242
12	3357	235	0.02709	0.00241
13	3615	234	0.02950	0.00240
14	4052	234	0.03190	0.00238
15	4306	233	0.03428	0.00237
16	4541	231	0.03664	0.00235
17	4773	230	0.03899	0.00233
18	5224	227	0.04132	0.00231
19	5453	226	0.04364	0.00230
20	5702	225	0.04593	0.00228
21	6127	223	0.04821	0.00226
22	6353	222	0.05046	0.00223
23	6575	221	0.05270	0.00221
24	7016	217	0.05491	0.00219
25	7236	216	0.05710	0.00217
26	7454	214	0.05927	0.00215
27	7671	213	0.06142	0.00212
28	10104	211	0.06354	0.00210
29	10315	210	0.06564	0.00208
30	10525	206	0.06772	0.00205
31	10734	204	0.06977	0.00203
32	11141	203	0.07179	0.00200
33	11344	201	0.07379	0.00198
34	11545	177	0.07577	0.00195
35	11745	176	0.07772	0.00193
36	12143	174	0.07965	0.00190
37	12340	172	0.08155	0.00188
38	12533	171	0.08343	0.00185
39	12724	167	0.08528	0.00183
40	13114	166	0.08711	0.00180
41	13302	164	0.08891	0.00178
42	13466	162	0.09068	0.00175
43	13651	161	0.09243	0.00173
44	14032	157	0.09416	0.00170
45	14212	155	0.09586	0.00168
46	14370	154	0.09754	0.00165
47	14544	152	0.09919	0.00163
48	14717	151	0.10082	0.00160
49	15070	147	0.10242	0.00158
50	15237	145	0.10400	0.00156
51	15405	144	0.10555	0.00153
52	15552	142	0.10709	0.00151
53	15715	141	0.10860	0.00149
54	16056	137	0.11008	0.00146
55	16216	136	0.11155	0.00144
56	16354	135	0.11299	0.00142
57	16511	133	0.11441	0.00140
58	16645	132	0.11581	0.00138
59	16777	130	0.11718	0.00135
60	17130	127	0.11854	0.00133
61	17257	126	0.11987	0.00131
62	17405	124	0.12118	0.00129
63	17532	123	0.12248	0.00127
64	17656	121	0.12375	0.00125

CNV=020000/(PI/4)) IS: 10430.3779297
CNV2=0.125/020000 IS: 0.00001526

SECTION 2

INTERACTIONS:
SENSES, VISION, AND VOICE

A robotic system is a computer system closely bound to the real world. Its interaction with the real world is evidence of this binding. A key part of its interaction, especially when humans are present, are the human-like attributes: tactile and kinesthetic senses, vision, hearing, and voice. We've grouped a series of past articles on these interactions in this section.

We start the exploration of interactions with Alan M. Thompson's "Introduction to Robot Vision." Following the introduction, we expand our vision theme in an application area through an article by Arnold G. Reinhold and Gordon VanderBrug entitled "Robot Vision for Industry."

After the general introductory articles, we turn to technical details of the problem of vision. An object or scene made of objects exists in the real world of three dimensions. Alan M. Thompson's "Camera Geometry for Robot Vision" provides us with specific optical engineering background on how to describe the mapping of the real world scene to the focal plane of some lens system. Now that we have an image on a focal plane, Raymond Eskenazi provides "Video Signal Input" so our attention can be focused on the problems of obtaining a digitized version of that image.

Then we come to the problems of figuring out what the digitized images mean. Robert Cunningham discusses "Segmenting Binary Images" as one approach to the analysis problem. Another approach is to somehow transform the image from a photographic representation into a line drawing—the process, edge detection, is nontrivial. Ellen C. Hildreth discusses the subject in "Edge Detection in Man and Machine." Then, once we've found our edges, we have to represent the scene in memory. Joel M. Wilf treats a method called "Chain Code" as one way of representing the line image form of a scene.

Now, the vision of optics and video input is a fundamentally expensive way to view the world with today's technology. In order to experiment with the concept of image digitization, you don't necessarily have to spend megabucks. In fact a few dollars can go a long way as shown by Don McAllister in his article "Build a Low-Cost Image Digitizer." Don continues on the subject of low-cost experimental sense inputs with his second article, "Multiple Sensors for a Low-Cost Robot." The ultimate minimum of vision input is the single-bit, single-pixel sensor. Martin B. Winston discusses this form of vision in "Opto 'Whiskers' for a Mobile Robot."

Finally, to conclude the section on interactions, we turn to the output side of the problem with Tim Gargaliano and Kathryn Fons' article "Teach Your Robot to Speak." The use of voice synthesizers is the proven method of low-cost interactive computer output. It is used in the Heathkit HERO-I robot as well as numerous experimenters' prototype robots.

The problem of 'understanding' sensory input is one of the central themes of robotics. This is the first in a recurring series on Robot Vision, written by professional researchers in the field.

Introduction to Robot Vision

Alan M. Thompson
NASA Jet Propulsion Laboratory
California Institute of Technology
Pasadena, CA

Few people stop to consider the complexities involved in converting the physical inputs to their senses into conscious recognition of the objects or phenomena causing them. A burst of acoustical energy distributed appropriately in frequency and duration over the audible spectrum can be changed by the action of ear and mind into the perception of the word, "Hello!," said by your best friend, seemingly without any delay after it is spoken. Similarly, a two-dimensional pattern of reflected light, with appropriate variation in brightness and color, is easily recognized as your friend's face within a fraction of a second of your first glance.

The volume of processing required to distinguish each of these particular patterns from all the other possible combinations of sounds and objects that you know has so far been greater than the resources of our largest and fastest computers, programmed by our best researchers. This may be surprising to those who have the impression that all that's required to make a computer "see" is to "plug in" a TV camera. Just the volume of "raw" data contained in a standard video signal is greater than the input capacity of all but the larger minicomputers, and is enough to easily carry several *thousand* telephone conversations simultaneously (thus easily proving the old

Reprinted from Robotics Age, *Volume 1, Number 1, Summer 1979*

Figure 1. Half-tone method of reproducing photographs (exaggerated).

proverb about the worth of one picture). To make the information contained in this signal accessible to a program requires suitable sampling hardware, and that is only the prerequisite for the complex process of recognition.

Humans have the ability to recognize familiar objects from almost any angle, over a broad range of distances and lighting. Most of the physiological and psychological processes that underly this ability are unknown, and, since most of recognition occurs beneath the level of consciousness, the phenomena can only be studied indirectly. With little evidence available from living systems, vision researchers in the field of artificial intelligence must develop theories that attempt to explain recognition in a way that can be duplicated in a computer. This involves deciding what knowledge is needed about the structure and appearance of objects, how it can be represented as a computer data structure, and designing procedures for using the knowledge to achieve recognition.

Numerous useful systems have been developed by restricting the recognition task to distinguishing between a limited number of objects in a carefully controlled environment. In these special cases techniques can be developed that take advantage of the simplification of the problem by being limited to the analysis of a few well chosen characteristics. These methods, some of which will be described later, have found successful applications in military and industrial systems, and offer researchers important clues to the structure of more effective general theories.

Video Image Input

To understand the process of setting a picture into a computer it is useful to refer to the "half-tone" technique used by printers for reproducing photographs. Photos are printed as a matrix of tiny dots; the size of the dot determines the brightness of the picture at that point. The photo in Figure 1 shows this process exaggerated, so that you may have to move back to recognize the subject. The size of a dot (black or white) determines the percentage of white area contained in that cell of the matrix. This percentage gives a measure of the brightness of that cell, allowing a digital representation of the image as the brightness number (referred to as the grey level) of each picture cell (referred to as a pixel) stored in a matrix that corresponds to the original picture.

Normal video, however, is an analog signal, consisting of

a series of horizontal "raster" scans of the image from top to bottom. Each line of the scan begins with a synchronization pulse of negative polarity, followed by a positive waveform that corresponds to the brightness of the image as a function of time as the scan moves across the image. (Color information is multiplexed into a higher band, and then decoded at the receiver.) Detection of the next horizontal synch pulse signals the end of the line and causes the scan to jump to the beginning of the next line down. (Figure 2.) When the scan reaches the bottom of the image, a vertical synchronization pulse occurs, and the scan returns to the top for the start of a new "frame". To convert this signal to a pixel matrix (for a black and white picture) a clock circuit is synchronized to a multiple of the horizontal scan frequency. The number of pulses of the "pixel clock" per scan line is the number of pixels per line in the digitized image. At each pulse, the video signal is sampled and converted to binary, and the results are

deposited into the computer's memory (see Figure 3). If a sufficiently slow scanning rate is used, a normal I/O channel may be used to read the converted grey levels, but for standard broadcast video rates, the use of a dual-port video memory is essential on all but the fastest computers. To avoid distortion in the converted image, the video scan should be linear, that is, the horizontal position of the scan is proportional to the elapsed time since the horizontal synch and the vertical position is proportional to the number of horizontal lines since the vertical synch. In the new solid-state imaging arrays using CCD or CID LSI technology, each pixel position is fixed in the silicon matrix, so linearity is guaranteed. On some models, however, the horizontal pixel spacing is not the same as the vertical, resulting in a linear distortion in the converted image that can easily be accomodated by the image processing program. Also, the drive circuitry for solid-state cameras produces a suitable pixel clock, simplifying the sampling circuit.

Once the image is represented in memory as a two-dimensional array of grey-levels, the task of enhancement and analysis can begin. Image enhancement usually refers to the processing performed to improve the quality of a picture for subsequent use by humans, (as done in the unmanned space program at JPL to process spacecraft images of planets) but may also be useful as the first step of automatic image analysis. Examples of such procedures are contrast stretch and normalization, in which a transformation is computed that maps the original range of grey levels into one that more fully utilizes the available range of the grey scale. In most automatic image recognition systems, however, a transformation of the entire image is usually not necessary, since the program can use operators that are insensitive to the actual range of brightness.

Figure 2.   A standard analog video signal. Each line of the raster scan is represented by a signal whose voltage is proportional to the brightness along the line.

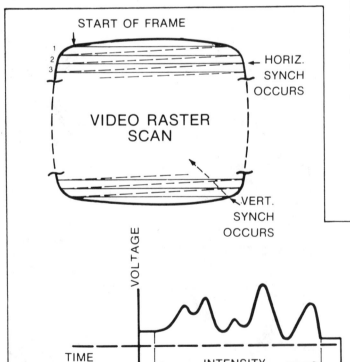

"Low-Level Processing", or Finding Primitive Features

The first step of image analysis is to locate areas in the image that correspond to whole objects or parts thereof. Ideally, there should be an easy way of isolating the image of each object from the background, (as in the case of a white egg sitting on black velvet) so that recognition techniques can be applied to just the object of interest. Unfortunately, in most environments this is not always possible. A trick used in some industrial applications is to paint the background a different color than the objects to be identified, but most research systems that operate in less constrained environments rely on methods that recognize objects without first having to determine their complete outlines. Two of the most common tools used to find objects are edge-detection and clustering. Both techniques attempt to locate the boundaries of objects or regions in the image so that their location, size, and shape may be computed as clues for recognition.

Edge detection is based on the fact that there is often a difference in brightness between object and background. Areas of high contrast can be found in the digitized image by looking through the pixel array for jumps in the grey level. If color information is available, this task is simplified because there could be a color change that might not be as noticeable based on intensity alone. Edges are located in the array by applying an operator that examines a small neighborhood of adjacent pixels and computes an "edge probability" value. If the value is above a set threshold, a notation is made in some data structure (possibly the pixel array itself) signifying the presence of an edge at the appropriate spot in the image.

The size of the neighborhood or "window" examined by the edge detector determines its sensitivity to noise. The grey level of any one pixel is subject to numerous noise sources that may result in considerable error. The accumulation of photons by the sampling element, whether vidicon tube or solid state array, is a statistical process subject to a normal error distribution based on intensity and sampling time. Signal transmission and the A/D conversion process all contribute electromagnetic noise. The presence of a noise 'spike' in the pixel array can lead to the detection of false edges. To help avoid false edges, adjacent pixels can be averaged by the edge detection function to reduce the effect of the noise. An edge in the image may be oriented at any angle in the neighborhood "window" around the point to be tested. A detector that looks for edges only horizontally or vertically loses much of the information in the picture. There is always a tradeoff between the reliability of an edge detector and its speed. Averaging over a larger window is slower and more accurate, and is capable of detecting fuzzy edges that a smaller window might miss. Detecting lines or spots requires special operators that measure the difference in brightness between a central region and its surroundings. A survey of popular edge detection techniques can be found in [1].

A typical edge detection function is the Sobel gradient operator. Given that the image is represented as an array, let the 3x3 window centered around one pixel be represented by:

A	B	C
D	E	F
G	H	I

Figure 3. Conversion from analog video to digital grey levels.

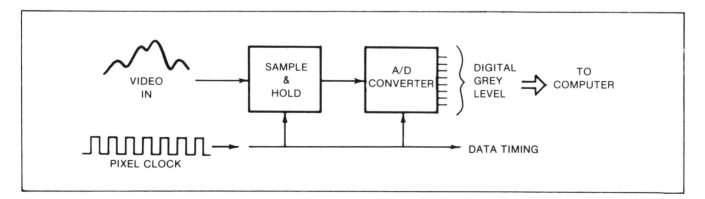

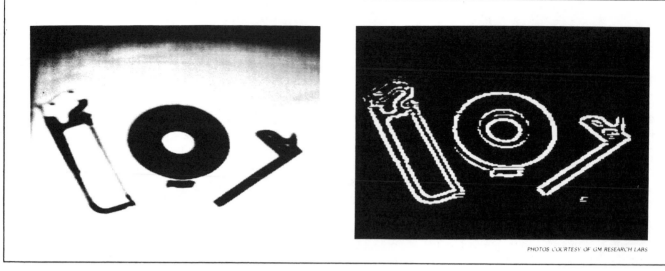

PHOTOS COURTESY OF GM RESEARCH LABS

Figure 4. Edge detection. The video image (left) is processed by an edge detection operator to produce an "edge picture". (right) Processing may be done by software or, as shown here, by digital hardware operating at full video rates.

Then the 'edge value' at pixel E is computed by the formula:

$$edge(E) = \left\{ \begin{array}{l} [\ (A+2B+C) \cdot (G+2H+I)\ ]^2 \\ +\ [\ (A+2D+G) \cdot (C+2F+I)\ ]^2 \end{array} \right\}$$

where, inside the square root, the letters stand for the grey levels of the corresponding pixels. Edge values above an experimentally determined threshold value are assumed to indicate the presence of the edge of some region in the picture such as an object or shadow.

This edge detection function has the advantage of being relatively fast, although the small sampling window (3x3) makes it more sensitive to noise and less sensitive to fuzzy edges. It is easily implemented in software and can be simplified further by using the approximation:

$$\sqrt{(a^2 + b^2)} \approx |a| + |b|$$

Summing the absolute values of the alternate differences avoids both the multiplications and the square root, and, because of its simplicity, can be implemented in a fast digital hardware module as part of a robot vision system. The JPL Robotics Project has such a hardware implementation, capable of producing an "edge picture" in "real time" at standard video rate, allowing the robot to track several moving objects simultaneously. [2] The basic operation of the edge detection function is shown in Figure 4. The output of the detector is displayed as the "edge picture" of the simple image shown. Note that even in this high-contrast image, there are still some gaps in the edges produced by the detector. A more complex edge detection function would produce a less fragmented outline at the expense of greater computation, but it is generally accepted that advanced object recognition programs should be capable of functioning even with incomplete, fragmented outlines of the objects to be recognized.

"Clustering", also known as "region-growing", refers to a process of collecting clusters of adjacent pixels that have similar properties. The decision whether or not to include a pixel in the region is made by a "discrimination function" which computes the desired property and compares it with a threshold value or with some value computed from the other pixels in the region. The simplest property of a pixel is its grey level, and many industrial robot vision systems locate objects merely by looking for regions whose grey level is sufficiently above (or below) that of a known background (conveyer belt, etc.) and produce a "binary" image in which each pixel is reduced to one bit, on or off according to the result of the threshold test. More complex clustering schemes use functions comparable to edge detectors that compute the property from a neighborhood, and, starting from a "seed" pixel in the interior of a region, expand the region to its boundary. Region growing has the advantage that the boundary of the region is always closed, so that it may be used to compute properties derived from its shape, such as various integral/statistical moments, etc. Noise in the image may result in inaccurate boundaries rather than in fragmented edges.

Image Recognition Methods

For many robot vision systems, the interpretation or classification of the processed outline is not necessary, since only the location and/or orientation of objects is needed (for grasping coordinates, etc.) If, however,

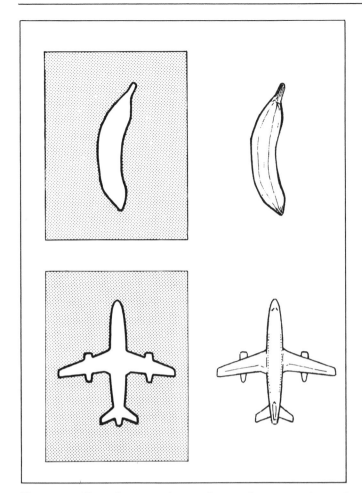

Figure 5. Template matching refers to the comparison of stored shapes with the observed image.

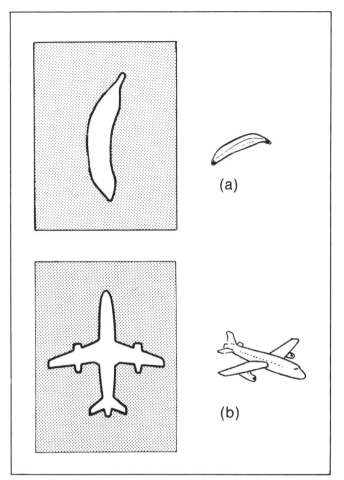

Figure 6. Problems with template matching: (a) Rotation and scaling, (b) perspective rotation.

recognition of objects is needed to achieve more "intelligent" behavior of the robot, then after objects in the field of view have been detected by one or more of the primitive operations described above, the resulting evidence is input to a cognitive process which (hopefully) determines the most likely interpretation of the image. The geometric structure of the edges, properties of the regions, and relationships between regions may be measured and compared to internal "models" of known objects. Models may take many forms, ranging from stored images, to statistical properties computed from the edges or regions, to abstract relations between the components of structures, or to various combinations of such methods. The design of a representation depends on the environment in which the system is to function, the number of objects in the database, and an analysis of their distinguishing features.

One simple recognition method that can work well in a restricted environment is "Template-Matching". This method works by comparing the image of an object (or its edge picture) with stored images (or outlines) of known objects. The stored "template" with the highest statistical correlation with the observed image is picked as the result. (Fig. 5) Clearly, this comparison is subject to many

constraints. Assuming for a moment that it is possible to isolate a complete object from the background in the image, the image of that object has to be "normalized" to a form suitable for comparing to the templates. If the object can appear anywhere in the scene, then a "window" around the object must be selected that matches the template size. If the object could be rotated in the image plane to a different orientation than that of its template, the image in the window may have to be rotated to the "normalized" orientation of the template, or the template transformed into the image. Since rotation of the image would be computationally expensive, requiring that the program perform a 2-dimensional coordinate transform for each pixel in the window, most systems that require such transformation do so by transforming a vector description of the outlines. Worse still, if the viewing area is too large, the image may have distortions due to the perspective transformation of the camera that reduce its similarity to the template. Worst of all, if the object could be rotated arbitrarily in three dimensions it might have a completely different appearance, rendering any single stored outline completely useless without complex, time-consuming 3-D surface rotational transforms. (See Figure 6.)

In circumstances where objects need not be recognized

from an arbitrary perspective and in which the field of view can be suitably restricted, template-matching and related methods can perform quite well. Since the template may tell the program where in the image to look for edges, it is possible for the program to expend more effort in looking for edges where they are expected. Alternatively, if edge fragments fall on or near the outline defined by a template, a reliable match may be obtained even in cases where the edge data are imperfect due to noise, poor contrast, or the limited ability of the edge detection operator. Templates can also be used to match outlines of different sizes by

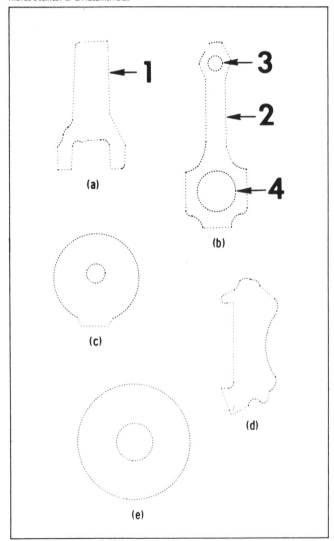

Figure 7. Five template models of automobile parts: (a) universal yoke, (b) connecting rod, (c) compressor body, (d) bracket, (e) gear blank.

including a scale transform in the comparison procedure, but this can add considerable computational overhead if the scale change is arbitrary.

Examples of applications meeting the requirements for successful use of template-matching include recognizing printed characters, the identification of navigation checkpoints by cruise missiles, and, significantly, recognizing industrial parts on conveyer belts. In the last case, it would seem that template-matching would have trouble, since parts could be placed on the belt in different orientations, i.e., with a different side up than the one expected in the template. The rotation is usually not arbitrary, however, because most parts have a limited number of stable resting positions on the belt. To apply template-matching it is necessary to have a template for each possible stable position. If the orientation of the part around a vertical axis is not controlled, it will still be necessary to perform the comparisons with the templates either by a 2-D rotation of the template or by basing the comparison on rotationally invariant properties of the templates.

To illustrate these techniques, let's examine a system for parts recognition developed at the General Motors Research Laboratories [3]. In this system the templates are represented as sets of connected curves (straight lines and circular arcs) called "concurves". Each set of concurves forms a two-dimensional model of the outline of some known object in a stable resting position. The outer boundary of an object can be represented by a single closed concurve, and each interior hole in the object requires a concurve to describe it. Figure 7 shows five models; the numbers indicate individual concurves in a model. Each concurve is represented as a set of orientation-independent properties such as length, number of arcs, total angular change of all arcs, etc. Orientation-dependent data is stored separately as a set of vectors perpendicular to the concurves at regularly spaced intervals along them.

To recognize parts in a scene the system analyzes the results of edge detection to form a set of concurves that most accurately fits the observed edge data (Figure 8). Since transforming the orientation-dependent model data into image coordinates is computationally expensive, the system first compares the more abstract concurve properties to compute estimates of the "likelihood" of a match before the detailed transformations are made. Each concurve in the image is compared to each concurve in each model to compute the average likelihood of finding each kind of object in the scene. This requires N (number of concurves in scene) times M (total number of different

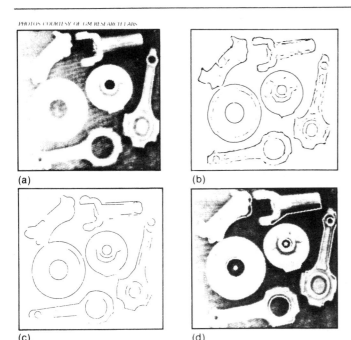

(a)   (b)

(c)   (d)

Figure 8. Recognition of parts on a conveyor belt: (a) Digitized scene, (b) Results of edge detection, (c) Concurves derived from edges, (d) Recognized models superimposed on the image.

concurves in the model database) comparisons, which can be large. However, this can be done fairly quickly since only the scalar properties are used. The vector description is transformed into the image only to verify the location of the most likely object, and to mark edges in the image as "used-up" so that they won't have to be considered in subsequent comparisons.

Because the actual templates are not used in the initial comparisons, this system is more accurately described as a form of "model-matching" based on the abstract properties of the concurves, but its restriction to two-dimensional patterns of a known size emphasizes its dependence on the templates. Also, since the template is used to verify a match and to mark edge data as "consumed", the system is able to work well even in the presence of visual noise (imperfect, fragmented edge data) and in cases of partial occlusion of the objects, as shown in Figure 9. This capability illustrates the power of template-related methods to deal with noisy images, since the template can tell the program what to look for and where to look for it. Systems such as this one could have wide application in manufacturing, both because of their ability to work well in the restricted environment and for the ease with which new objects may be added to its recognition repertoire. An example of a new object can be "shown" to the system under favorable conditions (high contrast, isolated object) and the program can compute and store both the template and its related properties. The speed of such systems (about 30 sec. for the complex scene in Figure 9) can be significantly increased by the use of special-purpose digital vision hardware such as that

described above, making them suitable for the rapid interaction with robot manipulators required for assembly-line applications.

Recognition of Objects in a Three-Dimensional World

The GM recognition system illustrates how orientation-independent properties associated with the system's model of an object can be used to greatly reduce the need for template rotations when matching against edge data in the scene. Without such techniques, the matching process would be much more time-consuming. In 3-D environments, the additional transformations of scale, rotation, and perspective must be considered, and we immediately realize that the complexity added by each new degree of freedom makes the literal comparison of the image of some object with stored object models even more impractical. To compute the projection of a 3-D model of an "archetypical" object into an image is almost too costly to be used for anything other than verifying the system's most likely "guess" about what it is seeing.

Even the possibility of storing "literal" models of the surface structure of objects becomes impractical if the number of objects is large. The surface models used in computer graphics systems to describe the shape of curved objects sufficiently to compute a projection from

Figure 9. A scene with occluded parts: (a) Digitized scene, (b) Derived concurves, (c) Recognized models superimposed.

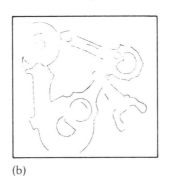

(a)

(b)   (c)

any perspective require thousands of bytes of storage. Since we do not know the capacity of human visual memory or the mechanisms behind its functioning, we cannot rule out the possibility that such literal surface models are somehow actually "stored", but merely the magnitude of the volume of information required makes the prospect unlikely. What seems more probable is that some form of encoding is used that compresses the unique characteristics of a particular kind of object into a highly condensed form.

Clues to the nature of the possible abstractions are suggested by examining drawings of familiar objects made by small children before they acquire the knowledge of perspective representation. These drawings often have the characteristic of showing structural relationships between the components of complex objects with little emphasis on the details of the components. (See Figure 10). The classic example is the stick figure representation of the human body. The fact that the child may be able to draw a more detailed representation of some individual part, such as an arm or a foot, suggests to some researchers that there may be a form of hierarchy in the understanding process. This may take the form of a pyramidal structure of definitions in which complex objects are described by successive layers of increasing detail, or in which recognition is accomplished by processes operating at

Figure 10. Drawings by children often show structural relationships without attention to actual appearances. The figure on the right is supposed to be a cube, and captures the significant feature of the squareness of the cube's faces.

different levels of detail. In each case, the top levels would be most concerned with the connections between the major components of the object and the lower levels with descriptions of primitive shapes. The ability of the human brain to "represent" and remember such structural knowledge would be intimately related to its aptitude for spatial relationships in general.

To translate this neuro-physiological theory of vision into a computer program, a researcher must decide what kind of knowledge is needed to accomplish recognition and how it can be represented and used effectively. For systems in which recognition procedures operate on descriptions of the objects, this may involve specifying a vocabulary and a grammar for expressing structural relationships within a particular level of detail and also for describing the connections between levels. In the GM recognition system, the top level of the description of an object contains the derived abstract numeric properties of the templates, with the description of its actual shape left to the lower level. In models of three-dimensional objects, the abstract structural relationships describing the object at the higher levels may have a fairly straightforward translation into English, using relational terms such as "Is-a-part-of", "Is-a-kind-of", "Is-connected-to", "Consists-of", etc., applied to terms denoting parts of the object at various levels of the description.

Consider the child's stick figure drawing of the human body. Such a drawing corresponds closely to a description of the connectivity of the limbs, and forms an appropriate top level description. The features of such a drawing can be translated into a computer data structure using relational terms, forming the description represented in Figure 11. From each named "node" in the description, other connectives would point to the description of that component in the next level of detail. These relations form a directed graph structure, in which the description of an individual object would form a "tree" with the name of the object as the top node. Note that the top-level description need not refer to any numeric properties of the image of the object, such as the shape of the limbs or the angle at which they join to the torso. Since the object may appear in a picture viewed from any perspective or with the limbs held at any angle (within their range of movement), such details would not be very useful in recognition. The abstract description captures the essential features that should be invariant in any image of the object.

Of course, at some level of the description numeric detail is essential. Many kinds of objects may have similar connectivity descriptions at the top level. For example, the connectivity graph for the nodes in the top level

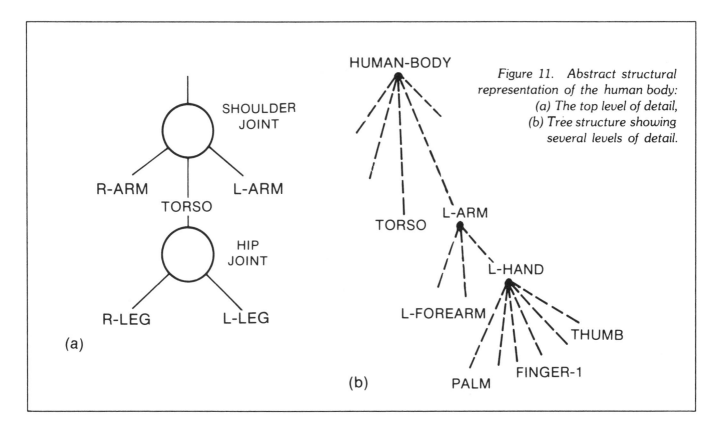

Figure 11. Abstract structural representation of the human body: (a) The top level of detail, (b) Tree structure showing several levels of detail.

description of the human body applies equally well to the description of many mammalian bodies. The only way to discriminate between each kind of mammal is to examine the sizes and shapes of the components. Comparison at this level of detail is computationally expensive, and should only be performed after descriptions of potential matches are selected at the top level, similar to the procedure used in the GM system.

The method of representing the shape descriptions is a major design consideration. The large amount of storage required for numeric surface models and the difficulty of comparing them suggests a need for a compressed representation that may be easily computed from an image and easily compared with the shape descriptions in the stored models. This is especially true for the curved surfaces typical of objects found in nature. For manufactured objects, modelling by geometric solids may be sufficiently concise and quite accurate.

One form of shape description that can work well for curved objects is the "generalized cone" representation introduced by Binford [4]. In this scheme, the shape of a component may be described by specifying an axis and successive cross sections of the object in planes perpendicular to the axis (Figure 12). For many objects, the choice of an axis may not be unique, but for elongated objects the obvious choice is one along the longest dimension that passes through the centers of the successive cross sections. The axis itself may be a curve, possibly represented as a series of line segments between the cross sections. The cross sections should be single closed curves; if the object branches, a joint should be

declared and new axes specified for the separate branches. This representation works well for objects whose cross section varies smoothly along the axis. Further data compression can be obtained by storing statistics and properties derived from the cross sections, such as the average area, its ratio to the length of the axis, whether the cross sections are more cylindrical or conical,

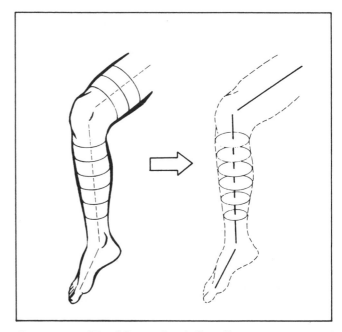

Figure 12. The "Generalized Cone" representation of curved solids.

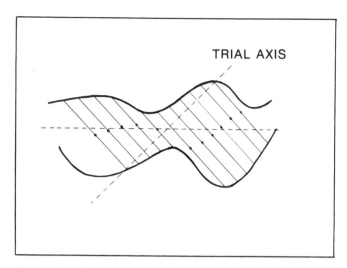

Figure 13. Finding the best choice for the axis of a region by iteration.

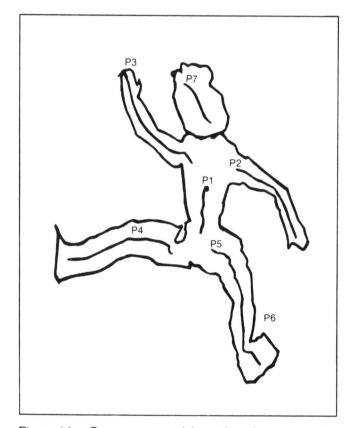

Figure 14. Cones computed from the edge picture of a toy doll.

etc., instead of storing the actual plane curves for each cross section along the axis.

Although restricted to two-dimensional recognition, the GM system demonstrates some of the characteristics essential for systems operating in the unrestricted 3-D world. To reduce the need for time-consuming coordinate transforms and literal image comparisons, a description of the image is formed in terms of the available vocabulary of

image primitives and relational connectives. Properties derived from the description are used as an index into the database of stored models and likely candidates are selected. Then, discrimination between the candidates is based on comparisons at levels of increased detail. In more elaborate systems, the program can use the more detailed levels of the model to tell it where to look for particular features in the image that may not have been in the first-pass description of the observed object. Similarly, a statistically good match between some region in the image and a component of some known object can be used to direct the search for the other components of that object in the image.

A working recognition system built on the concept of model descriptions of objects in terms of generalized cones has been described by Nevatia and Binford [5]. Although the image input to the system is from a laser rangefinder that returns the three-dimensional coordinates of object surface points, the same techniques could be applied to 2-D images, though discrimination based on the absolute size of parts of objects would not be possible (unless stereo vision input is available). Edge detection in a laser range image involves filtering the image array for discontinuities in range, rather than in contrast, and results in an analogous edge picture in which the 3-D location of the edge points is also known. Features of the top level description of an object in an image are used to index into the model database, selecting model descriptions for more detailed comparison.

The system works by first computing a generalized cone description of all the objects in the scene, using the edge picture. Since the axes of parts of the objects are not initially known, producing the description is an iterative process in which axes in several directions are proposed. For each trial axis, 2-D "cross-sections" of the object are constructed that are normal to the axis and at regularly spaced intervals along it (Figure 13). By fitting a new axis to the midpoints of the cross sections, a better choice may be found for the next iteration. If the process does not converge in a few steps, then the region may have no well-defined axis. If a good choice of axis is found, the axis is extrapolated at either end to cover as much as possible of the region. The extrapolation is terminated if the end of the region is encountered or if a discontinuous jump in the size of the cross section indicates that a joint with some different region may have been reached. Figure 14 shows the result of computing generalized cones to describe the image of a toy doll.

The next step in the recognition process is to construct a symbolic description of the generalized cones for each object. Regions described by a single axis are characterized as conical or cylindrical, and shape statistics are computed. The areas between well-defined axes are identified as joints, and may be described according to the number of axes they connect and the spatial configuration of the connected axes. The result is a connectivity graph (Figure 15) which (hopefully) is similar to the graph for the stored model of that kind of object. If there are disconnected regions adjacent to some object in the scene, potential joints are hypothesized that connect them to the object and may be verified or rejected during the recognition process.

Finally, the resulting description must be matched against the stored models. Since the computed descriptions may differ from the models due to imperfect edge evidence, missing parts, extra parts, distortion due to the particular perspective in the scene, etc., this matching must be statistical as well as literal. A key problem is finding the smallest possible number of candidates to compare. The solution taken in this system is to use features of the object description to compute an index into the model database. Similar indices are computed for each stored model, so that given an index computed from an object in the image, a list of models with the same index is immediately available. Several such indices may be computed for a single model, based on features of the largest components in the model. Once the smallest set of candidates is found, the actual matching of descriptions can take place. The match with the best score (based on the statistical matches of the individual pieces) is then verified to determine if differences between the object and model descriptions can be satisfactorily explained. In this case, the object description of Fig. 16(b) matched the graph of Fig. 12(a) better than the other candidates in the program's database, despite the extraneous joints J4 and J5 observed in the image.

If pieces of a candidate model are not matched in the object model, potential joints identified in the description phase may be examined to locate the missing pieces. Figure 16 shows another view of a toy doll in which, due to image noise, the regions for an arm and a leg are disconnected. The matching process correctly joined the two regions with the larger description and concluded that the object was a doll, the next most likely candidate being a toy horse, whose connectivity graph is the same as the doll's except for the addition of a tail. The match with the toy horse was rejected based on the differences in the statistical matches of the individual pieces. (Inclusion in the model of the possible range of movement of the joints could have helped rule out the match with the horse as

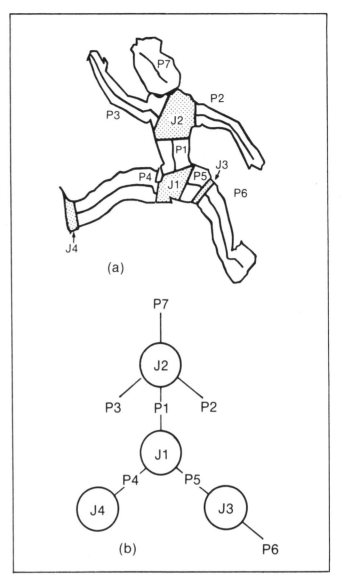

Figure 15. Resulting top-level description of the doll: (a) Cones connected by joint regions, (b) Resulting connectivity graph. Note the extra joints (J3 & J4) derived from the cones.

well.)

This system requires several minutes of computer time (on a DEC KA-10 computer) to process an image and recognize the objects, even with only five stored models. Most of the time is spent computing the generalized cone description from the edge data, a process which, like the description phase of the GM recognition system, can be greatly speeded up by the use of hardware edge detection and simple parallel processing schemes. The effective use of parallel processing may ultimately be the key to fast vision systems. The processes of description generation, model comparison, and verification all involve repetitive application of standard procedures to different sets of independent image and model data. If these procedures could be applied in parallel, the results could be made available to the next step in the recognition process immediately upon completion (with suitable process synchronization). This would result in a speed increase roughly proportional to the degree of parallelism.

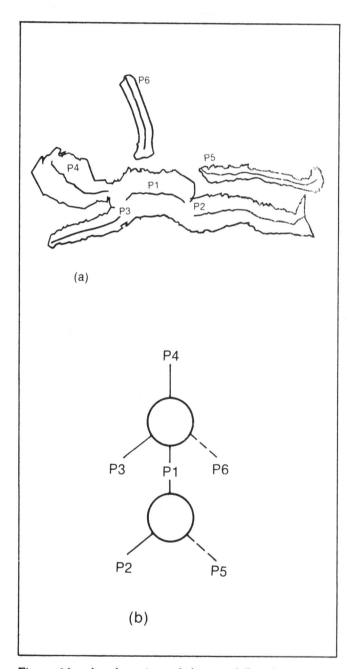

Figure 16. *Another view of the toy doll with increased image noise and a different orientation of the limbs and body. (a) Edge data and derived cones, (b) Resulting connection graph. A potential joint connects the left arm and leg to the rest of the graph.*

Summary

This article has attempted to give the reader some insight into the complexity of machine recognition. The intent was not to provide an exhaustive survey of the field of automatic perception, but rather to concentrate on examples that illustrate the variety of processing involved and an indication of the capabilities of state-of-the-art systems. Discussion of the many successful systems dealing with restricted domains, such as those limited to the recognition of toy building blocks and other geometric solids, or image segmentation based on texture descriptions, or statistical pattern classification, is beyond the scope of any single article. Hopefully, by considering some representative systems, the character and methodology of computer vision research may be conveyed.

The processes that underly image understanding extend beyond the problem of recognition. Indeed, the challenge of representing knowledge and its effective application to solving problems underlies all of automatic cognitive inference and is a central theme of Artificial Intelligence. Computer vision in particular offers immediate benefits for increased manufacturing productivity as well as many other areas of automation. Research in this area, as well as in other aspects of cognition, will almost certainly have a profound impact on the use of computers in industry, space exploration, and domestic applications. ⓝ

References:

[1] R. Duda and P. Hart, *Pattern Classification and Scene Analysis,* New York, Wiley and Sons. (1973).

[2] R. Eskenazi and J. Wilf, "Low-level Processing for Real-Time Image Analysis", IEEE Workshop on Computer Analysis of Time-Varying Imagery, Philadelphia, Pa. (April 1979).

[3] W. Perkins, "Model-Based Vision System for Scenes Containing Multiple Parts", Proc. of the Fifth International Joint Conference on Artificial Intelligence (IJCAI5), Cambridge, Masschusetts, (Aug. 1977).

[4] T. Binford, "Visual Perception by a Computer", IEEE Conf. on Systems and Controls, Miami (Dec. 1971).

[5] R. Nevatia and T. Binford, "Description and Recognition of Curved Objects", Artificial Intelligence 8, (1977), North-Holland Publishing Co.

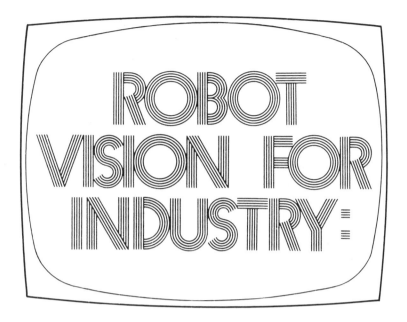

The Autovision™ System

Arnold G. Reinhold and
Gordon VanderBrug
Automatix Incorporated
Billerica, MA

Advanced robot technology such as computer vision is rapidly moving out of the laboratory and into practical applications in the demanding (both physically and economically) factory environment. With the introduction of the Autovision system, Automatix joins the growing number of high-technology firms engaged in the marketing and application of these developments.

A Brief Review of Machine Vision

Computer processing of visual images has been the subject of an enormous amount of research in the past 25 years. Almost since the invention of the computer, people have tried to couple computers with television cameras to develop a machine that could see. This technology has had some limited success in applications such as interpreting aerial photographs, guiding "smart" bombs, and counting red blood cells and chromosomes, not to mention broadening our understanding of the perceptual process.

In light of the hopes and effort invested in it, however, the spread of this technology must be considered disappointingly slow. There are two main factors holding back machine vision: First is the enormous amount of data represented by a TV image. A typical black-and-white home television picture requires about 8 million bits of information for faithful reproduction, and a frame arrives every 30th of a second! This amount of data exceeds the memory capacity of most of the minicomputers that are responsible for the rapid influx of computer technology in so many other fields.

The second reason is the lack of algorithms or methods of processing this information that are rapid enough and "robust" enough for general use. The difficulty of building a fast algorithm is easy enough to understand in light of the huge amounts of data that must be processed for one image. The difficulty of building a robust algorithm is perhaps harder to understand for someone familiar with the tremendous successes of computers in other fields. (A computer program is called *robust* when it can work successfully in a wide variety of situations, as opposed to a few laboratory demonstrations.) Most people would consider guiding a spaceship to a precise landing on the moon far more difficult than identifying a pencil in a photograph of a cluttered desk. The very opposite is true. The moon landing problem can be reduced to very precise equations. These equations are well suited to the numerical orientation of computers in existence today, and can be programmed and solved rapidly. By contrast, no programming language today contains a statement like "find the pencil" or even "look for something long and skinny, with a sharp point at one end and a pink lump at the other." The human brain, which finds the moon landing problem challenging, has evolved over billions of years to be extremely adept at finding pencils on tables.

The SRI Algorithm

A major breakthrough in the field of machine vision took place in 1974 when Gerald Agin and R. O. Duda [1,2] of Stanford Research Institute developed an algorithm for a simple version of the vision problem that provided useful answers in many situations, and was at the same time robust and efficient to compute.

The inherent difficulty of the vision problem is illustrated by how many simplifying assumptions were necessary to make the SRI algorithm work. To begin with, the algorithm assumes that the picture has been reduced to a very simple form—called a binary image. This reduction, which is done either by high contrast lighting or by some other—as yet unperfected—preprocessing technique presents the picture to the computer in the following simple form: The picture is stored as an array of dots or pixels (short for picture elements) which are either on or off (1 or 0). The 1 pixels represent the object while the 0 pixels represent the background. Thus, in order for the algorithms to work at all, useful information about the object must be entirely represented by the object's silhouette. (A sample binary image as seen by the computer is shown in Figure 1.) Of course no color information is understood; but, in addition, information that might be derived from texture, shading or three-dimensional perspective is also lost. The algorithm further requires that the object be entirely contained within the field of view of the camera, and, if more than one object is in the field of view, they may not overlap.

Given all these restrictions, it is surprising that this algorithm is any more than a laboratory curiosity. Yet, its power derives from its success in meeting the two criteria

mentioned above that have defeated other vision algorithms in the past. It is fast and it is robust, in that it can operate on any image meeting the above specificiatons. In addition, the algorithm is easy to use, and the information it provides is useful in a wide variety of industrial circumstances. In terms of speed, the simplicity and deterministic nature of the algorithm, together with recent advances in microcomputers, allow operating rates of six images per second or more. In terms of useful information, the SRI algorithm can reliably report on the position and orientation of an object, and such fundamental properties of an object as its area, moments of inertia, perimeter, ratio of perimeter to area (a measure of irregularity), maximum radius and a variety of other powerful geometric invariants. Furthermore, the algorithm will produce these invariants repeatably and reliably for a wide variety of objects and conditions. Its one principal drawback is the need to reduce information to a binary image, which generally requires high contrast lighting and a carefully structured viewing area.

One important aspect of the SRI algorithm is ease of use. The key parameters about an object are entered into the computer by a simple training operation. The object is shown to the camera in several positions and the computer statistically accumulates all the information it needs. This process, called training, can be carried out in a matter of a few minutes by an unskilled operator.

Much further research has gone into the SRI algorithm

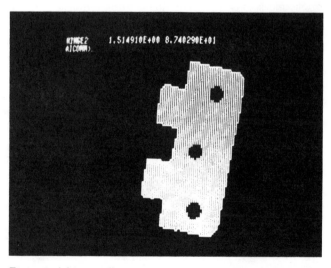

Figure 1. A binary silhouette image of a door hinge part, as seen by the computer.

since 1974, principally by the Vison Group at Stanford Research Institute, to further perfect this algorithm and remove some of the limitations. [3,4]

One thing the original SRI algorithm did not do is handle shades of gray in an image. If, instead of storing one bit of information per picture element, there was provision for storing 4 bits per pixel, these bits could be interpreted as a binary number representing 16 different levels of light intensity or gray tone. While hardware capable of converting video signals to gray scale and storing them has been available for some time, processing hardware that is fast enough and algorithms that are robust and general purpose have been scarce. [5]

Automatix Enters the Vision Scene

In the introduction of new technology to industrial use, the successful completion of a research project is only the beginning. The performance of the algorithm must be

Figure 2. The Autovision system, consisting of a TV camera, processor, and control console, shown in a typical application.

improved to where it can operate at the rates encountered in factory situations. It must be installed on a computer that can survive in the factory environment and work reliably for years with a minimum of maintenance. Training, applications support and repair service must be provided. The very largest corporations, with their own research laboratories, are sometimes able to bring new technology from the laboratory to the factory floor by their own efforts. Most companies, however, do not have the specialized resources to do this, and for them the adoption of new technology is strongly dependent on the availability of reliable commercial vendors.

Automatix Incorporated (AI), formed in January 1980 by a group with extensive experience in robotics, minicomputers and CAD/CAM technology, has as its charter the commercialization of advanced robotic technology in the form of complete modular "systems." AI believes that vision technology, as represented by the SRI algorithm and a number of its recent developments, is ready for such commercialization. The first AI product is a system called Autovision™(Figure 2). It is a vision module that combines the SRI-developed vision technology with the latest 32-bit microprocessor design, industrial packaging, advanced digital camera technology, and specific applications software. The SRI vision software serves as a nucleus around which further innovations—developed at AI and elsewhere—will be integrated with advanced concepts in robotic control. [6,7]

The Makeup of a Vision Module

For an advanced computer-based product like a vision module to be industrially viable, its manufacturer must provide three critical elements: hardware, software, and services. Of the three, hardware is probably the most tangible. In the case of the vision module the hardware may be further subdivided into five major subsystems: the camera, the vision preprocessor, the computer, the factory interfaces, and the enclosure. (A block diagram of the AI Autovision System is shown in Figure 3).

In its simplest form, the camera system is nothing more than an ordinary closed-circuit vidicon television camera of the type widely used for surveillance and videotaping. Such cameras are available from many manufacturers along with a large number of accessories, including lens, iris, mounting system and rugged enclosure. For many purposes such vidicon-based cameras are perfectly adequate. Where precisely repeatable measurements are necessary, however, vidicons have some drawbacks.

Because they scan a photosensitive surface with an electronically deflected electron beam, vidicons are sensitive to small changes in the deflection signals, causing a distortion of the picture. While this distortion is not bothersome to a human viewer, it can cause as much as a 10% error if measurements are made from the image, particularly at points near the outer edges. Recently, integrated circuit engineers have developed arrays of photosensitive cells on a single integrated circuit that can be scanned by digital means. Cameras based on such chips are commercially available and are smaller, more accurate and more rugged than the vidicon-based system. Because of the difficulty of building integrated circuits with hundreds of thousands of elements, all perfect, such cameras are also substantially more expensive.

One major aspect of the camera subsystem is the lighting, which can be a key factor in the success of a vision application. A variety of lighting techniques have been employed successfully in various industrial vision applications. Specific lighting system recommendations are extremely application dependent.

The second major subsystem in a vision module is the video preprocessor. The enormous amount of data contained in a digital camera image can easily swamp even the largest computer. Using high-speed dedicated electronics to carry out processing steps that reduce this data to a more manageable amount therefore makes sense. On the other hand, if the electronics is too carefully tuned to one particular algorithm it may not be usable on other algorithms, and the system will be unable to take advantage of future developments. A sensible compromise is to use advanced microprogramming techniques to build a vision preprocessor that can be adapted to a wide variety of algorithms. Such a microprogrammable preprocessor is one of the major advances in the Autovision II system from Automatix.

The Automatix vision preprocessor is also equipped to handle gray-scale images. Indeed, all images are acquired using the gray-scale process and the resulting image is thresholded to a binary picture under program control. This allows the program to try several different thresholds on the same image to get the best results. In addition, the microprogrammed arithmetic unit in the preprocessor can perform calculations on the gray-scale image, making normally ponderous gray scale algorithms more efficient.

Industrial Applications

What are some typical applications of this algorithm? Surely one of the most exciting is its use to guide heretofore blind robots. In almost all applications where robots are used today, the object must be in a precisely located, previously known position and orientation so that the robot can pick it up. If the object is displaced at all from the location, the robot will grasp at empty air or, even worse, damage the object or itself.

Needless to say, in many industrial applications objects are not accurately registered or fixtured ahead of time. In such cases a vision system must give the robot two pieces of information to pick up the object. First is the object's position. The SRI algorithm provides that information by computing the object's orientation. Here the situation is a bit more complex. For an object that is basically planar in nature, such as, say, a monkey wrench, a robot need only know the principal axis of the object. In the SRI algorithm this is determined by computing the axis of inertia. A more complex object might have several stable positions while lying on a work surface. Imagine a large hex nut, which might be lying hole vertical or horizontal. The SRI algorithm treats such objects by storing key information about each of the stable states and using its recognition capability to determine which of the stable states the object is in. Thus, the algorithm is able to tell the robot the object's position, its orientation and its stable state. The robot, in turn, would have separate programs for picking up the object in each of the several states.

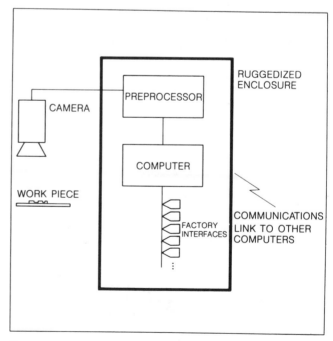

Figure 3. Autovision™ block diagram.

Inspection

Another major use of the SRI algorithm is sorting and part inspection. The geometric invariants calculated by the algorithm, in addition to providing direct information about the part, form a signature that helps distinguish different parts from each other. During the training process the algorithm automatically identifies the most useful sorting features or "discriminant" for a given set of parts.

Thus, if a number of different part types were coming down a conveyor belt, a vision module using the SRI algorithm could sort those parts into different bins—either by commanding a robot or by turning on or off diverter vanes or blasts of air. Perhaps of wider importance is the ability to distinguish good parts from bad. In many processes it is important to know whether all manufacturing operations have taken place successfully. A mounting plate, for example, might have a number of holes, brackets, and studs, each created by a different operation. One hundred percent inspection of such mounting plates by a vision system could avoid costly rework when an installed defective mounting plate has to be replaced. Another example is injection molding, where an inadequate plastic charge can result in misshappen features that can be recognized by the vision module. One thing the algorithm is not capable of is very high-precision measurement. Precision machined parts must often be inspected to better than one-tenth of the manufacturing tolerance—often a tenth of a mil (0.0025 mm). Such problems await other solutions.

Figure 4. An illustration of a typical inspection task.

The Computer

The third subsystem in a vision module is the central computer. No other part of the system has been so profoundly affected by the enormous advances in solid-state technology. The first microcomputers, introduced less than a decade ago were 4-bit systems. While these were, and still are, used to replace complex special-purpose logic in a wide variety of applications, it was only with the introduction of 8-bit devices that microprocessors took on the flavor of real computers, with operating systems, higher-level languages, and mass storage. Sixteen-bit microprocessors have been recently introduced and are meeting with wide success. However, Automatix has chosen to use the most advanced microprocessors, with 32-bit internal design, in Autovision. Automatix felt that the large processor requirements represented by vision processing, robot control and other demanding factory tasks require the very best in microcomputer performance. Not only does such a microprocessor offer very high processing speeds but its wide address architecture permits the marshalling of large amounts of high-speed memory to further improve system performance. Therefore, application development at AI should not be computer limited.

The fourth key subsystem is the industrial interface. No matter how clever or efficient the vision system is, it does no good unless it is able to affect other operations in the factory. In addition, the vision process itself must be synchronized with other processes. Finally, the vision system may have to coordinate with other computers in an integrated factory information system. Autovision meets these needs by providing discrete opto-isolated signal lines that can be used to drive solenoids, relays and motors and to sense signal levels and contact closures in coordinated machinery. It also provides serial and parallel interfaces to other computers.

The final key subsystem is an enclosure that can protect the sophisticated electronics of the vision module from various hazards, including the dust, fumes and temperature extremes often encountered in factories. At the same time the total package must facilitate reliable operation and rapid restart in the event of external power failures. The Autovision System cabinet is being designed to meet NEMA 12 standards, thus assuring suitability for installation in most factories.

Software

Of the three major elements of a viable vision product—

hardware, software, and service—software is the most complex and intricate, and, if not done properly, can be the most obscure and error-prone component. It has become a cliché that sofware costs are increasing at the same time hardware costs are decreasing. Perhaps one reason is that hardware people tend to adopt new technologies universally while software continues to cling to old methods. No engineer today would design with 10-year-old chips, yet the most popular software tools and languages of today were developed in the vacuum tube era. The persistence of obsolete techniques, such as writing entire systems in assembly language, is in part due to the limited memory capacities of 8- and 16-bit microcomputers. Automatix has chosen to base its software development on a modern well-structured high-level language in widespread use, namely Pascal. Selected critical performance subroutines are rewritten in assembly language to assure high-speed execution once they have been debugged in Pascal. These routines usually comprise less than 5% of the total code.

It is uncommon for the largest engineering investment in a product such as Autovision to be in the software area. Most companies, including Automatix, consider the details of their software to be valuable trade secrets and guard them closely. However, some potential users of advanced systems will need to modify or expand parts of the software to work in special circumstances. Automatix has taken the unusual step of providing arrangements whereby customers may have access to the Autovision source code upon signing necessary licensing and trade secret protection agreements.

An Application Language—RAIL™

Modifying the Pascal software programs would be undesirable for most customers, however, since modifications require a computer programmer and incur some risks of introducing system errors. By giving a user without special training in computers an easy way to tell the vision module what actions it should take under various circumstances, and what criteria it should use in processing images, much of the need for system modification can be avoided. Automatix provides this capability with a simple yet powerful and flexible language called RAIL™ (Robot Automatix Incorporated Language). RAIL makes it possible to sequence the operations of the Autovision System so that the desired task is performed.

The flexibility and power of RAIL are illustrated by the example inspection task shown in Figure 4. The shaded rectangular region represents a heat sink and the smaller square represents an I.C. chip that has been bonded to it. If the chip is not too close to the edge of the heat sink, and is not tilted too much, the assembly is a good one. This example is motivated by an actual inspection task described in [8] and in [9].

The RAIL program for this inspection task is shown in Figure 5. We assume that the parts are moving down a conveyor and that there is a part detector which signals when a part comes into view. After declaring I/O ports for the conveyor-running sensor, part detector, and part-reject solenoid, the program asks the user for tolerances for the chip offset and the chip orientation. It then waits until the conveyor is on, and, while it remains on, processes each part that comes into view.

The actual inspection procedure is entirely contained in the single IF-THEN-ELSE statement. It simply states that if the difference between the maximum X-values of the heat sink and the chip (which are the rightmost points on the boundary of the respective components in Figure 4) is greater than or equal to the offset tolerances, and if the orientation of the chip is within the specified tolerance of 90°, then the part is good—otherwise, it is bad. The striking similarity between the RAIL code and the manner in which the inspection procedure might be described in English illustrates both the power and the simplicity of RAIL.

```
INPUT PORT 1 : CONVEYOR
INPUT PORT 2 : PART_DETECTOR
OUTPUT PORT 1 : BAD_PART

WRITE "ENTER CHIP OFFSET TOLERANCE: "
READ OFFSET_TOL
WRITE "ENTER CHIP TILT TOLERANCE: "
READ TILT_TOL

WAIT UNTIL CONVEYOR = ON

WHILE CONVEYOR = ON DO
 BEGIN
    WAIT UNITL PART_DETECTOR = ON
 PICTURE
 IF XMAX ("HEAT SINK") - XMAX ("CHIP") > = OFFSET TOL
       AND
    ORIENT ("CHIP") WITHIN TILT_TOL OF 90
    THEN
       BAD_PART - OFF
    ELSE
       BAD_PART - ON

 END
```

Figure 5. The RAIL™ program for the example inspection task of Figure 4.

ROBOTICS AGE: IN THE BEGINNING

XMAX and ORIENT are features calculated in software. RAIL can reference any feature of any component of the picture, simply by stating the name of that feature along with the name, assigned during training, of the desired component. This is extremely powerful because it allows for the use of any feature in either an arithmetic or logical expression. Additional capabilities of RAIL will make it possible to write programs to control robots and to integrate sensors with robots.

Services and Product Support

To guarantee the success of a high-technology application such as a vision system in the demanding environment of a factory, the potential user of such a product must evaluate the full range of services provided by its supplier. These include pre-installation applications support, thorough user training, documentation, spare parts, and field service support. No factory can tolerate frequent breakdowns by any piece of equipment that can stop the production line. Many customers will first choose to place one unit in a manufacturing development laboratory, where it may be supplied with periodic software updates reflecting advances in the state-of-the-art of vision processing technology in order to advise the factory on promising applications. The system must also be able to coordinate with other advanced technologies such as robots.

In addition to these features and services, a key element of the Autovision system's "Availability Assurance" program is the inclusion of thorough self-test diagnostics that permit rapid correction of problems by on-site user personnel using only basic electrical maintenance techniques such as module switching.

Conclusion

Early work in vision research, particularly the development of the SRI algorithms along with large scale improvements in microprocessor technology, has brought vision technology to a plateau from which practical industrial vision systems for inspection and robot control can be developed and successfully applied.

Automatix, along with other companies in the field, will deliver vision systems in 1980 as modules for use in inspection and in sensor controlled robot systems. Application software, fast image processing and extensive user support are required for commercial success. However, there is no doubt that in the decade of the 80's, the introduction of vision systems and other sensor modules will revolutionize factory automation.

References

[1] G. J. Agin and R. O. Duda, "SRI Vision Research for Advanced Industrial Automation," Proceedings of the Second USA-Japan Computer Conference, Tokyo, Japan, pp. 113-117 (1975).

[2] G. J. Gleason and G. J. Agin, "A Modular Vision System for Sensor-Controlled Manipulation and Inspection," *Proc. of 9th International Symposium on Industrial Robots*, Washington, D.C., pp. 57-70 (March 1979).

[3] D. Nitzan and C. Rosen, et al., "Machine Intelligence Research Applied to Industrial Automation, Ninth Report, SRI International, Menlo Park, CA (August 1979).

[4] R. C. Bolles, "Symmetry Analysis of Two Dimensional Patterns for Computer Vision," Proceedings of 6th International Joint Conference on Artificial Intelligence, Tokyo (August 1979).

[5] S. Barnard, "Automated Inspection Using Grey Scale Statistics," IEEE Transactions on Pattern Analysis and Machine Intelligence (to appear).

[6] VanderBrug, G. J., and R. N. Nagel (1979). "Vision systems for manufacturing." To appear: Proc. of 1979 Joint Automatic Control Conf., IEEE Inc., New York.

[7] VanderBrug, G. J., J. S. Albus, and E. Barkmeyer (1979). "A vision system for real time control of robots." Proc. of 9th International Symposium on Industrial Robots, SME, Dearborn, MI. pp. 213-222.

[8] Baird, M. L., "An Application of Computer Vision to Automated I.C. Chip Manufacture." Third International Joint Conference on Pattern Recognition, IEEE Computer Society, Coronado, CA, pp. 3-7 (1976).

[9] Kashioka, S., Ejiri, M., and Sakamoto, Y., "A Transistor Wire-Bonding System Utilizing Multiple Local Pattern Matching Techniques," IEEE Trans. on Systems, Man, and Cybernetics, vol. SMC-6, no. 8, Aug. 1976, pp. 562-570.

CAMERA GEOMETRY

for Robot Vision Relating what the camera looks at to what the camera sees.

Alan M. Thompson
Robotics Age Magazine

The Image-Object Relationship

Most articles dealing with robot vision are primarily concerned with the problem of detecting, describing, and recognizing target objects in an image, usually the digital representation of an image taken by a TV camera. In this article, we will take a look at the geometric relationships between objects in the real world outside of the robot and their images in the digitized picture. Clearly, knowing these relationships is essential for tasks that require the robot to *physically* interact with its environment, as opposed those where it just *passively* derives information by looking at a scene. Even in the latter case, it is often necessary to describe the location of objects quantitatively.

In general, robots must deal with multiple coordinate systems. Manipulator control programs typically describe the location of the robot's hand, as well as target points, relative to the base of the arm. The location of machine tools or part-feeders, though, may be expressed in terms of an absolute *workspace* frame. For a mobile robot, the navigation program must be concerned with the location of the robot in some universal frame. For each of these problems, the robot designer is faced with providing coordinate transforms that allow location measurements made in one frame to be expressed relative to another.

For a vision-equipped robot, the relationship between the real world and the two-dimensional world of an image presents special problems due to the nature of the imaging process. Not only can distortions of perspective complicate the problem of recognizing objects in the general case, but these distortions also complicate the problems of automatic range measurement and of accurately determining spatial relationships from visual information.

One of the critical pieces of information needed by a vision subsystem is a description of the camera—for only with this knowledge can the location of targets, whether landmarks or workpieces, be derived from their position in the image that the camera sees. Assuming that the robot has identified a target point in an image (by some perceptual or cognitive process such as those I described in an earlier article [1]), the *geometric transformation* between the object world and the image can be used to help compute the target's real-world coordinates. Similarly, if we know the target's coordinates relative to the camera, we can use the transformation to tell us where to find the target in the image. As I will show, some methods of automatic range measurement by binocular vision require a combination of these operations, applying the transformation in both directions.

102

Reprinted from Robotics Age, Volume 3, Number 2, March/April 1981

The tools needed to derive the image-object relationship come from the broader field of *projective geometry*, which finds practical application in a number of important areas, including mapmaking and computer graphics. It is used here to find the projection that maps points in three-dimensional space into a plane by projecting them through a single point. The 3-space is, of course, the real world; the plane, the focal plane of the camera. The point used for the projection is the camera's *focal center*, defined below, and the resulting mapping is called the *perspective transform*.

The Camera Model

Before we can start writing equations to describe the perspective transform, however, we must first have a mathematically precise description, or *model* of the structure of a camera. To avoid having to deal with the complexities introduced by lenses, we'll start by limiting our description to the simplest type of camera, the pinhole camera. (Figure 1)

In a pinhole camera, the image of a point in 3-space is made by a ray of light that leaves the point, passes through the pinhole, and intersects the rear plane of the camera, presumably leaving an image on the film. For an "ideal" pinhole, the ray of light from the point is an ideal line, so that the image of the space point is itself a point, eliminating the need to focus. The pinhole, of course, is the camera's focal center, which serves as the projection

point. Let the vector **C**, measured in some convenient external coordinate system, describe the location of the focal center.

The next thing we do is define the location of the image plane relative to the focal center. In a film camera, this is the location of the film that records the images of all the points in the field of view of the camera. For a video camera, it is the sensitive surface of a vidicon tube or solid-state imaging array. The orientation of the image plane can be described by a direction vector orthogonal to it, shown in the figure as the camera's *aiming vector*, **A**. The *focal distance, f,* measured from the focal center to the image plane along the vector **A**, is an important factor in determining the size of an object's image.

To locate points in the image plane, we must define two coordinate axis vectors **H** and **V**, orthogonal to **A**, that correspond in this case to image horizontal and vertical directions, respectively. For a film camera, the orientation of these axes is rather arbitrary, but for a video camera, the vectors' orientations are determined by the manner in which the image is scanned and digitized, as will be explained.

By using the three direction vectors **A**, **H**, and **V** to define a coordinate system origined at the focal center **C**, our model of the pinhole camera is complete. (Figure 2) We can use our camera-center frame to provide an exact description of the location of the image plane in space: the plane can be defined as

$$\mathbf{C} - f\mathbf{A} + u\mathbf{H} + v\mathbf{V},$$

where u and v are scalar parameters that can be used as the coordinates of points in the image plane. Of course, for

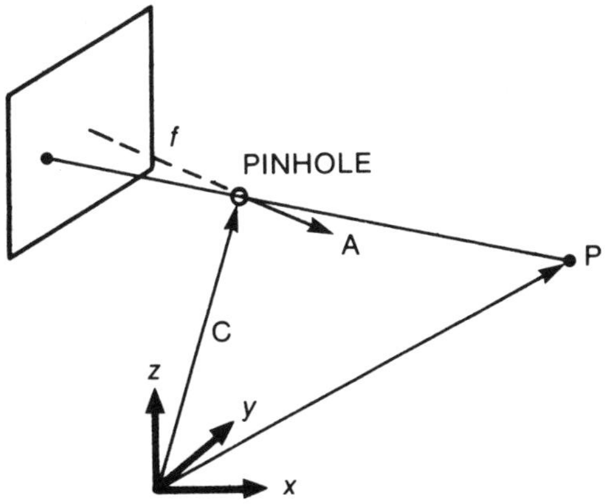

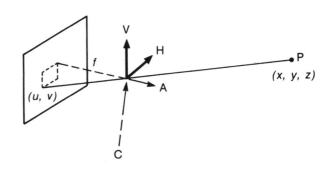

Figure 1. A pinhole camera. The image of point **P** is formed by a ray of light that passes through the pinhole at **C**. The aiming vector **A**, and the focal distance f, determine the location of the image plane relative to **C**.

Figure 2. Image horizontal (**H**) and vertical (**V**) vectors serve as coordinate axes that permit the location of points in the image plane to be measured. Here, the image of point **P** is located at image coordinates (u, v).

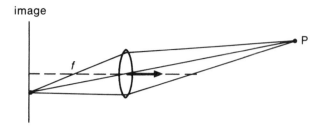

image

Figure 3. With the addition of a lens, a cone of light from **P** is focused on the image plane, producing a brighter image. Although the focal distance f needed for focusing varies with the distance to the target, the basic pinhole camera model holds.

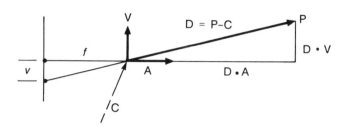

Figure 4. A side view of the imaging process. The components of **D** along the **A** and **V** axes are found by taking the scalar product, forming a right triangle as shown. One side of a similar triangle is given by the focal distance, so that v, the image coordinate in the direction of **V**, may be computed.

a real camera the sensitive part of the image plane is bounded in uv space, thereby defining the field of view of the camera. With this model we can accurately say that for every point **P** in the field of view, there is a corresponding image of the point located in the image plane at some *image coordinates* (u, v). We are now equipped to examine the geometry in detail to find the actual relation between the target point **P** and its unique location in the image.

Although our camera model was derived by assuming a pinhole camera, it turns out, fortunately, that it also holds quite well for most general purpose lenses, too. Figure 3 shows a side view of a camera with a larger aperture that uses a lens to focus the image of the point **P** onto the image plane. In this case, all the light from **P** that falls on the lens is concentrated on the image point, producing a much brighter image. For fixed-focus cameras, the camera model remains identical, and the focal distance f corresponds to the focal length of the lens. In variable focus cameras, however, the lens can be moved further away from the image plane to focus on points close to the camera, thereby moving the focal center and changing the focal distance. Lenses with multiple elements are complex to describe, and with them, the images of straight lines in space may appear curved (wide-angle, "fish-eye" lenses, etc.). For these lenses, this simple *linear* camera model does not hold.

The Camera's Perspective Transform

Let's start by taking a side view of the camera geometry and examining just the vertical component of the projection. In Figure 4, the camera's aiming vector **A** and image vertical vector **V** lie in the plane of the paper. First, we form the displacement vector **D** from the camera's focal center **C** to the target point **P**. We can *project* this vector onto the **A** and **V** axes by taking the scalar (dot) product of **D** with the respective unit vectors. The resulting scalars give the lengths of two sides of a right triangle as shown. Since **V** is parallel to the image plane, the solution for the vertical component v of the image of the point is obtained by noticing that a similar right triangle is formed by v and

the camera's focal distance f. Thus, since the sides of similar triangles are proportionately scaled, we can say that:

$$v = f \frac{\mathbf{D} \cdot \mathbf{V}}{\mathbf{D} \cdot \mathbf{A}} \quad , \text{ where } \mathbf{D} = \mathbf{P} - \mathbf{C}$$

By a similar process, taking a top view of the geometry, substituting the image horizontal vector **H** for **V**, we can solve for the image horizontal coordinate of the image of **P** as:

$$u = f \frac{\mathbf{D} \cdot \mathbf{H}}{\mathbf{D} \cdot \mathbf{A}}$$

These two equations are all it takes to find the image of a known target point in the image plane, provided we know the 3-space coordinates of the target and the camera parameters f, **C**, **A**, **H**, and **V** that define the location of the camera center and the image plane in the same 3-space. Determining these parameters requires a process of camera *calibration* that we'll return to later.

Note that, although we assumed that **A**, **H**, and **V** were unit vectors to aid the derivation, the resulting expressions for v and u, and, in fact, the actual derivation, are independent of their respective lengths, as long as their lengths are the same. This fact will become significant later when we discuss the calibration procedure.

So far we haven't said anything about the fact that for robot vision, the image from the camera must be digitized and made available to a computer. The relations we derived for the image coordinates of a target point measure the coordinates in the same units used to measure the camera's focal distance, f. This may be fine for film cameras where the negative itself provides a record of actual image measurements, but for a digitized picture we must consider the additional mapping from image plane to the representation of the image provided by the digitizer.

Describing the Digitized Image

Typically, an image digitizer partitions the image plane

into a rectangular array of picture elements, or *pixels*. The number of bits of information per pixel is not important here; all we are concerned with is the correspondence between the rows and columns in the digitized picture and points on the imaging surface. For a solid-state imaging array, the correspondence between the imaging surface and the row and column is fixed geometrically by the device's pattern of light-integrating cells, providing the most stable mapping. For a vidicon tube, however, the scanning of the image is accomplished by deflection coils, and the mapping is subject to drift. The digitizer counts the number of video lines, and the column number is obtained by counting a "pixel clock" that also triggers the sampling and conversion of the video signal. (*See article in this issue.*) The row and column numbers provided by the digitizer are typically used to index the memory address at which the digitized pixel information is stored, and the computer program keeps track of the correspondence between the stored data and the original pixel address in the image.

The actual mapping from the pixel address to the physical image coordinates is determined by four quantities: the pixel address (row and column) of the element nearest the origin of the uv coordinate frame (i_o, j_o), and the sampling *resolution*, the number of pixels per unit length in the image horizontal and vertical directions, given by n and m, respectively. The mapping can then be written as:

$$u = n(j - j_o) \quad \text{and} \quad v = m(i - i_o).$$

It is highly desirable for n and m to be the same, that is, that the pixels should be "square" instead of "rectangular". If this is the case, then some pattern-matching processes are simplified because a silhouette will appear to have the same proportions when rotated to a different orientation in the image plane. The scan characteristics of a vidicon camera can be altered to achieve this, but not all solid-state cameras have this feature.

It turns out to be extremely simple to combine this additional transformation from image to pixel coordinates into our original mapping from 3-space into the image. Equating the expression for v in terms of the row i with the vertical term of our original transform and solving for i, we get:

$$i = \frac{f}{m} \frac{\mathbf{D} \cdot \mathbf{V}}{\mathbf{D} \cdot \mathbf{A}} + i_o, \text{ which can be written as}$$

$$i = \frac{\mathbf{D} \cdot \mathbf{V}'}{\mathbf{D} \cdot \mathbf{A}}, \text{ where } \mathbf{V}' = \frac{f}{m} \mathbf{V} + i_o \mathbf{A}$$

By redefining $\mathbf{V}$ we can still use the simple equation for the projection and completely account for the origin and resolution of the image digitization. A similarly redefined image horizontal vector $\mathbf{H}'$ is used in an expression for j. For the rest of this discussion, $\mathbf{H}$ and $\mathbf{V}$ will refer to these redefined vectors.

It is worth pointing out that due to the slight diagonal slope of the horizontal scan in a vidicon tube, caused by the continuous vertical motion of the beam, the horizontal and vertical axes of the image are not necessarily orthogonal. While this does result in some shape distortion since the pixels aren't exactly square, it doesn't affect our equations for the perspective transform.

Locating Targets in 3-Space

The perspective transform is a valuable tool with many applications in computer graphics and image processing. With it, any three dimensional model of an object or scene can be projected into a plane for graphical display, allowing it to be viewed from different angles. For robot vision, the expressions we have derived so far can be used to tell us where in an image to look for targets when we know their location in some external 3-space frame.

More typically, however, we are concerned with giving the robot the capability to see something in an image and then compute its location in the surrounding environment. Unfortunately, this is not as simple as computing the projection was, because the image only supplies two pieces of information, the row and column coordinates, for any real-world target point whose projection it contains. More specifically, target points lying on the same line through the camera center, but at different ranges from it, will all appear at the same image coordinates. Without additional information, one camera alone cannot determine range.*

Even so, by supplying a somewhat arbitrary constraint we can use the image coordinates to solve for a direction vector that points to the target from the camera center. This pointing vector can then be used for various triangulation schemes. The length of the vector $\mathbf{D}$ used in the equations for u and v is the range r from $\mathbf{C}$ to the target. Since we don't know this range, we can make the assumption that $\mathbf{D} \cdot \mathbf{A} = 1$. This allows one component of $\mathbf{D}$,

---

*Actually, monocular ranging schemes have been devised that rely on restricting the depth-of-field of the camera's focus and then writing software that attempts to decide when the image of the target is in focus. Calibrating the focus control of the camera can then be used to obtain an approximate range measurement.

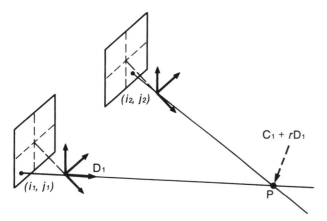

Figure 5. Finding the image of a target at (i_1, j_1) in camera 1 allows a direction vector, D_1, to be computed. Seeing the same target in camera 2 at (i_2, j_2) allows its range from either camera to be determined.

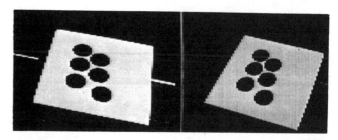

Figure 6. An actual stereo pair taken by adjacent TV cameras. The horizontal shift of a target point, between its coordinates in the left and right images, is the primary range clue.

to be written in terms of A and the other two components. Knowing the two image coordinates i and j of a target point allows us to write two equations for the two remaining components of D, which can then be solved algebraically.

If the same target point can be identified in different calibrated cameras, then the (approximate) intersection of lines emanating from the camera centers in the respective directions can be computed to solve for range. Due to various errors (round-off, digitization, etc.) an exact intersection may not exist, so this computation can be cumbersome. A simpler, though less general method, can be used if the two cameras are displaced in the approximate direction of the image horizontal axis.

In Figure 5, two cameras are shown with a lateral separation along H. Target point P has an image at (i_1, j_1) in camera 1, which is used to solve for direction unit vector D_1. The location of P in 3-space can be written as $C_1 + rD_1$, where r is the undetermined range parameter. Rather than solving for the direction vector D_2 from the image of P in camera 2, (i_2, j_2), since the cameras are displaced laterally it is easier just to plug the indeterminate expression for P into the equation for the horizontal (column) coordinate of its image:

$$j_2 = \frac{D_2 \cdot H_2}{D_2 \cdot A_2}, \text{ where } D_2 = rD_1 - C_2 + C_1$$

There is only one unknown, r, and the equation can be solved algebraically for the range from camera 1.

I would hasten to point out that it is not always easy to find the same target point in images from two different cameras. However, since the image of the ray $C_1 + rD_1$ (as r varies from 0 to infinity) in camera 2 is readily determined, the area of the image to be searched is gratefully constrained. Even so, the techniques of performing the match can become quite complex and will be the subject of future articles. Many methods rely on statistical matching of regions of the image around the target points,

as in [2]. Another clever approach is to use just one camera and monitor the shift of the target in the image as successive pictures are taken while the camera is moved along a lateral track, as in [3]. Of course, a mobile robot may be able to accomplish the same result by watching the apparent motion of targets as it moves along.

Figure 6 shows a stereo pair of images taken by adjacent TV cameras having a baseline separation of approximately .5m. From examining the two views of the scene, it is apparent that the lateral shift of the position of a target point relative to the centerline between the two images increases the closer it is to the cameras, whereas a target further away is shifted less. It is this *parallax* shift that is the primary range clue in a side-by-side camera configuration such as this one.

The Advantages of a Structured Environment

Often it is possible to obtain range information by taking advantage of some additional constraint imposed on the robot's environment. If the objects or target points the robot looks at can somehow be confined to a known two-dimensional surface such as a plane, then the two equations provided by the image of the target are again sufficient for a completely determined solution.

Figure 7 (a) shows a camera positioned above a plane. If we know where the plane is in 3-space, then any point P in the plane can be described as $P = P_o + mX + nY$, where P_o is the 3-space vector locating some point in the plane, X and Y are some coordinate axes defined in the plane, and m and n are the parameters used to define the coordinates in the plane of the point in question. Here again, if we see some target point that we know lies in the plane, the system reduces to two equations in two unknowns and is readily solved. (A straightforward exercise left the the enterprising reader.)

The greatest advantage is obtained when the camera is oriented so that the image plane is parallel to the object plane, as shown in Figure 7 (b). Here, since the camera's aiming vector A is orthogonal to the plane, the value of $D \cdot A$ computed from any point in the object plane equals the same constant: the height of the camera above the

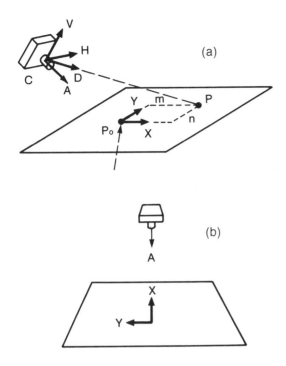

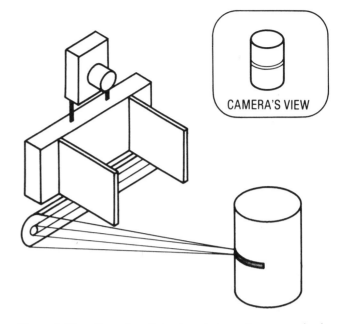

Figure 8. *The effect of confining targets to a plane may also be obtained by projecting a plane of light, as done in this system developed at NBS. [5] The timing of the flash is synchronized with the camera scan, so that it sees a bright band across the target. The location of the light plane is known, allowing range measurement to points on the target it illuminates.*

Figure 7. *When target points are confined to a known plane, so that their location is completely specified by two coordinates (m, n) in the plane, there is a one-to-one correspondence between the location of points in the image and their position in the plane. In (a), the ray **D**, computed from the image coordinates, intersects the plane at a unique, easily determined, m and n. When the camera is aimed directly at the plane (b) rather than obliquely, the mapping from image coordinates to plane coordinates is linear, and may be further simplified by aligning the axes in the plane with the image axes **H** and **V**.*

plane. Since the denominator of the perspective transform equations is now constant, the mapping becomes linear, and there is no shape distortion due to perspective. This permits the use of simple object recognition strategies based on matching silhouettes. Industrial robot vision systems based on this configuration are commercially available. [4, 5]

Constraining the environment isn't the only way of providing the structure necessary for monocular range measurement. At the National Bureau of Standards, researchers devised a light source using a flash tube and a cylindrical parabolic reflector that is capable of projecting a thin plane of light. [6] (Figure 8). With both the projector and a solid-state camera mounted on the robot's gripper as shown, the equation of the projected plane of light relative to the camera center frame is known. The triggering of the flash tube is synchronized with the vertical drive, so that the band of light visible where the projected plane intersects a target object is easy to spot. Since a point in the projected plane of light can be described with two coordinates, finding the location of its image permits accurate range measurement. Robot vision using this technique will soon be commercially available as part of a robotic welding system. [5]

The Problem of Camera Calibration

Our discussion so far has not dealt with the question of how to find the vectors **C**, **A**, **H**, and **V** that define the location, orientation, focal distance, and digitizer parameters for our camera. Clearly, it would be quite difficult to measure these values mechanically with enough accuracy to yield a camera model useful for automatic range measurement. In this section, I will describe a method that can be used to compute the camera calibration vectors from data collected during a training session in which the robot records the image coordinates of target points whose 3-space location is known.

All of our 3-space measurements, both for the locations of points in space and the components of the camera orientation vectors, must be made in a suitable external coordinate frame that is fixed relative to the camera. If the camera is stationary, then a coordinate frame attached to some reference point on the floor can be used. If the camera is mounted on a pan/tilt platform so that it can be aimed, then the calibration must be performed with the platform set in one position. Later, the pan/tilt platform itself can be calibrated so that measurements made in a camera frame defined on the platform can be transformed into some other frame as necessary, using the azimuth and elevation angles of the platform as parameters.

For a mobile robot, the camera calibration should be performed in a coordinate frame attached to a point on the body of the robot that has a fixed relation to the camera mounting. Once performed, the calibration will permit the position of targets to be determined relative to the robot, so that knowledge of the robot's location and heading (maintained by its navigation system) can be used to refer the relative locations of targets into some stationary frame. Alternatively, the robot may use the relative locations of landmarks it sees to determine its absolute location visually, as in [3].

Let's consider the case of a stationary camera viewing a grid pattern placed on the floor, as shown in Figure 9. Using the stationary coordinate frame as a reference, the location of easily visible reference marks on the grid can be accurately specified. The first operation of the camera calibration program is to identify successive reference marks on the grid, recording for each mark m its image coordinates (i_m, j_m) and its location in 3-space $\mathbf{P_m}$. Thus, each reference point will have five numbers associated with it. It is *essential* that data be collected with the horizontal grid placed at two or more different elevations, or once with it on the floor and again with it on the wall, otherwise there will be insufficient data for a solution to be obtained. Likewise, the total number of data points, n, must be greater than the number of unknowns. To find the four 3-space vectors we seek, a minimum of twelve data points should be taken, and $n>20$ is preferable.

The object of the calibration procedure is to perform a least-squares fit of the parameters of the camera model to the observed calibration data. Before we can write a program to do this, however, we must first examine the system analytically and formulate a solution algorithm.

We want to find the calibration vectors that produce the best fit in the relations

$$i_m = \frac{\mathbf{D_m} \cdot \mathbf{V}}{\mathbf{D_m} \cdot \mathbf{A}} \quad \text{and} \quad j_m = \frac{\mathbf{D_m} \cdot \mathbf{H}}{\mathbf{D_m} \cdot \mathbf{A}}$$

where $\mathbf{D_m} = \mathbf{P_m} - \mathbf{C}$ and $m = 1, \ldots, n$.

If our system were perfect, then each of the 2n equations would hold exactly. Sadly (but not unexpectedly) this is not the case. Round-off and digitization errors, camera nonlinearities, etc., combine to prevent it. Therefore, we have to formulate an expression for the "slop" in the fit. Starting with the equations for i_m, we can write this as

$$S = \sum_{m=1}^{n} (\mathbf{D_m} \cdot \mathbf{V} - i_m \mathbf{D_m} \cdot \mathbf{A})^2$$

This is the sum of the squared error terms we want to minimize by proper choice of calibration vectors. The solution to this system can be obtained by applying

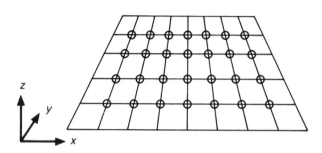

Figure 9. Camera calibration given a coordinate system defined on the floor (or wherever), the 3-space coordinates of reference marks on the grid may be measured and associated with their corresponding images as seen by the camera. Then, a least-squares fit of the camera calibration vectors may be obtained.

standard linear algebra techniques, provided it is rewritten in the correct form. Let's call the individual error terms inside the parenthese e_m. We can rewrite each e_m as

$$e_m = (\mathbf{P_m} \cdot \mathbf{V} - i_m \mathbf{P_m} \cdot \mathbf{A} - C_V - C_A), \text{ where}$$
$$C_V = \mathbf{C} \cdot \mathbf{V} \text{ and } C_A = \mathbf{C} \cdot \mathbf{A}.$$

Now we can more easily see the system as a set of n linear terms in 8 unknowns, 3 unknowns each for $\mathbf{V}$ and $\mathbf{A}$, and two more for C_V and C_A. The next step is to recall that our choice of $\mathbf{A}$ as a unit vector was arbitrary. To aid the solution, we can asume that one coordinate of $\mathbf{A}$, say $\mathbf{A_1}$, equals 1. This reduces the number of unknowns in each term to 7, in place of the eighth a constant term $i_m \mathbf{P_{m, 1}}$. We can always scale the solution for $\mathbf{A} \cdot \mathbf{A} = 1$ later if desired.

If we think of each term e_m as an element of an n-dimensional vector $\mathbf{E}$, then we can write an expression for $\mathbf{E}$ as an n by 7 matrix $\mathbf{Q}$ times a 7 by 1 column vector $\mathbf{W}$ containing the unknowns, with an n-vector $\mathbf{B}$ for the constants:

$$\mathbf{E} = \mathbf{QW} - \mathbf{B},$$

and the expression for the slop can be rewritten as

$$S = \mathbf{E} \cdot \mathbf{E} = \|\mathbf{QW} - \mathbf{B}\|^2$$

where the bars indicate the magnitude of the n-vector. Expressed this way, we are looking for the 7-vector $\mathbf{W}$ that minimizes S. As described in [2], the expression for the error n-vector $\mathbf{E}$ can be converted to a standard 7 by 7 inhomogeneous linear system whose solution is the calibration vector $\mathbf{W}$ that we seek. The conversion is accomplished by multiplying by the transpose of the coefficient matrix and forming the expression

$$\mathbf{Q}^T\mathbf{E} = \mathbf{0}, \text{ or } \mathbf{Q}^T\mathbf{Q}\mathbf{W} = \mathbf{Q}^T\mathbf{B}.$$

Linear systems in this simple form can be solved by a variety of means: Cramer's Rule, Gaussian elimination, the Gauss-Seidel interative method, etc. The latter two approaches are more suitable for computerized solution; flow charts and FORTRAN programs for these are available in [7].

At this point, we know $\mathbf{A}$, $\mathbf{V}$, C_A and C_V. Knowing the values for $\mathbf{A}$ and C_A permits us to build an n by 4 expression for the slop in the fit for the n column coordinates j_m. The 4 unknowns are C_H and the three components of $\mathbf{H}$. If you write the solution procedure to operate on any two dimensional array, then the same procedure may be used to solve both systems. (You may have to write your own array access functions if the language you used does not permit passing arrays of variable dimension to subroutines.)

After we have C_A, C_H, C_V, the 3-space location of the camera, $\mathbf{C}$, is obtained by solving the 3 by 3 linear system

$$\mathbf{C} \cdot \mathbf{A} = C_A$$
$$\mathbf{C} \cdot \mathbf{H} = C_H$$
$$\mathbf{C} \cdot \mathbf{V} = C_V$$

and our model of the camera is completely determined.

It's a good idea to substitute the results into the original expressions for the slop of the fit to see how good your solution is. If the predicted location of the image of a known reference mark is more than a few pixels away from where it actually appeared in the image, then the solution is not accurate and should be repeated with a better set of data. The broader the distribution of your data points in 3-space, the more likely a good solution will be obtained by this method.

If your program should fail to obtain a solution (overflow, underflow, divergent iterations or whatever), then perhaps your implementation lacks precision or your original data collection was insufficient. In the former case a reformulation of your arithmetic expressions may prevent the loss of information due to round-off errors, while in the latter you should collect a new set of calibration data with a broader spatial distribution. Another potential problem is if the arbitrary choice of $A_1 = 1$ is inconsistent with your choice of the external coordinate system, i.e., that $A_1 = 0$ in that frame for your camera mounting. Without moving the camera, this can be remedied by redefining the correspondence between the 3-space coordinate indices and the actual x, y, z axes.

Concluding Comments

Although big "number-crunching" programs like the camera calibration procedure described here can place a burden on a microcomputer based system, the perspective transform requires only a few floating point operations to perform. Also, you should only need to run a calibration when the camera position or some digitizer characteristic is changed. Analog drift in the scan circuitry of a vidicon camera may necessitate more frequent calibration runs, however.

With the calibrated camera model, your robot vision system is equipped to make accurate measurements of spatial relationships in the surrounding world. Although a more complete presentation of binocular vision and other generalized range measurement methods must be delayed for future articles, the tools provided here are an essential foundation, and serve quite well in environments where targets are constrained to a two-dimensional space like a plane. ▣

References

[1]   Alan Thompson, "Introduction to Robot Vision," *this book*.

[2]   Y. Yakimovsky and R. Cunningham, "Extracting 3-D Measurements from Stereo TV Cameras," *Computer Graphics and Image Processing*, No. 7, 1978, Academic Press.

[3]   Hans Moravec, "Obstacle Avoidance and Navigation in the Real World by a Seeing Robot Rover," Report CS-80-813, Dept. of Computer Science, Stanford University.

[4]   VS-100 Vision System, by Machine Intelligence Corp., Palo Alto, Calif.

[5]   Autovision I and Robovision II, by Automatix, Inc., Burlington, Mass.

[6]   R. Nagel, *et al*, "Experiments in Part Acquisition Using Robot Vision," National Bureau of Standards Report MS79-784.

[7]   S. D. Conte, *Elementary Numerical Analysis*, McGraw-Hill Series in Information Processing and Computers, New York (1965).

A TV digitizer that reads in binary images
at video frame rates.

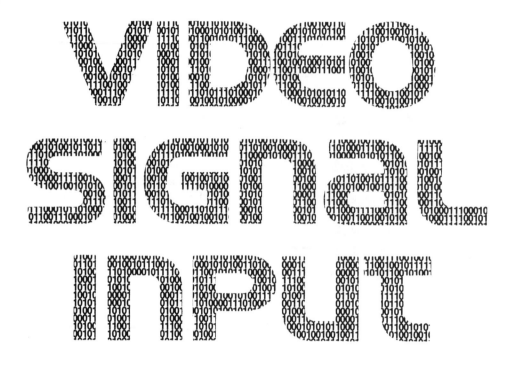

VIDEO SIGNAL INPUT

Raymond Eskenazi
Culver City, CA

Introduction

Television cameras have the potential to provide for a robot what eyes provide for a human: the richest source of sensory information. But before a computer can get useful information from a camera, the analog video signal must be converted to a digital representation in the computer's memory. This is the function of a video image *digitizer*.

The conversion process begins by partitioning the image into a rectangular array of picture elements, or *pixels*. At each pixel, the digitizer measures the average brightness, assigns an integer *gray-level* that corresponds to that brightness, and stores it in computer memory. The digitized image, therefore, is typically stored as a two-dimensional array of gray-levels. Each gray-level's location in the array—its row and column position—corresponds to the pixel's location in the image (or is directly related to it). Figure 1 illustrates this scheme.

In this article I will describe the design of an inexpensive digitizer. For about $50 in parts, it will interface a television camera to a standard microcomputer DMA (direct memory access) port and store an array in the computer's memory that is derived from the video image. First, though, let's explore some of the considerations that influence the design.

Digitizer Design Considerations

The spatial *resolution* of a digitized image is related to the number of pixels in each image row and column defined by the digitizer. The theoretical resolution of an image refers to the volume of information available. When more pixels are sampled in an image, the image is usually represented with more accuracy, up to the theoretical limit. 35mm motion picture film, with a resolution of about 2000 × 2000, has a far greater image quality than standard broadcast television, whose resolution is on the order of 480 × 320 pixels. The human eye detects an image with a resolution of more than 4000 × 4000 pixels.

Clearly, increasing the spatial resolution of the digitizer beyond the theoretical resolution of the camera provides no increase in image information. Image digitizers that operate by scanning film can be designed with a great deal of flexibility, as the spatial resolution can be set rather arbitrarily. For a video digitizer, however, this is not the case, since the serial signal transmission requirements dictate that the image must be partitioned into rows (lines) at the camera. Thus, our row resolution is limited by the scan circuitry in the camera and is either difficult or impossible to change.

The horizontal (column) resolution of a video signal, however, is determined by the quality of the camera. More expensive cameras have a higher data rate, or bandwidth. Studio cameras, costing thousands of dollars, have the greatest bandwidth. An inexpensive monochromatic camera for home video or closed circuit television (CCTV) surveillance typically has less than a 3 MHz bandwidth, for which a 256 pixel column resolution is reasonable. Between these extremes lie numerous other types, including the commercially available solid-state CCD or CID cameras whose resolution is fixed by the lattice of light-integrating cells in the silicon substrate.

These solid-state cameras, by eliminating the analog drift intrinsic in vidicon scan circuitry, provide the most accurate calibration for automatic range measurement. However, their relatively high cost ($2000-5000), places them beyond the budget of most schools and individual experimenters. For our design, therefore, we will assume

that a cheap CCTV camera will be used, and select a column resolution of 256 pixels per line.

The Electronics Industries Association (EIA) has defined a standard format for video signals used in the U.S., designated RS-170. In this standard, the image has 480 lines and is divided into two *fields*—each consisting of 240 lines scanned from top to bottom. One field contains all the odd numbered lines; the other field contains the even numbered lines. Each field is scanned in 1/60 second, and scanning is alternated between fields. Between each field, the scan has several line periods to return to the top, bringing the total number of horizontal cycles per complete vertical cycle to 525. If we sample one complete field, the spatial resolution of the image will be 240 × 256 pixels. Figure 2 diagrams the EIA video signal format.

Gray-scale resolution is another important characteristic of the digitized image. This is the number of possible brightness values that can be assigned to any pixel. In many image processing applications, an 8 bit gray-scale is used, for a resolution of 256, where zero represents a completely dark pixel and 255 represents one of maximum brightness. With one byte of storage required per pixel, a 240 x 256 pixel image requires 61,440 bytes of computer memory—practically enough to consume the 16 bit address space of most micros!

One approach that is becoming more economical as memory prices come down is to dedicate an entire separate address space to storing the image. With the memory alternately accessible from the computer and the

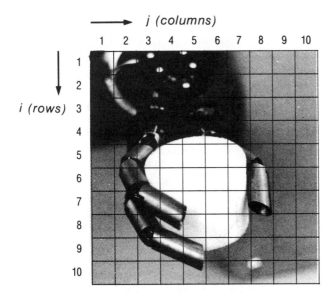

Figure 1. The first step of digitization is to partition the image into cells (pixels) addressed by row and column, shown (exaggerated) on the left. Within each pixel, the digitizer measures and assigns a number corresponding to the image brightness (right).

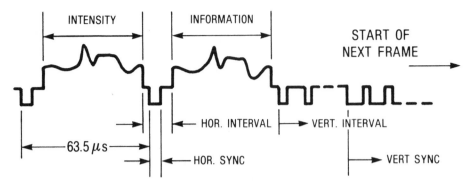

Figure 2. An EIA RS-170 composite video signal. The period between successive horizontal intervals represents one horizontal scan line in the image. The start of the horizontal interval terminates the intensity information on the line, and the horizontal sync pulse triggers a "flyback" to the start of the next line down. A full frame of lines is terminated by the start of the vertical interval, and vertical sync causes the scan of the next frame to begin at the top.

digitizer, the latter can deposit a digitized image in the memory at the full video rate, later making individual pixels available to the computer at memory access speeds. Because of its similarity to taking a "snapshot" of the video scan, this technique is called *frame grabbing*. Many types of frame grabbers are commercially available, some even for the S-100 bus (the CAT-100 by Digital Video Systems, for example), but the prices are still relatively high ($1000-$10,000), and constructing one from parts would be an ambitious project for the amateur.

Fortunately, there are many ways we can reduce the amount of image memory needed, making it feasible to use the computer's main memory for image storage. We could, for example, store only a portion of the image. With two pixel addresses, we can define the upper left and lower right corners of a rectangular region in the image. If this "window" is programmable, we can read only the rectangular portion of the image that contains useful information or that we have room to store and process.

We can further reduce the amount of memory the image needs by sacrificing spatial resolution. But this throws away valuable information—for many applications it would be quite useful to have the full 240 x 256 pixel resolution. It makes more sense to lower gray-scale resolution. If we allow each pixel only two possible gray-levels (0 or 1), we reduce our storage needs eightfold, to less than 8k bytes. Using this "binary image" has the added advantage of making our analog-to-digital conversion (ADC) circuitry much simpler and less expensive.

In addition to the problem of allocating memory for an entire image, the video input system must deal with the high data rate, or bandwidth, of a video signal. If we want to read in a 256 pixel line at the full video rate, we would need a data rate of over 5MHz—beyond the capability of most microcomputers and high enough to effectively stall

computation on most minis. By accepting a lower gray-scale resolution and buffering several pixels per data word—in our case 8 per byte—we do not need as high a data rate.

If our camera were looking at an unchanging scene, we would not have to grab the whole image in 1/30 second. We could lower the data rate by sacrificing digitizing time rather than resolution. There are two common ways of doing this: with *point* and *column* digitizers. A point digitizer keeps track of the pixel row and column address in the scan, and when it matches the address of the desired point, it samples the video signal, converts it to a gray level, and presents the result to the computer. The computer may have to wait up to 1/30 second for the desired point to "come around" in the scan. This is the scheme used in the "Digisector," by Microworks, Inc.

A column digitizer is able to read in an image much faster. Given the number of a desired column, it digitizes the pixel for that column on each successive line in the image. At the end of 1/30 second, when every line has been scanned, a complete column of pixels has been stored in memory. Reading only 30 columns/sec (max.), a column digitizer would take over 8 seconds to convert an entire 256 column image. Since a robot vision system usually has to react to a changing environment, I chose to discard both these low-speed input schemes in favor of frame grabbing—reading the image in one 1/30 second scan, but with only one bit per pixel to reduce the data rate and storage requirements.

We can design the digitizer so that it grabs one video field of 240 x 256 pixels every second and stores it as a binary image in memory. Adding a few more requirements, though, would make it a much more powerful device. First, the actual data transfer should be performed by a standard DMA card. For image transfer using this digitizer design on an 8 bit machine, the DMA must be able to handle one byte every 1.6 microseconds. A 16 bit machine would need to handle only a 3.2 microsecond DMA cycle.

Another attractive feature is *windowing*. Sometimes only a portion of an image is worth considering. The digitizer should be able to select any rectangular window in the image, and read in only that part. The window should be set under computer control.

Variable thresholding is another function that would

improve the digitizer's power. Instead of having just one fixed analog level as the boundary between the digitized values of 0 and 1, we would like to set an arbitrary level under computer control. Furthermore, we can extract more information from the picture by setting *two* thresholds and having the digitizer output a "1" only if the brightness falls *between* them and "0" otherwise. I have provided this feature, in addition to windowing, in the digitizer design presented here.

The Question of Synchronization

Most commonly available CCTV cameras produce a *composite* video signal (Figure 2) which contains the image intensity information as well as the horizontal and vertical sync signals that terminate a line and field, respectively. This poses a problem in the design of a video digitizer, since the digitizer must use these signals separately to reset its row and column counters. A sync separator circuit could be added to extract these signals (and this is the approach used in some commercially available systems), but these circuits are not exactly simple, and adding them increases both the cost of the digitizer and the effort required to build it. They also do not solve the other problem that arises when we use composite sync from the camera—the accuracy of the pixel address.

The horizontal and vertical oscillators used in inexpensive video cameras are usually entirely analog, and therefore subject to drift and various other instabilities. Attempting to synchronize the pixel clock, with a frequency 256 or more times the horizontal rate, with such an external horizontal sync reference will result in the column addresses near the end of a line becoming increasingly less accurate, with concomitant loss of accuracy when these addresses are used to compute the 3-space coordinates of targets from their position in an image. This inaccuracy can occur, though to a lesser extent, even if a solid-state camera with geometrically precise pixel locations is used—unless you have access to the same pixel clock used by the camera (which fortunately is the case with most of these cameras).

A much better solution, and the one selected for use in this digitizer, is to produce all video timing signals, both for the camera and the digitizer, from a single crystal-controlled master pixel clock. The only problem with this approach is that you must either obtain a camera that is capable of running on externally provided sync signals or get the schematics and modify one so it can. The work involved in modifying one is actually much easier than

building a sync separator and results in a much cleaner and more accurate system.

The key to this approach is that there are inexpensive (about $10) TV sync generator ICs readily available from numerous sources that generate all necessary video timing signals. The one I selected for this circuit is the MM5320N (or the MM5321), by National Semiconductor. Clocked at a multiple of the pixel clock frequency, this chip produces either composite sync or separate horizontal and vertical drive signals for your camera, plus some other essential information.

One important signal that the 5320 gives the digitizer tells it where to look on the video signal for the DC level that corresponds to black. Since video signals are usually capacitively coupled, you cannot count on this "black" level to be at the signal ground, where it should be for the digitizer to make accurate conversions. As part of the RS-170 video signal standard, the EIA has specified that a portion of the signal immediately following the horizontal sync pulse should provide the black DC reference. The 5320 provides a signal that is active only during this interval, called the "Color Burst Gate" (CBG), since the interval may also contain color reference information. Of course, we shall be concerned here only with black and white signals.

The Digitizer Design

Given all the design considerations, we are ready to look at the functional architecture of the digitizer, shown in Figure 3. The pixel clock, shown in the lower left, provides a reference for the sync generator and also serves as the clock for the pixel column counter and for the shift register that buffers eight pixel values prior to their output to the computer. The signals produced by the sync generator are transmitted to the camera by 75ohm coaxial line drivers. Horizontal and vertical drive signals are also used to preset the column and row counters, respectively.

Synchronized with the reference drive signals, the camera returns its video output to the digitizer. The CBG signal from the sync generator triggers a DC restoration circuit that clamps the black reference to ground, resulting in a normalized analog video signal with image intensity levels across a line ranging from 0 to a maximum of .7 Volts. This referenced video is made available to the threshold comparators for conversion to binary gray levels.

The threshold levels used in the comparisons are set by the computer. Eight data lines from the CPU bus are

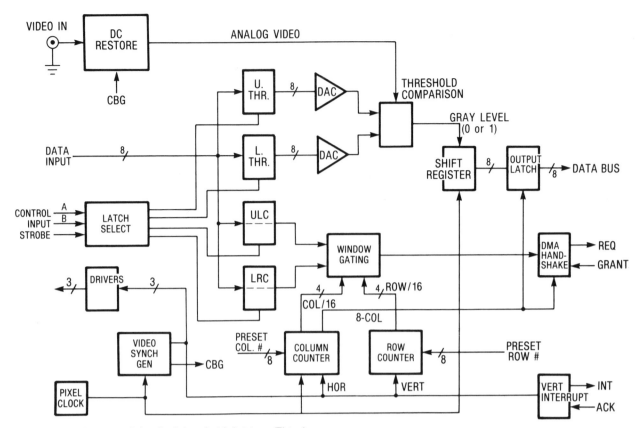

Figure 3. Block diagram of the dual-threshold digitizer. This design presumes that a DMA interface to the computer will be used. Two control bits (A and B) are required to select threshold or window registers for loading. Standard DMA and interrupt handshaking are provided.

latched into 8 bit upper and lower threshold registers. Two additional 8 bit latches hold the two window corner coordinates. The choice of which of the four registers is loaded by the CPU is made by a 1-of-4 selector circuit that requires a two bit control word. The number present on the two control lines when the data strobe from the bus occurs is used to select the appropriate register.

Each threshold register has at its output an 8 bit digital to analog converter (DAC), whose output is scaled so that the digital threshold range of 0 to 255 results in an analog output of 0 to .7V. Since these converters need only respond to changes in the threshold level set by the computer, inexpensive slow converters can be used. The dual threshold comparator outputs a pixel value of "1" only if the normalized video signal level lies between the two thresholds, and zero otherwise. The results of the comparison are clocked into the output shift register by the pixel clock.

Image windowing is achieved by a digital comparator that ensures the current pixel address lies within the rectangle defined by its upper left corner (ULC) and its lower right corner (LRC). To simplify matters, the row and column numbers used in corner coordinates must be a multiple of 16. Since the highest row or column number is always less than 256, this permits each corner coordinate

to be expressed in a single 8 bit byte, the four most significant bits (MSBs) determining the row and the least significant bits (LSBs) the column.

The final function to be performed is the transfer of the converted pixel values to the computer. A signal taken from column counter corresponding to one eighth the pixel frequency and is used to deposit in parallel the contents of the pixel shift register into an output latch once every 8 columns (on a multiple of 8). The same signal is used to latch the result of the window comparator onto the DMA request line of the interface, so that no request will be made unless the pixels come from within the window. The handshaking is completed by the interface granting the request, resetting the latch. The request must be granted and the data read from the output register before the next output byte is loaded into the register 8 columns later. For the 5.040MHz pixel clock used in this circuit, a data byte must be read by the computer every $1.6\,\mu s$ during the window. The digitizer requests an interrupt at the start of each field to call the program to prepare for an image transfer.

Circuit Description

Dealing with each of the functions in turn, let's refer now to the actual circuit design to see how each is realized in hardware. I recommend using low-power Schottky TTL chips (74LS series) throughout the circuit, due to their superior time-power performance. The exception is the

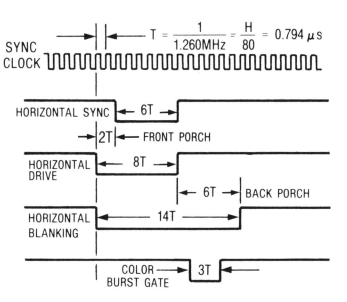

Figure 4. Detail of the horizontal interval. The horizontal drive (HD) signal is provided by the sync generator, lasting 8 sync clock periods. Blanking extends an additional 6 periods (back porch) during which the CBG signal defines the video DC reference.

pixel clock, which is made with 74S04 hex inverters for the sharpest (fastest) clock transitions. The inverter feedback configuration shown makes a reliable clock, but you may also use a standard clock chip or other clock designs, if desired. Although the 74LS (and 74S) devices are preferrable, ordinary TTL will work just as well, but a much stronger 5V supply will be needed if the 74LS chips are not used.

IC11, a 74LS163 synchronous counter, is used to divide the pixel clock by 4 to produce the 1.260MHz sync clock needed by the MM5320N sync generator, IC1. With pin 2 at –12V, the 5320 expects this clock frequency, which is 80 times the EIA standard horizontal rate (horizontal sync every 63.5µs). The relationship between the sync clock and the various signals in the horizontal interval is shown in Figure 4, which gives a precise description of the EIA "front porch" and "back porch" intervals and the CBG interval used for DC restoration.

The back porch, between the rising edge of the horizontal drive and that of the horizontal blanking signal, is of particular significance. When the blanking signal is high, the video signal contains valid intensity information. Since the 5320 only provides composite (horizontal and vertical) blanking, the digitizer uses the horizontal drive (HD) signal and the pixel clock to measure and skip this interval before starting to count the columns of the image. The timing diagram gives the duration of the back porch as 6 ticks of the sync clock, which translates to 24 ticks of our pixel clock. We will use this number later when we choose the preset value of the column counter.

Depending on the type of camera or how you modified it, you will need either composite sync (CSYNC) or both HD and VD signals at the camera. The circuit shows all three of these references converted to signals suitable for 75ohm coax by 74128 drivers (IC0). Another useful piece of information provided by the 5320 is the field index signal on pin 9. This signal goes low for one sync clock period at the start of the vertical drive (VD low) interval, but only on *odd* fields. This information will be made available to the CPU to identify which field a particular window of converted video data came from.

The DC restoration function is realized by transistors Q1 and Q2 (2N2907) and their associated passive network. Although the specs of the 5320 label the signal defining the reference interval as CBG and not $\overline{CBG}$, the signal is actually an active low. Therefore, an inverter is needed to produce an active high pulse to trigger the DC restoration circuit. When this "$\overline{CBG}$" is high, Q2 is turned on, clamping its emitter to (near) ground. Once it turns off, the time constant of the input capacitor and the 100K resistor is sufficiently long to hold the DC reference esentially constant until the next update.

The four 8 bit data registers are made using eight 74LS175 4 bit latches, IC3-10, assigned in pairs to the respective functions. Input data lines from the CPU bus are mapped appropriately to their inputs. (The input pinout shown on IC10 is used on all.) The clock (load) inputs of a given pair of latches are connected to one of the four outputs of a 74LS139 selector, IC2, which pulls one output low when strobed by the CPU, loading the selected register. Depending upon your choice of DMA interface, the two input bits that control the selector may be available as control or status bits from the interface, otherwise two bits from some other parallel port must be used.

There are a variety of choices available for the DACs used in the digitizer. Since the threshold values held in the latches should be changed no more than once per frame, practically any 8 bit DAC can be used. To keep the cost low, I selected the DAC-IC8BC by Datel (about $5). Two identical converters, configured as shown in Figure 5, are used as DACs 1 and 2 in the digitizer circuit. This converter provides an output current *sink* on pin 4 proportional to the digital input times an externally supplied referece current. This current is translated to a voltage ranging from 0 to .7V by the resistor network shown, but the inverting nature of the current output requires that the *complement* of the digital threshold value held by the latches be used as the input to the DAC. As the value in the latch varies from 0 to 255, the DAC input varies from 255 to 0, so that the translated voltage from the resistor divider varies from 0 to .7V, the standard range of normalized video. If the converter configuration you use

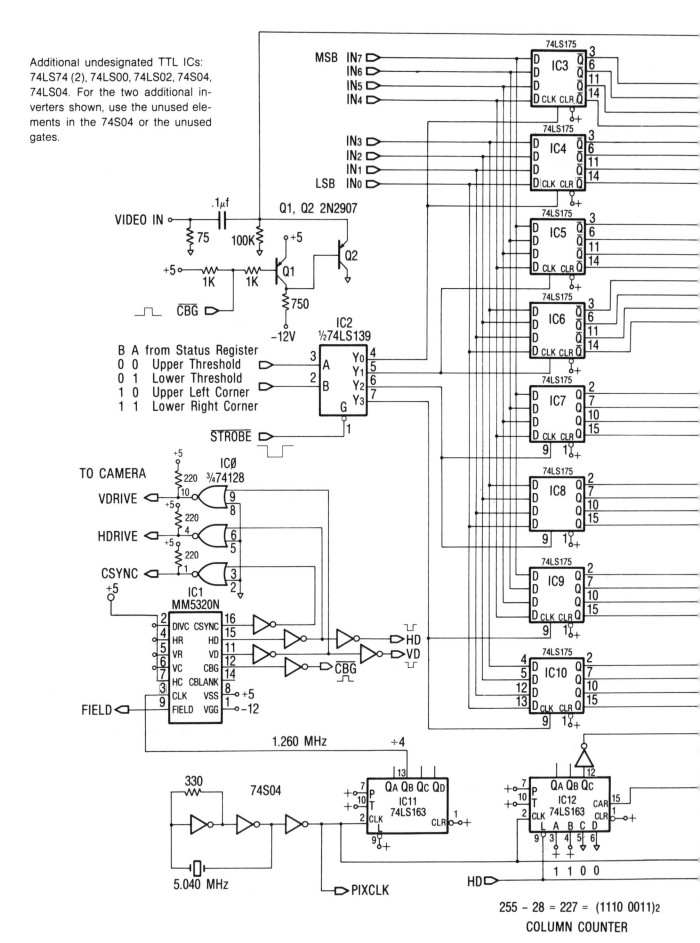

Additional undesignated TTL ICs:
74LS74 (2), 74LS00, 74LS02, 74S04,
74LS04. For the two additional inverters shown, use the unused elements in the 74S04 or the unused gates.

MSB IN7
IN6
IN5
IN4

IN3
IN2
IN1
LSB IN0

VIDEO IN

.1μf

75   100K

Q1, Q2 2N2907

+5

Q2

+5
1K   1K   Q1
750

CBG

−12V

IC2
½74LS139

B A from Status Register
0 0 Upper Threshold
0 1 Lower Threshold
1 0 Upper Left Corner
1 1 Lower Right Corner

STROBE

TO CAMERA

IC0
¾74128

VDRIVE
+5
220

HDRIVE
+5
220

CSYNC
+5
220

+5

IC1
MM5320N

DIVC CSYNC
HR   HD
VR   VD
VC   CBG
HC   CBLANK
CLK   VSS   +5
FIELD VGG   −12

FIELD

HD
VD

CBG

74LS175 IC3
74LS175 IC4
74LS175 IC5
74LS175 IC6
74LS175 IC7
74LS175 IC8
74LS175 IC9
74LS175 IC10

1.260 MHz

÷4

330   74S04

5.040 MHz

QA QB QC QD
P
T   IC11
CLK 74LS163
CLR

QA QB QC
P
T   IC12   CAR
CLK 74LS163
CLR
L A B C D

1 1 0 0

PIXCLK

HD

255 − 28 = 227 = (1110 0011)2

COLUMN COUNTER

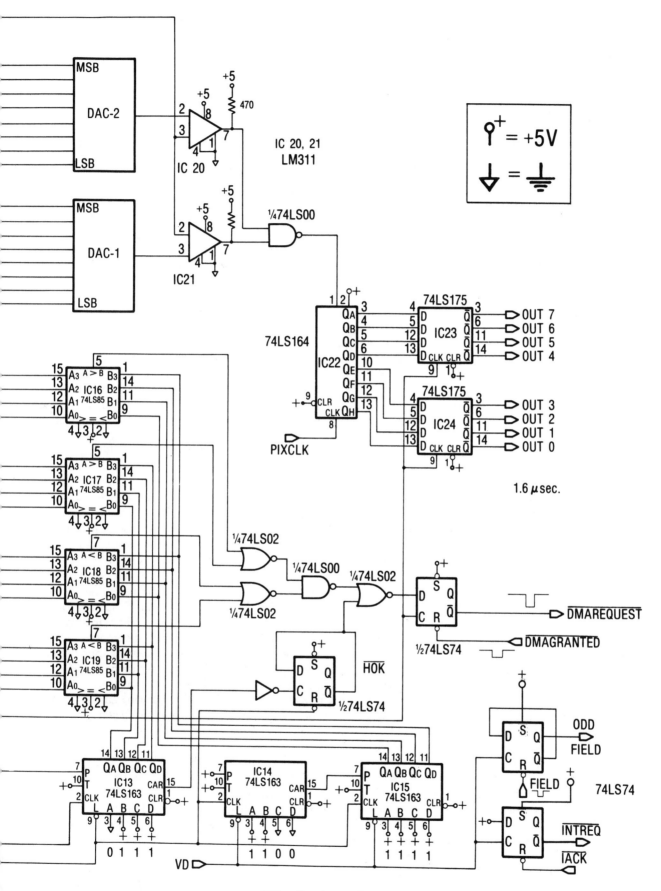

255 – 12 = 243 = (1111 0011)₂

LINE COUNTER

INTERACTIONS: SENSES, VISION, AND VOICE

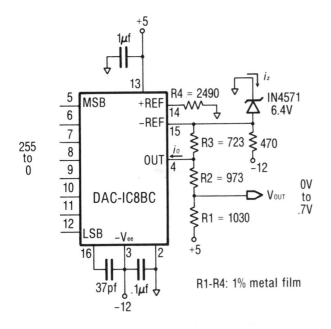

Figure 5. *Configuration of the DAC-IC8BC digital to analog converter. Digital input varying from 255 to 0 results in analog output from 0 to .7V. Metal film resistors with 1% tolerance are recommended. Other 8 bit DAC's may be used.*

needs the normal binary number (0-255) as input, then use the "Q" outputs of the latches, whose pin numbers are shown on the window latches, IC7-10.

If your robot vision application depends on the absolute accuracy of the thresholds, then the values of resistors R1-R4 in the DAC circuit are critical. R4 and the zener diode establish the reference current for the DAC, whose output is proportional to their values. R1-R3 form the divider that scales the output voltge to video levels. I recommend using metal film resistors with 1% tolerance, both for accuracy and temperature stability. If threshold accuracy is not critical, then carbon resistors may be used.

The output of each DAC is an analog threshold level, which is used as input to a common LM311 comparator (National Semiconductor), ICs 20 and 21, respectively. The output of DAC-1 is on the non-inverting input of IC20, so that the comparator's output is high only if the video level is *below* the threshold. Conversely, the lower threshold level from DAC-2 is on the inverting input of IC21, whose output is therefore high only when the video is *above* the threshold. A NAND gate is used to conjoin the two outputs, so that the input to the pixel shift register, IC22, is the *complement* of the actual pixel value. This is remedied later by using the complement outputs of the output register. The last 8 pixel values clocked serially into the shift register are simultaneously available in parallel on its output pins.

Counters IC12 and IC13 hold the LSBs and MSBs, respectively, of the pixel column number, which is determined by counting the pixel clock. The horizontal

drive signal HD is used to preset the counter at the beginning of each new video line. As mentioned earlier, the actual image information does not start at the end of the HD interval, but rather at the end of the blanking interval, 24 pixel periods after HD goes high. Also, note that since the sync clock used by the LM5320 is 80 times the HD rate, there are actually 320 pixel periods between horizontal sync pulses. As shown in the 5320 timing diagram (Figure 4), the horizontal blanking interval is 14T = 56 pixels, leaving 264 pixels of actual image, 8 more than we can use. By discarding the first 28 pixel counts following the HD interval, we skip the back porch and center our 256 columns in the valid video data. This is accomplished by presetting the column counter to 255 − 28 = 227. The period between the two overflows of the column counter (following HD high) define our valid column numbers.

The appropriate preset value for the line counters, IC14-15, is determined similarly. The vertical blanking period lasts 12 horizontal periods longer than the end of the vertical drive (VD low) interval. Since the line count is obtained by counting HD pulses directly, the VD signal is used to preset the line count to 255 − 12 = 243. Since there are only 240 image lines in a field, however, you should restrict your vertical windowing range to line numbers numbered from 0 to 239, inclusive.

The comparison of the 4 bit "nibbles" that define the MSBs of the corner coordinates is performed by four 74LS85 4 bit comparators, IC16-19. The column number MSBs are bussed to ICs 17 and 19, which compare them with the left and right column coordinates, respectively, and the row MSBs go to ICs 16 and 18, for the top and bottom row comparison. To get the windows to range over all the available rows and columns, the comparisons must be *inclusive*. Since the LSBs of the counters don't matter, a zero in a latch must cover a counter range of 0-15, and a latch value of 14 must cover a counter range from $14 \times 16 = 224$ to 239, etc. This inclusive windowing can be obtained with the comparator outputs gated as shown. The output of the NAND gate is high when any window extremum is exceeded and low otherwise.

The carry output of the column counter (suitably inverted) is used to to toggle a flip-flop (FF) whenever the column number rolls over from 255 to 0. Resetting the FF with HD (which presets the counter) guarantees that the first rollover following the HD interval will define the start of valid image column numbers. A signal to this effect ($\overline{HOK}$) is gated with the result of the window comparison. The output, high whenever valid video data lies within the desired window, is used to enable a DMA request to the interface. This request, however, must be initiated only

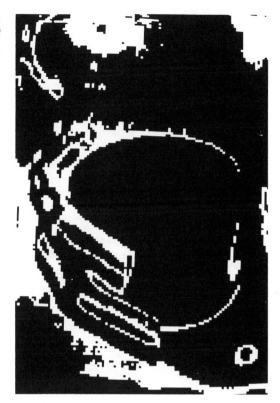

The results of a bi-threshold digitization. White pixels in this photo had grey levels between the two thresholds in the original scene. The use of binary images for robot vision usually requires specially lighted high contrast scenes.

after 8 pixel values have been buffered. Inverting the "4" bit of the column yields a signal that goes from low to high every 8 pixels on a multiple of 8—exactly the moment to latch the contents of the shift register into the output latch and initiate the DMA request.

The digitizer is completed with the addition of one more 74LS74. Both its FFs are clocked by the vertical drive signal VD. One is used to post an interrupt request at the beginning of each field, and the other is toggled by VD to identify the odd field. The field index signal from the LM5320 resets this latter FF to synchronize the odd field signal. These FF's, as well as the output data latches and DMA request FF, may be unnecessary if your DMA interface provides latches that can be used for these functions.

Some other digitizer design options are worth mentioning, too. If you're willing to spend more cash to save some assembly time, you can use a DAC that has data latches built in, such as Datel's DAC-UP8BC (about $14 each), and eliminate latches IC3-6. Refer to the Datel reference manual for details. Another option is to not use dual thresholding at all. If you left out DAC-2 and IC21, connecting the output of IC20 directly to the shift register, then the resulting image in memory will have "1"s for pixels *above* threshold 1 and zero for those below it. If you have access to separate horizontal and vertical sync signals, from a sync separator or other external source, you may use these to synchronize the LM5320 to external sync. (You will also need to preset the sync clock, IC11, appropriately.) Refer to the 5320 data sheet for further information.

The Digitizer in Action

To use the digitizer, it is first necesary to load the threshold and window registers. To load the threshold registers, first set the control bits to select the register you want, then output the desired threshold value to the memory or I/O address assigned to the digitizer interface. The window corner registers are not as simple, since both the row and column numbers of a corner must be encoded into a single 8 bit data byte for output to the digitizer. (If you have an additional control bit available to the digitizer, you may consider altering the design to select the 4 bit latches for the corner coordinates individually.)

Since only the four MSBs of the pixel row and column address are used for window comparison, the row and column coordinates of the ULC and LRC must both be multiples of 16. The *grid coordinates* (m, n) of a corner are

the two 4 bit numbers obtained by dividing a corner's image coordinates (row, column) by 16. Since the LSBs are not compared, grid coordinates (m, n) refer to all rows and columns between (16m, 16n) and (16m+15, 16n+15), inclusive. Thus, if the ULC and LRC have identical grid coordinates (m, n), a minimum 16 × 16 pixel window starting at (16m, 16n) will be transferred to the computer. To load a window corner register, place the grid coordinates (m, n) into the left and right (MSB/LSB) halves, respectively, of a single byte, select the appropriate register, and output the byte. Do the same for the other corner. Remember that since the last line of valid image information is on row 239, the highest meaningful row grid coordinate for the LRC is 14, since 14 × 16 + 15 = 239 (conveniently!).

Conclusion

Robot vision was once possible only on large, expensive computers, using either slow sampled conversion strategies such as point or column digitizers, or expensive frame grabbers and video image memories. Now, inexpensive digitizers, such as the one just described, are making vision systems affordable to many more schools, groups, and individuals. For example, the Dithertizer II™ by Computer Station is a complete, assembled digitzer with an Apple interface, employing some of the same principles as the one described here, for about $300—within reach of all but the most constrained budgets. Whether you build the digitizer described here or decide to buy one, one message is clear: microcomputer-based robot vision has arrived! ®

References

[1] *The TTL Data Book*, Texas Instruments Inc., Dallas, Texas. This is the professional design engineer's complete TTL reference.

[2] *TTL Cookbook*, by Don Lancaster. H. W. Sams & Co., Indianapolis, Ind. If you have never built TTL projects before, *this is essential reading*. Although a little dated (1974), it provides an excellent introduction to logic design and proper circuit layout and construction techniques.

[3] *MOS/LSI Databook*, National Semiconductor Corp., Santa Clara, Calif. Contains LM5320-5321 specifications and timing diagrams.

[4] *Engineering Product Handbook*, Datel-Intersil, Inc., Mansfield, Mass. Contains specs and application information for 8 bit DACs DAC-IC8B and DAC-UP8B.

SEGMENTING BINARY IMAGES

Robert Cunningham
Robotics Research Program
NASA Jet Propulsion Laboratory
California Institute of Technology
Pasadena, CA

Generalized image interpretation is still one of the most challenging and important research areas in robotics and machine intelligence. Although most current research is concerned with understanding gray scale pictures taken from an arbitrary 3D perspective, earlier results dealing with *binary* images, usually of objects confined to a plane, are being used in practical robot vision systems for a variety of applications. The main reasons for this are the compactness of representation and the simplification of the image analysis algorithms that result from structured scenes with only one bit of intensity information at each point in the picture. A 256 by 256 element binary image stored as a packed bit array requires only 8192 bytes of memory. Thus even a modest microcomputer can store a complete binary image of average spatial resolution and still have plenty of memory left over for image analysis software.

The algorithms that analyze a binary image are simpler than those which analyze gray scale images. With a binary image, the computations used in such functions as locating the edge of an object are reduced to logical operations involving only a few picture elements (pixels). As a result, binary image analysis programs are generally much faster than comparable programs that process gray scale images.

The process of producing a binary image significantly reduces the volume of data available from a standard TV camera. Of course, this reduction results in a certain amount of information loss as well, and this limits the applicability of binary image vision. However, in many robot vision tasks, particularly industrial ones such as locating parts on a conveyor belt, the designer can structure the environment to control the viewing conditions. By carefully choosing the illumination, viewing angle, and background composition, we can make it possible for vision software to distinguish target objects from the background on the basis of contrast alone. Under these circumstances, binary image vision methods can be quite effective.

One of the basic goals of a computer vision system is to determine the position and orientation of a target. A robot can use this information to guide tasks such as grasping an object, positioning a tool (as in welding or grinding), or mating parts in an assembly. In addition to locating objects, some jobs require the robot to identify what *kind* of object it is seeing or to inspect it for defects. In a controlled environment, the efficiency with which binary image vision can perform these functions has led to its use in several successful industrial systems. [refs. 1, 2]

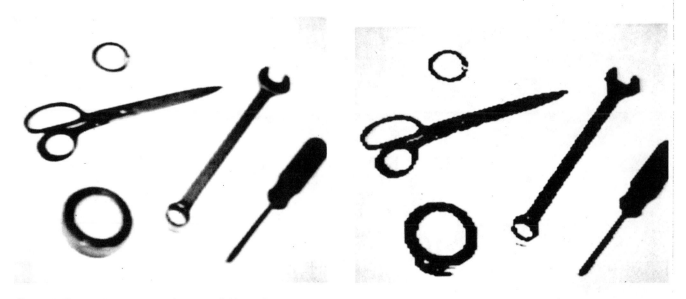

Figure 1. Converting a gray scale image (left) to a binary image (right) is done by setting to a 1 all pixels whose brightness lies above a threshold and to 0 all those below it.

Regardless of the ultimate application, the first thing any binary image vision system must do is to partition the image into regions, or *blobs*, that correspond to objects, holes, or background in the scene. After partitioning, it can analyze the blobs to compute information needed for the task at hand.

In this article, I will present two different algorithms that produce descriptions of the outlines of blobs in a binary image. One approach, called *connectivity analysis*, is most suitable for tasks where location information and shape *statistics* are sufficient. The other, called *boundary tracking*, produces a precise vector description of the outlines in addition to the other results. Following this, I will discuss some of the issues involved in using the blob descriptions for object recognition, tracking moving targets, and stereo vision for rangefinding.

Getting a Binary Image

Binary images are obtained by a thresholding process that converts a continuous tone or gray scale image to one with only two "colors," black and white, much like a silhouette. There are many ways of accomplishing this conversion, either in the video input hardware or, if a gray level image is available, in software. The simplest approach, called *global thresholding*, is to compare the brightness of each pixel to a common threshold level, assigning a value of 1 (white) wherever the threshold is exceeded and 0 (black) otherwise, as shown in Figure 1. A common extension of this approach is to use two thresholds, setting to white on only those pixels whose brightness falls between the two levels.

Global thresholding with either single or dual thresholds can be built directly into the video digitizer as a one bit analog to digital converter. *(See "Video Signal Input" [ref. 3] in our previous issue.)* With global thresholding, you must experimentally determine the threshold level(s) that give the best results in a particular environment. If, however, a gray scale image is available, the vision software can determine the best threshold to use by looking at the intensity *histogram* of the image.

The histogram (Figure 2) shows the number of pixels with a given gray level, plotted as a function of gray level. (This is represented in our program as an array, N(G), indexed by gray level.) Typically, to find a good global threshold, analyze the histogram to locate the valley between the peaks that correspond to object and background brightness levels, respectively, then set the threshold to the gray level at the valley.

Another approach is to use not a global threshold for the entire image but a *local threshold*. The threshold used at each pixel is determined by examining the neighboring pixels in a small *window* of the image. See Weszka [ref. 4] for a more complete discussion of threshold selection techniques.

The Blob-Finding Algorithms

Assuming we have acquired a binary image in a two-dimensional bit array, let's now consider how to describe its contents. Formally, a blob is a simply connected cluster of pixels which are all the same color (i.e., black or white). Since a picture may be composed of a number of blobs, including one for the background, we can most easily

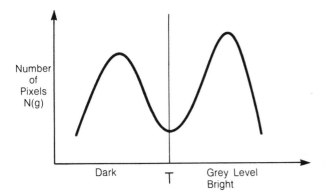

Figure 2. The histogram above represents an image with bright objects on a dark background similar to Figure 1. One method of selecting a threshold to obtain a binary image is to analyze the shape of the histogram, setting the threshold at the gray level corresponding to the valley (local minimum) which separate object pixels and background pixels.

represent it as a linked list of blob *descriptors*. Descriptors are records that contain everything we know about a blob: its area, centroid, perimeter, number or holes, color, and possibly other shape or type information. In addition, there are links which superimpose a hierarchical structure on the blob list to represent *nesting* relationships. These relationships are described by the words *parent*, *child*, and *sibling*.

In Figure 3, blob 1 has two children, blobs 2 and 4; the parent of blobs 2 and 4 is blob 1; and blobs 2 and 4 are siblings. Including a descriptor for the background in the blob list is convenient because it unifies the nesting relationships: blobs 1 and 3 are siblings whose common parent is the background. The parent-child relationship applies to *immediate* containment only, so that blobs 2 and 4 are not children of the background. Note that in a binary image, if two blobs are adjacent (touching), then one must contain the other, since adjacent blobs must have different colors.

Our first method of finding blobs and building the blob list description of an image is *connectivity analysis*. As its name suggests, this algorithm examines the local connectivity around each pixel of the image, building up individual blob descriptions on a pixel-by-pixel basis. Our second method is *boundary tracking*. This algorithm infers the connectivity of pixels by following the edge of a blob to determine its complete extent and outline. Connectivity analysis is the more elegant approach of the two in that all processing is done in a single raster scan in the image. The boundary tracker alternates between a raster scan search for new blobs and following the boundary, which requires random access to the image.

However, the boundary tracking algorithm has an important feature that connectivity analysis lacks—it produces a *chain-code* description of the boundary of each blob as one of its primary outputs. This is an encoded sequence of unit direction vectors that define the exact shape of the blob's outline. *(See the article "Chain-Code" [ref. 5] in our previous issue)* This important product can make the boundary tracking algorithm the most useful of the two partitioning methods for applications that require an explicit shape analysis of object boundaries.

Before going into the details of the two approaches we must consider some problems with the definition of connectivity that impact both algorithms. Adjacent pixels are *connected* if they are the same color, but we need a convention to specify *which* pixels are adjacent. The two most natural conventions in a rectangular grid are *4-connectivity* and *8-connectivity*. (See Figure 4) 4-connectivity allows a pixel to be connected to any of its four nearest neighbors in the cardinal directions—up, down, left, and right. 8-connectivity allows it to be connected to all eight of its immediate neighbors by including the diagonal directions as well. This is where a problem can arise: if, for example, we use 8-connectivity to cluster pixels into white blobs, then we must use 4-connectivity to define black blobs, or *vice versa*. This is necessary to avoid a paradox that arises in images where a blob meets itself at a single point, such as point *A* in Figure 5a or 5b.

If all blobs are 8-connected, then from the object's point of view, the central black square is disconnected from the

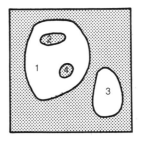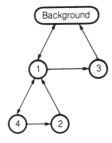

Figure 3. The blob list forms a tree-like structure to represent nesting relationships. The scene at the left contains four blobs. Blobs 1 and 3 are contained in the background. Blobs 2 and 4 are contained in blob 1. The diagram at the right indicates the linkage information stored in the blob records. The double arrow from blob 1 to blob 4, for example, indicates the CHILD link of blob 1 and the PARENT link of blob 4. The arrow from blob 4 to blob 2 is the SIBLING link of blob 4.

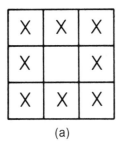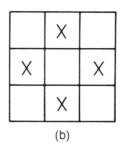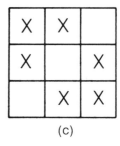

(a)   (b)   (c)

Figure 4. The x's indicate which pixels in a 3 by 3 neighborhood are connected to the center pixel using (a) 8-connectivity, (b) 4-connectivity, and (c) 6-connectivity.

surrounding background and is thus a hole is the object. However, from the central square's point of view, it is connected to the background and thus is not a hole. If all blobs are 4-connected, the object views the square as part of the background while the square views itself as a hole in the object. Either way, it is impossible to generate a consistent hierarchical blob description of the scene.

One solution to this problem is to adopt a double standard which applies 4-connectivity to one color and 8-connectivity to the other. This complicates the blob finding algorithms to a certain extent, but not prohibitively so. Another solution is to adopt a 6-connectivity convention [2] which treats black and white pixels equally and avoids the paradox of Figure 5.

Unfortunately, however, 6-connectivity is not a perfect solution either. In Figure 5a the central square is 6-connected to the background, while in Figure 5b, which depicts the same object rotated 90°, the square is not connected to the background. Thus, while any image may be described consistently, the description of a particular object may vary from scene to scene. Actually, due to quantization errors in digitized images, it is unlikely that an object will consistently appear as only one of the Figures 4a or 4b. Instead, it is more likely that pixels in the neighborhood of point A will appear as either black *or* white depending upon the location and orientation of the object in a particular scene. The object could appear as it does in Figures 5c or 5d as well, so that we can't expect a consistent description of such an object from one scene to the next regardless of the connectivity convention we use. We are therefore left with an essentially arbitrary decision as to which connectivity convention to adopt. In the interests of simplifying the algorithms presented here, I consistently use 6-connectivity. Now, with that out of the way, let's take a look at the blob-finding algorithms.

The blob descriptor record used by the algorithms is shown in Figure 6. The fields for the hierarchical links and the statistical features are common to both the connectivity analysis and the boundary tracking algorithms. The

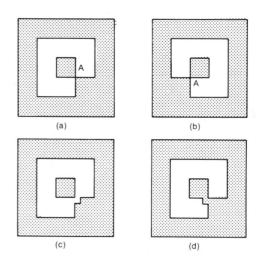

(a)   (b)

(c)   (d)

Figure 5. The blob in (a) illustrates the connectivity paradox that arises if both black and white pixels are exclusively 8-connected or 4-connected. An unambiguous description of the image, in which (a) is topologically equivalent to (c), is obtained using 6-connectivity. Note that if the object is rotated 90 degrees as in (b), the central square is a hole, topologically equivalent to (d). However, as illustrated in (c) and (d), the number or holes in this object is an unreliable feature since fluctuations in pixel values around point A due to noise and quantization errors make it unlikely that the object will be imaged consistently as in (a) or (b).

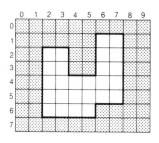

fields FLINK and RLINK are used to link the blob records into a doubly linked circular list that simplifies searching the entire collection of blob records. They are needed by the connectivity analysis algorithm, since it must insert or delete records in the list at random. In the boundary tracking algorithm, blob records are allocated sequentially as each blob is encountered so these fields are not needed. The boundary tracking algorithm has one extra field, CCB, which points to a chain-code block that contains the chain-code representation of the blob's boundary.*

As we shall see, both algorithms need a way to tell which blob a previously processed pixel has been assigned to. This information cannot be stored in the image array itself, since there is only one bit per pixel. Instead, a program must use some other representation, one that can be indexed by the pixel coordinates without requiring as much room as an entire image. One compact way of representing this knowledge is called *run-length encoding*. In the coding scheme, used by the connectivity analysis algorithm, each unbroken run of 0's or 1's in a raster line of the image is described by a record that tells its starting (leftmost) column, its length, and which blob the run belongs to. Each line of the picture can be described by a list of these run-length records.

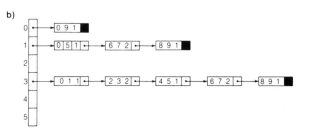

Figure 7. The blob finding algorithms generate a run-length list which encodes each run of consecutive 0's or 1's on a line as a three word record. The run-length list data structure for the connectivity analysis algorithm is shown in (a). Assuming the background is blob 1 and the object is blob 2, the three numbers in each record are interpreted as the start of a run, the number of pixels in the run, and number of the blob these pixels belong to, respectively. Part (b) shows the data structure for the boundary tracking algorithm. In this case, the second word is the right end of the run rather than its length.

As with the blob list, the structure of the run-length list differs in the two algorithms. In the connectivity analysis algorithm, run-length records are allocated sequentially as the image is scanned. In boundary tracking, run-length records are inserted and updated at random, so a linked list structure is necessary; it is also easier for this algorithm to keep the ending column of the run in the record instead of its length. In both cases, a table of pointers to the first record of each line is maintained. Figure 7 shows the run-length list structure for both algorithms. It's interesting to note that since the connectivity analysis algorithm operates strictly in a raster scan, it is very easy to modify it to accept a run-length description of the image as its only input. In this case the blob assignment fields would initially indicate only the color of each run.

Some final comments about the data structures: It is convenient to assume that there is a margin of at least one background pixel around the entire image. This greatly simplifies the algorithms, since there are then no special boundary conditions to consider. We are now ready for the details of the algorithms. In the following discussion,

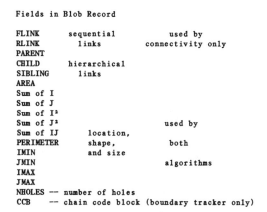

Figure 6. The contents of a blob record, sharing those fields unique to either the connectivity analysis algorithm or the boundary tracking algorithm, and the fields common to both algorithms.

*Milgram [6] describes an algorithm that determines chain-code boundary descriptions from a single raster scan of the image. It requires somewhat complex data structures since the boundary is constructed in several pieces.

Figure 8. *This is a complete PASCAL procedure implementing the connectivity analysis procedure described in the text. The calling program is not shown. The constants at the top, defining various array lengths and the image size, are representative. The actual values you choose will depend on your video input hardware and the particular application.*

```
CONST   MAXBLOB=25;  MAXRUN=1000;  MAXNUM=200;
        NROW=240;  MAXROW=239;  NCOL=192;  MAXCOL=191;

TYPE    BLOB=RECORD
                FLINK,RLINK,PARENT,CHILD,SIBLING:  INTEGER;
                COLOR,NUMBER,PERIM,NHOLES:  INTEGER;
                IMIN,IMAX,JMIN,JMAX: INTEGER;
                AREA,SUMI,SUMJ,SUMI2,SUMIJ,SUMJ2:  REAL
              END;

        RUN=RECORD
                START,LENGTH,NUMBER: INTEGER
              END;

        LINE=ARRAY [0..MAXCOL] OF INTEGER;

VAR     B:  ARRAY [1..MAXBLOB] OF BLOB;
        BP: ARRAY [1..MAXNUM] OF INTEGER;
        BN: INTEGER;
        R:  ARRAY [1..MAXRUN] OF RUN;
        RP: ARRAY [0..MAXROW] OF INTEGER;

PROCEDURE QUIT; (* this is a dummy — its supposed to be an error exit *)

PROCEDURE GETLIN(J:INTEGER; VAR L: LINE);
(* this is implementation dependent — it stores line J in array L *)

FUNCTION LOOKUP(N: INTEGER): INTEGER;
BEGIN
      · WHILE BP[N] < 0 DO N:=-BP[N];
        LOOKUP:= N;
END;

PROCEDURE BLOBSCAN(IL,JL,IR,JR: INTEGER);

VAR     I,J,WSTATE,CRUN,PRUN,NEWFLAG,LEFTEND: INTEGER;
        CURREG,REGABV,HOLREG: INTEGER;
        FREE,AVAIL: INTEGER;
        CLINE,PLINE: LINE;
        TMP,TMP2: INTEGER;

    PROCEDURE INIBLOB(N,L,R,J: INTEGER);
    BEGIN
      IF BN > MAXNUM THEN
        BEGIN WRITE('**BLOBNUM OVERFLOW'); QUIT  END
      ELSE
        BEGIN
          BP[BN]:=N;  B[N].NUMBER:=BN;  BN:=BN+1
        END;
      B[N].PARENT:=-1;
      B[N].CHILD:=-1;
      B[N].SIBLING:=-1;
      B[N].COLOR:=CLINE[R];
      B[N].AREA:=0;
      B[N].SUMI:=0;
      B[N].SUMJ:=0;
      B[N].SUMI2:=0;
      B[N].SUMIJ:=0;
      B[N].SUMJ2:=0;
      B[N].IMIN:=L;
      B[N].IMAX:=R;
      B[N].JMIN:=J;
      B[N].JMAX:=J;
    END;


    PROCEDURE ALLOC;
    BEGIN
      NEWFLAG:=0;
      IF AVAIL < 0 THEN
        BEGIN
          IF FREE > MAXBLOB THEN
            BEGIN  WRITE('**BLOBLIST OVERFLOW'); QUIT END
          ELSE
            BEGIN
              CURREG:=FREE;
              FREE:=FREE+1;
            END
          END
        ELSE
          BEGIN
            CURREG:=AVAIL;
            AVAIL:=B[AVAIL].FLINK;
          END;
      INIBLOB(CURREG,R[CRUN].START,I-1,J);
      TMP:=B[REGABV].FLINK;
      B[REGABV].FLINK:=CURREG;
      B[CURREG].FLINK:=TMP;
      B[TMP].RLINK:=CURREG;
      B[CURREG].RLINK:=REGABV;
    END;

    PROCEDURE UPDATE(EOL: INTEGER);
    VAR    L,K: REAL;
    BEGIN
      L:=I - R[CRUN].START;
      R[CRUN].LENGTH:=I - R[CRUN].START;
      R[CRUN].NUMBER:=B[CURREG].NUMBER;
      B[CURREG].PERIM:=B[CURREG].PERIM + 2;
      K:=R[CRUN].START;
      B[CURREG].AREA:=B[CURREG].AREA  + L;
      B[CURREG].SUMI:=B[CURREG].SUMI  + L*((L-1)/2 + K);
      B[CURREG].SUMJ:=B[CURREG].SUMJ  + L*J;
      B[CURREG].SUMI2:=B[CURREG].SUMI2 + L*((L-1)*(K + (2*L -1 )/6) + K*K);
      B[CURREG].SUMIJ:=B[CURREG].SUMIJ + L*((L-1)/2 + K)*J;
      B[CURREG].SUMJ2:=B[CURREG].SUMJ2 + L*J*J;
```

```
      IF R[CRUN].START < B[CURREG].IMIN  THEN   B[CURREG].IMIN := R[CRUN].START;
      IF I-1 > B[CURREG].IMAX THEN  B[CURREG].IMAX := I-1;
      B[CURREG].JMAX:=J;
      CRUN:=CRUN+1;
      IF CRUN > MAXRUN THEN
        BEGIN WRITE('**RUNLIST OVERFLOW'); QUIT END;
      IF EOL = 0 THEN
        R[CRUN].START:=I
      ELSE
        BEGIN
          R[CRUN].START:=IL;
          PRUN:=PRUN+1;
        END;
    END;

    PROCEDURE STATE7;
    BEGIN
      UPDATE(0);
      LEFTEND:=I;
      NEWFLAG:=1
    END;

    PROCEDURE STATE4;
    BEGIN
      LEFTEND:=I;
      PRUN:=PRUN+1;
      REGABV:=LOOKUP(R[PRUN].NUMBER);
    END;

    PROCEDURE STATE3;
    BEGIN
      UPDATE(0);
      PRUN:=PRUN+1;
      REGABV:=LOOKUP(R[PRUN].NUMBER);
      CURREG:=REGABV
    END;

    PROCEDURE STATE2;
    BEGIN
      IF NEWFLAG <> 0 THEN ALLOC;
      UPDATE(0);
      B[CURREG].PERIM:=B[CURREG].PERIM + I - LEFTEND;
      B[REGABV].PERIM:=B[REGABV].PERIM + I - LEFTEND;
      CURREG:=REGABV
    END;

    PROCEDURE STATE6;
    BEGIN
      IF NEWFLAG <> 0 THEN ALLOC;
      UPDATE(0);
      B[CURREG].PERIM:=B[CURREG].PERIM + I - LEFTEND;
      B[REGABV].PERIM:=B[REGABV].PERIM + I - LEFTEND;
      CURREG:=REGABV;
      PRUN:=PRUN+1;
      REGABV:=LOOKUP(R[PRUN].NUMBER);
      LEFTEND:=I;
    END;

    PROCEDURE STATE1;
    BEGIN
      HOLREG:=REGABV;
      PRUN:=PRUN+1;
      REGABV:=LOOKUP(R[PRUN].NUMBER);
      IF NEWFLAG <> 0 THEN
        BEGIN
          NEWFLAG:=0;
          CURREG:=REGABV;
          B[CURREG].PERIM:=B[CURREG].PERIM + I - LEFTEND;
        END
      ELSE IF CURREG = REGABV THEN
        BEGIN
          TMP:=B[CURREG].CHILD;
          B[HOLREG].SIBLING:=TMP;
          B[HOLREG].PARENT:=CURREG;
          B[CURREG].CHILD:=HOLREG;
          B[CURREG].NHOLES:=B[CURREG].NHOLES + 1;
          B[CURREG].PERIM:=B[CURREG].PERIM - B[HOLREG].PERIM
        END
      ELSE
        BEGIN
          B[CURREG].PERIM:=B[CURREG].PERIM + B[REGABV].PERIM + I - LEFTEND;
          IF B[REGABV].IMIN < B[CURREG].IMIN THEN
            B[CURREG].IMIN:=B[REGABV].IMIN;
          IF B[REGABV].IMAX > B[CURREG].IMAX THEN
            B[CURREG].IMAX:=B[REGABV].IMAX;
          IF B[REGABV].JMIN < B[CURREG].JMIN THEN
            B[CURREG].JMIN:=B[REGABV].JMIN;
          B[CURREG].AREA:=B[CURREG].AREA + B[REGABV].AREA;
          B[CURREG].SUMI:=B[CURREG].SUMI + B[REGABV].SUMI;
          B[CURREG].SUMJ:=B[CURREG].SUMJ + B[REGABV].SUMJ;
          B[CURREG].SUMI2:=B[CURREG].SUMI2 + B[REGABV].SUMI2;
          B[CURREG].SUMIJ:=B[CURREG].SUMIJ + B[REGABV].SUMIJ;
          B[CURREG].SUMJ2:=B[CURREG].SUMJ2 + B[REGABV].SUMJ2;
          IF B[REGABV].NHOLES <> 0 THEN
            BEGIN
              B[CURREG].NHOLES:=B[CURREG].NHOLES + B[REGABV].NHOLES;
              TMP:=B[REGABV].CHILD;
              WHILE TMP > 0 DO
                BEGIN
                  B[TMP].PARENT:=CURREG;
                  TMP2:=TMP;
                  TMP:=B[TMP2].SIBLING;
                END;
              B[TMP2].SIBLING:=B[CURREG].CHILD;
              B[CURREG].CHILD:=B[REGABV].CHILD;
            END;
```

```
          B[B[REGABV].RLINK].FLINK:=B[REGABV].FLINK;
          B[B[REGABV].FLINK].RLINK:=B[REGABV].RLINK;
          BP[B[REGABV].NUMBER]:= -B[CURREG].NUMBER;
          B[REGABV].FLINK:=AVAIL;
          AVAIL:=REGABV;
          REGABV:=CURREG;
        END;
    B[HOLREG].PERIM:=B[HOLREG].PERIM + I - LEFTEND;
  END;
BEGIN (* BLOBSCAN *)
  BN:=1;
  FREE:=2;
  AVAIL:=-1;
  FOR I:=IL TO IR DO PLINE[I]:=0;
  INIBLOB(1,0,MAXCOL,0);
  B[1].FLINK:=1;
  B[1].RLINK:=1;
  RP[JL]:=1;
  R[1].START:=IL;
  R[1].LENGTH:=NCOL;
  R[1].NUMBER:=1;
  PRUN:=1;
  R[2].START:=IL;
  CRUN:=2;
  FOR J:= JL+1 TO JR DO
    BEGIN
      NEWFLAG:=0;
      WSTATE:=0;
      RP[J]:=CRUN;
      GETLIN(J,CLINE);
      REGABV:=LOOKUP(R[PRUN].NUMBER);
      CURREG:=REGABV;
      FOR I:= IL+1 TO IR DO
        BEGIN
          WSTATE:= (WSTATE DIV 4) + 4*PLINE[I] + 8*CLINE[I];
          PLINE[I]:=CLINE[I];
          CASE WSTATE OF
            7,8:    STATE7;

            4,11:   STATE4;

            3,12:   STATE3;

            2,13:   STATE2;

            6,9:    STATE6;

            1,14:   STATE1
          END;
        END;
        UPDATE(1);
      END;
  END;
```

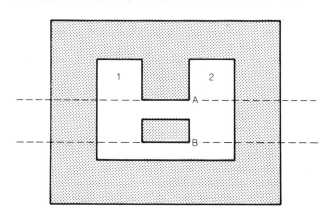

*Figure 9. In the connectivity analysis algorithm, separate blob
records are maintained for the components labeled 1 and 2 until
the raster scan reaches the point labeled A, at which point the
two records merge into one. When the scan reaches point B, it
discovers that the same blob has surrounded an interior hole.*

the image size is NROW lines by NCOL columns. Pixels
are addressed by (I,J) = (column,line) where I can vary
from 0 to NCOL–1 and J from 0 to NROW-1. The upper
left corner of the image is (0,0).

Connectivity Analysis

A version of the connectivity analysis algorithm written
in PASCAL is given in Figure 8. Blob records are allocated

from an array B, indexed by blob number, and run-length
records from array R. Array RP, indexed by image line
number, holds pointers to the first run-length record of
each line. The constants define the particular limits of my
system—yours may differ. Procedure GETLIN is the
interface to the bit array of the binary image. I don't show
the actual method it used to unpack an image line into an
integer array, since it is highly machine dependent for
optimum efficiency.

The connectivity analysis algorithm scans the image
from left to right, top to bottom, updating the descriptions
of each blob which intersects the current scan line. At the
end of each run of 0's or 1's, the run-length list is updated
and blob statistics are accummulated. If any pixel of the
run just completed is 6-connected to a pixel of the same
color on the previous line, an existing blob is extended to
include this run. Otherwise, a new blob is record is
allocated. Each blob is assigned a unique number. Pixels
belonging to the blob are labeled with this number in the
run-length list.

A blob may initially have more than one label associated
with it, as shown in Figure 9. It is not until the scan reaches
point A that the two components labeled 1 and 2 are
recognized as being part of the same blob. At this point
they are merged—the record for blob 2 is released after
combining the statistics of blobs 1 and 2.

However, there are still references to blob 2 in the run-
length list, so we need a method of making labels 1 and 2
equivalent. The method I use involves a table of blob
pointers (BP in the program) which is indexed by blob
number. If the entry in the table for a given blob number is
positive, then it is a valid pointer to an active blob record. If
the entry is negative, then the original blob record has
been merged and released, and the corresponding positive
value of the entry is the number of the blob that was
merged with the original blob. It may be necessary to chain
through several locations if more than one merge has
occurred in the same blob.

The hierarchical relationships are updated whenever a
blob is complete, as detected by the surrounding blob
closing on itself. When this occurs, as shown at point B in
Figure 9, the surrounded blob is inserted into the list of
children of the current blob by manipulating the PARENT,
CHILD, and SIBLING pointers in the two blob records.

The program could compute the moments and other
blob statistics on a pixel-by-pixel basis. However, it's
better to update blob statistics at the end of each run since
this requires far fewer computations. Also, new blob
records are not allocated until the first run of a new blob
has ended. Whenever a new blob record is allocated

Updating Blob Statistics

Area, Sums of I, J, I², J², and IJ

Since the I values of a run of L pixels starting with with I_0 are I_0+0, I_0+1, ..., $I_0+(L-1)$, and since

$$\sum_{k=0}^{L-1} k = L(L-1)/2,$$

then the contribution of the run to the sum of I is $L(I_0+(L-1)/2)$.

Similarly, since $\sum_{k=0}^{L-1} k^2 = L(L-1)\cdot(2L-1)/6$,

and $(I+k)^2 = I^2 + 2Ik + k^2$, then the contribution of the run to the sum of I² is

$$\sum_{k=0}^{L-1} (I_0+k)^2 = L[I_0^2 + I_0(L-1) + (L-1)(2L-1)/6].$$

Since J is constant over the run, the contribution of the run to the sum of IJ is $LJ(I_0+(L-1)/2$, the sum of J is incremented by LJ and the sum of J² is incremented by LJ^2.

Figure 10. The formulae for updating moments at the end of each run in the connectivity analysis algorithm.

(procedure ALLOC in the program), its moments, perimeter, and hole count are zeroed. The blob's minimum enclosing rectangle is initialized by setting JMIN to J (the current line number), IMIN to the starting column number of the run, and IMAX to the last column in the run.

As new runs are added to the blob on successive lines, the statistics have to be updated (procedure UPDATE). The moments are incremented using the formulae given in Figure 10. These take advantage of the fact that J, the line number, is constant over the length of the run, and, since the run covers consecutive columns, the moments involving I can be collected using the closed forms for sums of powers of consecutive integers. The enclosing rectangle is updated by checking if the left or right endpoints of the new run exceed the old values of IMIN or IMAX, respectively. Since the blob grows steadily downward, JMAX is simply set to J, and there is no need to change JMIN. JMIN is only updated when two blobs merge, in which case the other extreme must be checked and changed as well to accurately represent the extent of the resulting merged blob. The perimeter is incremented by two to account for the vertical boundary segments at each end of the run. The contributions to the perimeter of the horizontal boundary segments are handled specially, as I will describe shortly.

The various actions described above take place whenever there is a transition from black to white or from white to black in either the current line or the previous line. To detect these transitions, the image is scanned with what amounts to a 2 by 2 "window" into the image whose lower right corner is always the current pixel. Thus, the previous pixel on the same line and the two corresponding pixels on the previous line make up the other elements in the window. Since each pixel can be colored either black or white (0 or 1), there are 2^4 or 16 possible patterns that can appear in the 2×2 window. These 16 patterns or window *states* are shown in Figure 11.

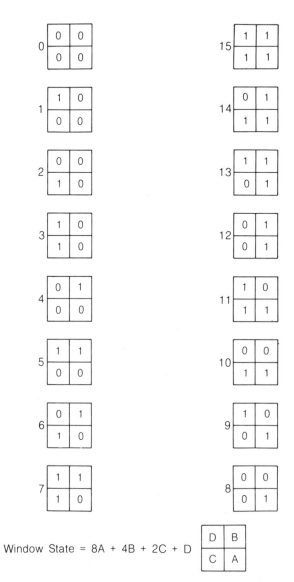

Window State = 8A + 4B + 2C + D

Figure 11. The 16 possible configurations of 0's and 1's are shown above, numbered according to the convention shown at the bottom. Note that they are arranged in complementary pairs.

These window states are numbered by treating the values of the pixels in the window as a four bit binary number, according to the formula shown in the figure. Although there are 16 distinct states, there are actually only eight cases to consider. Since we want to treat transitions from black to white the same as those from white to black, the binary complement of a state number (1's and 0's interchanged) represents the same transition and should be handled identically. Of the eight remaining cases, two (0,15 and 5,10) require no action at all, since there is no horizontal transistion. Let's now consider the other six cases in detail.

In the discussion below, as well as in the program, I is the position of the scan corresponding to the column coordinate of pixel A in the 2 × 2 window. CURREG and REGABV are initially pointers to the blob records corresponding to the pixels C and D, respectively. At the beginning of each line, CURREG and REGABV are both set to the background color (since there is one pixel margin of background around the entire image) and the scan actually starts in the second column from the left. In the program, variable WSTATE, derived using the state labeling formula in Figure 11, identifies which window state has occurred. (The DIV operator used in the code produces the same results due to the iteration.) Here, then, are the six transitions we still need to consider:

Case 7,8: This is possibly the start of a new blob. Set the variable NEWFLG to true. It is also the beginning of a new horizontal boundary segment (which will contribute to the perimeter) and a new run. Save the I coordinate in the variable LFTEND for later use in the update. Update the statistics of the run just completed.

Case 4,11: Although this is not the start of a new run, it does mark the beginning of a new horizontal boundary segment—therefore LFTEND is again set to I. Set REGABV to the blob assignment of pixel B, obtained by looking it up in the run-length list.

Case 3,12: Update the run-length list and blob statistics for the run just ended. REGABV and CURREG are both set to the blob assignment of pixel B.

Case 2,13: If NEWFLG is true, allocate a new blob record for the run just completed. Set CURREG to the number of this record and NEWFLG to false. Update the statistics of the new blob, in including the perimeter, which is incremented (from zero) by the amount I-LFTEND. Set CURREG to REGABV.

Case 6,9: If NEWFLG is true, allocate a new blob and update it as in Case 2,13. Set CURREG to the current value of REGABV to the blob label of pixel B.

Case 1,14: There are three possibilities here. If NEWFLG is true, then what was originally assumed to be a new blob has merged with an existing blob. If NEWFLG is false, then either two distinct blobs have merged or a blob has closed on itself surrounding the blob containing pixel D. Save REGABV (pixel D's blob) in variable HOLREG, and then set REGABV to the label of pixel B. If NEWFLG is true, then set CURREG to the new REGABV and NEWFLG to false. This will include the new run in the old blob.

If NEWFLG is false and CURREG is *not* equal to the new REGABV, then merge the two blobs by releasing REGABV, deleting it from the sequential blob list and combining both blobs' statistics into CURREG. Update the blob number table BP[REGABV] to make REGABV equivalent to CURREG. If, however, NEWFLG is false and CURREG *is* equal to REGABV, then add HOLREG to the list of children of CURREG, since it is indeed a hole in CURREG (increment the hole count as well).

Regardless of NEWFLG, this case also terminates a horizontal boundary segment of HOLREG and CURREG. If HOLREG is not really a hole (CURREG not equal to REGABV), then subtract the perimeter of HOLREG from that of CURREG (since we only want the outer perimeter) and increment the perimeter of HOLREG by I-LFTEND.

To perform a connectivity analysis of a scene, the computer program calls the procedure BLOBSCAN with parameters that define the upper left and lower right corners of a rectangular window into the image bit array. When BLOBSCAN returns, the blob records can be examined by following the linkages from the background blob record, B[1], either sequentially (following FLINK or RLINK) or hierarchically (the CHILD field points to the first child blob, whose SIBLING link points to the next, etc.).

Boundary Tracking

The boundary tracking algorithm traverses the boundary of each blob in the image when it is first encountered in a raster scan of the image. As the boundary is traversed, blob statistics are computed and the run-length list is updated to incorporate the new blob. In addition, the algorithm produces a chain-code representation of the boundary [5]. A new blob is recognized whenever the raster scan encounters a color transition between (I–1, J) and (I, J) which is not yet represented in the run-length list. When this occurs, the raster scan is suspended and the boundary tracker is invoked to immediately get a complete description of the new blob. When the boundary tracker has completed, the raster scan is resumed at (I+1, J).

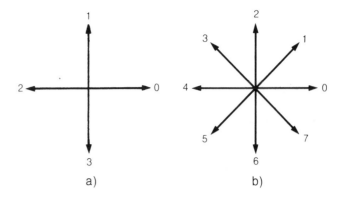

a)   b)

Figure 12. The boundary tracking algorithm uses a four direction chain-coding scheme with the four directions numbered as shown above in (a). The numbering for eight direction chain-coding is shown for comparison in (b).

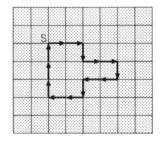 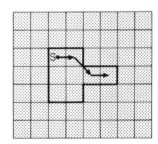

(a)   (b)

Figure 13. Starting at the point labeled S and traveling clockwise, the four direction chain-code for the object in (a) obtained by following pixel edges is 00330010033322322211111. The eight direction chain-code for the same object obtained by connecting pixel centers in (b) is 06701066445442222. Notice that the area and perimeter of the object derived from the two chain-code representations is different.

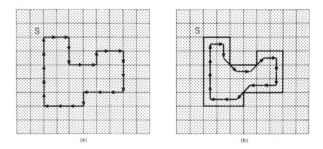

Figure 14. The boundary tracking algorithm uses four direction chain-coding (left) to avoid situations like that shown at the right which may occur if eight direction chain-coding is used. In the latter example the boundary must retrace itself to form closed loop.

Obviously, the boundary of each blob will be encountered several times during the raster scan, but the boundary tracker is invoked only once per blob, since all encounters after the first are represented in the run-length list.

There are three distinct elements of the boundary tracking algorithm. The first is the tracking procedure itself. The second is the procedure that computes the blob statistics as the boundary is traversed. The third is the procedure that updates the run-length list. Let's look at each of these in turn.

The boundary tracker views each pixel as a unit square in the image plane, bounded by four edges. A boundary pixel is one which has one or more edges in common with a background pixel (for the remainder of this section, the word *background* is used in a generic sense to indicate the *parent* of a blob). These edges are called *boundary segments*. The boundary of a blob is a sequence of directed boundary segments represented using a four-direction chain-coding scheme. The four directions are left, up, right and down, numbered 0, 1, 2 and 3, respectively (see Figure 12a). Note that in the eight-direction chain-code (Figure 12b), these directions are labeled 0, 2, 4 and 6, respectively.

Each boundary segment in the chain code representation corresponds to one edge of a boundary pixel. Thus, associated with the boundary as we have defined it is a sequence of boundary pixels. An alternative way to represent the boundary is with a chain-coded list of vectors that connect adjacent boundary pixels. This requires the eight direction chain-coding scheme. Figure 13 shows both boundary representations for the same blob. While the two representations are equivalent in the sense that one can be uniquely derived from the other, there are some differences. First, some of the statistics, particularly the area and perimeter, have different values, depending on which representation is used. This is not a serious problem since the values will be consistent if the same representation is always used. Note that the statistics obtained from the boundary segment representation are identical to those obtained from the connectivity analysis algorithm described earlier. The second difference is illustrated in Figure 14. In this case, the blob has a component that is only a single pixel wide. In the boundary pixel representation, we cannot obtain a closed boundary unless we allow the boundary to backtrack over itself.

Here, we avoid this situation by using four direction chain-code.

The boundary tracking procedure operates as follows. Suppose the raster scan has just encountered a new blob at (I,J). This becomes the first boundary pixel and the coordinates are saved in variables (I0,J0). The chain code list is initialized with a "right" (0) boundary segment code corresponding to the top edge of the initial pixel, and this value is also stored in the variable CC. The procedure BTRACK, with alterable (VAR) arguments CC, I, and J, is now called repeatedly to locate successive boundary segments around the blob. Each call to BTRACK replaces the old values of its arguments with new ones. It returns in CC the next boundary code, which is added to the chain code list, and in I and J the coordinates of the next boundary pixel. The tracking procedure terminates when CC=0 and (I,J)=(I0,J0). The field CCB in the blob record is set to point to the completed chain-code list. (The chain-code is usually stored in packed form, i.e., four codes per byte for four direction code.)

The boundary traversal is clockwise with respect to the interior of the blob, or, equivalently, the blob is always on the righthand side of a directed boundary segment. At each step, BTRACK decides whether to turn to the left or right or to continue straight ahead. This requires examining the two pixels, labeled A and B in Figure 15, that lie ahead and on either side of the current boundary segment. The figure assumes that a white blob is being tracked.

The 6-connectivity convention results in slightly different

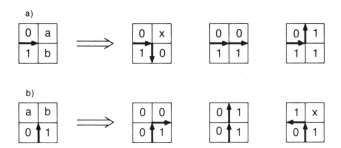

Figure 15. The rules for determining the next boundary segment in the boundary tracking algorithm depend on the colors of pixels A and B in the 2 by 2 windows shown at the left for the cases when the last segment was "right" (a) and "up" (b). It is assumed that the object is composed of 1's. The rules for proceeding from a "left" or "down" boundary segment are obtained by rotating (a) or (b), respectively, by 180 degrees.

decision rules for vertical than for horizontal boundary segments. In all cases, if pixel A (Figure 15) is black and B is white, the boundary continues in the same direction. The differences arise in deciding to turn left or right. In the case where the current segment is horizontal (left or right), pixel A is not 6-connected to the current boundary pixel. Thus the boundary turns left only if pixels A and B are *both* white. If pixel B is black, then the boundary turns right regardless of the color of A.

In the case where the current segment is vertical, pixel A *is* 6-connected to the current boundary pixel, so the boundary turns left if pixel A is white regardless of the color of pixel B. The boundary turns right only if pixels A and B are both black. Each of these cases is shown in the figure. In general, when the boundary turns left or continues in the same direction, the next boundary pixel is A or B, respectively. When the boundary turns right, the

a) Moment Formulae

$$\text{Area} := P_1$$

$$\sum I \quad := P_2/2$$

$$\sum J \quad := P_3/2$$

$$\sum I^2 \quad := P_4/3$$

$$\sum IJ \quad := (2P_5 - P_6)/4$$

$$\sum J^2 \quad := P_7/3$$

b) Accumulation of P terms

If boundary code = 0 ($\rightarrow$)

$$X := X+1$$
$$P_3 := P_3 - Y^2$$
$$P_7 := P_7 - Y^3$$

If boundary code = 1 ($\uparrow$)

$$Y := Y+1$$
$$P_1 := P_1 - X$$
$$P_2 := P_2 - X^2$$
$$P_4 := P_4 - X^3$$
$$P_5 := P_5 - X^2Y$$
$$P_6 := P_6 + X^2$$

If boundary code = 2 ($\leftarrow$)

$$X := X-1$$
$$P_3 := P_3 + Y^2$$
$$P_7 := P_7 + Y^3$$

If boundary code = 3 ($\downarrow$)

$$Y := Y+1$$
$$P_1 := P_1 + X$$
$$P_2 := P_2 + X^2$$
$$P_4 := P_4 + X^3$$
$$P_5 := P_5 + X^2Y$$
$$P_6 := P_6 + X^2$$

Initially, $X := I_0 - .5$ and $Y := J_0 - .5$

Figure 16. In the boundary tracking algorithm, moments are calculated by the formulas shown in (a). The terms P_1 and P_2, etc. are accumulated according to the formulas shown in (b), where the new value of each term is computed from its previous value and the current pixel edge coordinates.

boundary pixel remains the same as the current one. The rules for tracking the boundary of a black blob are identical, with the roles of black and white reversed.

Next, let's consider the accumulation of blob statistics as the boundary is traversed. The perimeter of the blob is simply the length of the chain-code list and thus requires a counter which is imcremented for each iteration of BTRACK. The moments (area, I, J, I^2, J^2, IJ) are computed according to the formulas described in [5]. Since we are using a four direction chain-code, we need only the formulas for horizontal and vertical segments. These are summarized in Figure 16.

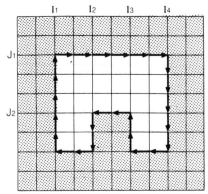

Evolution of
line J₁:

a) `[ 0 | NCOL | 1 | 0 ]`

b) `[ 0 | ? | back | • ]` → `[ ? | I4 | obj | • ]` → `[ I4+1 | NCOL | back | 0 ]`

c) `[ 0 | I1-1 | back | • ]` → `[ I1 | I4 | obj | • ]` → `[ I4+1 | NCOL | back | X ]`

Evolution of
line J₂:

a) `[ 0 | NCOL | back ]`

b) `[ 0 | ? | back | • ]` → `[ ? | I4 | obj | • ]` → `[ I4+1 | NCOL | back | 0 ]`

c) `[ 0 | I3-1 | back | • ]` → `[ I3 | I4 | obj | X ]`

d) `[ 0 | ? | back | • ]` → `[ ? | I2 | obj | • ]` → `[ I2+1 | I3-1 | back | X ]`

e) `[ 0 | I1-1 | back | • ]` → `[ I1 | I2 | obj | X ]`

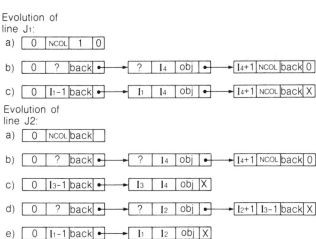

Figure 17. The evolution of the run-length list as the boundary of the object at the top is traversed is illustrated for line J₁ and J₂. Question marks indicate unknown values which will be filled in later when the line is crossed.

Taking I as an example, the value computed by these formulas is equivalent to summing the I coordinates of each pixel in the blob, as is done implicitly in the connectivity analysis algorithm. However, we can not use the pixel coordinates directly in the boundary traversal formulas since the boundary segments do not pass through pixel center, but rather lie on the edges of pixels. Thus the values of x and y in the formulas of Figure 16 must be the coordinates of a corner of the current boundary pixel, and are those of the form $x = I \pm .5$ and $y = J \pm .5$. Recalling that the boundary is initialized with the top edge of the first boundary pixel, the coordinates of the starting point of the boundary for the purposes of computing moments becomes $(I-.5, J-.5)$.

The enclosing rectangle represented by IMIN, JMIN, IMAX, and JMAX is initialized by setting IMIN and IMAX to I0, and JMIN and JMAX to J0. By the nature of the algorithm's raster scan search for new blobs, J0 *must* be the value of JMIN, so JMIN need never be changed. JMAX is updated whenever the next boundary segment is down. Likewise, IMIN or IMAX are only updated when the next boundary segment is left or right, respectively. The natural place to do this is in the procedure BTRACK

where the current value of CC, holding the code for the current boundary segment, must be tested to determine which pixels it examines to compile the next boundary segment.

Finally, the algorithm must maintain the run-length list. Each run-length record has four fields, labeled IL, IR, BLOBID, and LINK, corresponding to the left and right endpoints (column coordinates) of the run, a pointer to the blob record for the pixels in this run, and a pointer to the next run-length record for this line. The LINK field is –1 for the last record of each line. The run-length list is initialized to represent the entire image as single blob of background pixels; i.e., there is one record per line of the form (0, NCOL-1, BACK, –1) where BACK is the number of the actual background blob record.

As a blob boundary is traversed, new run-length records are created for each run of pixels contained in the blob. Assuming that for some line J the blob extends from I1 to I2, the run-length list can be updated as follows:

1. Find the background record line J that contains the run, i.e., IL, I2, I2, IR.
2. Create a new run-length record (I1, I2, NEW), where NEW is the current blob number. (The LINK field is set in step 4.)
3. Replace IR in the background record by I1–1 and set the LINK field of this record to point to the new record created in step 2.
4. Create a new background record (I2+1, IR, BACK, —), set the LINK field of this record to the old value in the original background record, and set the link of the record created in step 2 to point to this record.

Unfortunately, the actual procedure for updating the run-length list is a little more complicated than the above description, since I1 and I2 must be determined by two separate crossings of line J in the boundary tracking procedure. Assuming that line J is crossed at I2 first, and noting that the right end of a run must be associated with a "down" boundary segment, the above procedure can be modified as follows:

1. Find the background record for line J that contains I2.
2. Create a new run-length record for the current blob (–1, I2, NEW, —).
3. Replace IR in the background record by –1 and fix its LINK.
4. Create a new background record (I2+1, IR, BACK, —) and set the LINKs of both new records.

The –1's in the new blob record and the original

background record indicate that the corresponding end-points are not yet known. These fields will be resolved by a later crossing of line J by an "up" segment which determines the left (right) end of the blob (background) run. This requires a different procedure that is called when line J is crossed the second time at I_1. In this case, the boundary intersects an adjacent pair of unresolved run-length records, the first one representing background pixels and the second one representing blob pixels. The update procedure under these conditions consists of simply replacing the –1's by the correct values. The IR field of the background run gets I_1–1, and the IL field of the blob run gets I_1. Figure 17 shows the evolution of a portion of the run-length list for a simple image.

The initial crossing of a run of pixels can also occur on the left, i.e., at I_1, instead of on the right as described above. If the boundary tracker is following an edge downward in the image (increasing J) and then the edge swings upward again (due to a concavity on the top edge of the blob), the pixel that runs on the rightmost projection of the boundary will be initialized on the left first. An example of this is shown at point A in Figure 18. Handling this case requires a set of procedures analogous to those described above, linking in run-length records that leave the right end undetermined initially and resolve it when the boundary recrosses the line on a downward segment.

Another situation can arise due to concavities in the boundary: the boundary may intersect a completed run of pixels that is already marked as part of the object being traced. This is essentially the same as the first case described, where a "down" segment intersects a background run, except now a piece of background has to be "cut out" of the object. The crossing can occur on the right with an "up" segment, as shown at point B in the figure (the completed object run would be from point C to D, or on the left with a "down" segment, as shown at E. In either case, a background run with one end unresolved is spliced into the object run, and later corrected at the second crossing of the line as before.

If the boundary tracker is invoked starting at a point *inside* a concavity of the boundary, then one other special case can occur. The boundary crosses between two unresolved runs, which would normally resolve them, but the crossing is in the "wrong" direction, i.e., the boundary segment orientation implies that the object is on the left, but the left run is a background run. By always starting the tracker on a convexity of the boundary, as in the case of the algorithm described here, which invokes the tracker on the first image line from the top that encounters the object, this problem cannot occur.

When the boundary tracking algorithm has finished processing the image, the resulting blob list will have all the hierarchical linkages between blobs in the scene, just like those produced by the connectivity analysis method (though the order of blob allocation, and thus the order of blobs in the various lists, may differ). In addition to the blob statistics, however, each blob record generated by the boundary tracker has a pointer to the chain-code list that gives an exact description of the boundary. The moments accumulated by both algorithms are sufficient for computing a blob's centroid and major axis (angle of minimum moment of inertia), but if higher-order moments are needed, the chain-code list can be traversed to derive them. Alternatively, of course, fields for additional statistics could be added to the blob record and their values accumulated during the first traversal of the boundary.

Regardless of which scene analysis algorithm you choose, the result you get is a complete description of the patterns of black and white in the image. I should point out that even if you are restricted to using connectivity analysis in a single raster scan of the image (due to the practical limitations of your image access method, for example), you can still get a chain-code description of the blob outlines if necessary by modifying the boundary tracking algorithm to accept the run-length list as input. If efficient random access to the image is available, however,

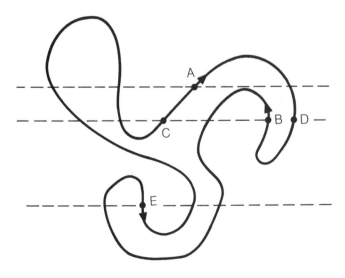

Figure 18. Concavities in the outline of a blob can give rise to cases that require special handling. At A, the initial crossing of a run occurs on the left. At B a background run must be "cut out" of run CD. A similar situation occurs at E.

it's better to use the boundary tracking algorithm first if you need chain-code.

Using the Image Description

Beyond simply locating targets in two dimensions, applications of robot vision usually involve three basic categories: object recognition and inspection, stereo vision for range measurement, and tracking moving targets. Within the particular limitations of binary images, the basic blob description approach described here can be used to advantage for each of these functions. Although a complete treatment of each of these topics will be reserved for future articles, there are a few comments pertaining to the use of blob descriptions that should be addressed here.

Object Recognition

In its simplest form, object recognition consists of matching a blob of unknown category against a finite set of possible objects. Each possible object is represented in a *model* in terms of features that can be derived from information stored in a blob record. There are two basic approaches to modelling, and hence to recognition. One is purely numerical. Objects are modelled by a list of *n* different features, such as area, or moment invariants, perimeter, each of which can be expressed by a single number. It is convenient to think of the values in the list as the coordinates of a point, or *feature vector*, in an *n*-dimensional feature space where each axis representing ideal descriptions of each object, perhaps with a small envelope of acceptable variations about each point, that is gained by a "training session" where the system is shown sample objects. A blob can then be recognized by finding its "nearest neighbor" among the set of model points in feature space, possibly with a minimum acceptable "degree of match." [refs. 1,7]

The other approach to modelling represents the shape of each object explicitly in terms of its boundary. This is usually done by approximating the chain-code description of the boundary as a sequence of straight line segments and possibly circular arcs. In this case, recognition is either based on template matching to find the boundary description in the model which most closely matches the boundary of a particular blob in the image, or else the boundary is used to verify or possibly disambiguate matches based on numerical features. [ref. 8]

Because the only information available is derived from the shape of the outer boundary of a blob, binary image vision using these techniques is particularly ill-equipped to recognize objects in three-dimensional scenes. The shape of an object's outline, and hence its blob statistics, can change dramatically when viewed from different perspectives, even if it is possible to provide sufficient contrast between object and background by suitably structuring the scene. Matters are made worse by occlusion, the apparent juxtaposition of objects from a particular point of view. In this case, the two objects would be merged into one binary blob in the scene.

Stereo Range Measurement

The location of an object in 3-space cannot be determined from a single view of the object unless there is some known constraint on one dimension. [ref. 9] Usually, the constraint is that the object lies in a plane whose position relative to the camera is known. In the absence of such a constraint, we need at least two views of the object from different camera locations. These can be provided by a stereo pair of images taken by two cameras mounted some distance apart.

The problem in making stereo measurements is to match points from the two images which correspond to the same point in 3-space. One of the classical approaches is *stereo correlation*, which relies on a statistical correlation between the distribution of gray level values in a neighborhood surrounding the target point in the two images. [ref. 10] This method is not particularly suited to binary images, since the correlation is undefined unless there is some variation in gray level. Thus, the only possible basis for correlation would be the shape of the boundary in the neighborhood of an edge point—a very poor basis for discrimination if there is more than one object in the scene.

Fortunately, we can remove the ambiguity in matching edge points if we initially perform *global* blob matching. For any point in one image, the stereo camera geometry model specifies a line segment in the other image which contains the matching point [9]. Taking as a target point the centroid of a blob from one image, the centroid of the corresponding blob in the other image should lie near this line, unless the difference in perspective is too great (which can happen if the target is too close to the cameras, for example). Once we have identified candidate matches on on the basis of camera geometry, then the blob statistics can be used to select the best match. You can then match edge points to provide several 3-D measurements of points on the border of the object. Knowing the 3-D location of three non-colinear known features of the object is sufficient to completely determine its position and orientation in 3-space.

Tracking moving targets

Provided that the target or some feature on it has sufficient contrast to be readily distinguishable from the background, binary vision using blob descriptors may be sufficient to measure its translational and rotational velocity in a succession of images. Here again, the centroid of the target blob is the primary source of information. Given the difficulties imposed by the need for high contrast and lack of occlusion, the major concern in the context of robotics is to execute a tracking algorithm in real time. If the results of object tracking are to be used as feedback in a control loop, as in visual servoing, for example, then the tracking should be continuous and there should be as little delay as possible from the time an image is formed to the time new target position and velocity measurements are available.

In most cases, real time tracking systems are designed to operate at 30Hz, corresponding to the 30 frame/sec. rate of a standard video signal. The first thing required to achieve this processing rate is a digitizer system that operates at the full video rate, such as the one mentioned earlier [3]. Secondly, the image analysis and tracking software is usually written in assembly/machine code optimized for speed. Thirdly, rather than attempting to analyze the entire image, the algorithms are modified (like BLOBSCAN in the listing) to operate within a window into the image which can be accurately placed where the target is expected to be in a given frame, based on its position, velocity and acceleration calculated form previous frames and allowing for uncertainty in the prediction. The higher the cycle rate of the algorithm, the less the appearance of the target is likely to change from one image to the next.

Conclusion

The techniques and applications I've described here should give you some idea of the potentials of robot vision using binary images—as well as an appreciation of its limitations. The algorithms also give you the basic tools for building your own binary image vision software. Under the right circumstances, these methods can give you a powerful tool for providing a robot with valuable sensory input at a very low cost in hardware and development effort.

References

[1]   G. J. Agin, "Computer Vision Systems for Industrial Inspection and Assembly," Advances in Computer Technology—1980, Volume 1 sponsored by ASME Century 2, San Francisco, Calif., August 12-15, 1980, pp. 1-7.

[2]   M. R. Ward, L. Rossol and S. W. Holland, "CONSIGHT: A Practical Vision-Based Robot Guidance System," 9th International Symposium on Industrial Robots, Society of Manufacturing Engineers and Robot Institute of America, Washington, DC, March 1979, pp. 195-212.

[3]   R. Eskenazi, "Video Signal Input," *this book*.

[4]   J. S. Weseka, "A Survey of Threshold Selection Techniques," **Computer Graphics and Image Processing 5,** 1976, pp. 382-399.

[5]   J. M. Wilf, "Chain Code," *this book*.

[6]   D. L. Milgram, "Constructing Trees for Region Description," **Computer Graphics and Image Processing,** Vol. 11, 88-89 (1979), 88-99.

[7]   R. Wong and E. Hall, "Scene Matching with Invariant Moments," **Computer Graphics and Image Processing,** Vol. 8, 1978, pp. 16-24.

[8]   W. A. Perkins, "A Model-Based Vision System for Industrial Parts," **IEEE Transactions Computers,** Vol. C-27, 1978, pp. 126-143.

[9]   A. Thompson, "Camera Geometry," *this book*.

[10] Y. Yakimovsky and R. Cunningham, "A System for Extracting Three-Dimensional Measurements from a Stereo Pair of TV Cameras," **Computer Graphics and Image Processing,** Vol. 7, 1978, pp. 195-210.

IN MAN AND MACHINE

Ellen C. Hildreth
Artificial Intelligence Laboratory
Massachusetts Institute of Technology
Cambridge, MA

Most vision systems, whether human or machine, begin by finding edges—the significant intensity changes in an image. How the vision system uses edges depends on its function. Some systems immediately begin interpreting edges as the boundaries of objects, shadow lines, and so on. [1, 14] With a representation of edges, we can analyze the texture, surface contours, range, and motion of objects in the image. All of these later analyses have one common starting point—the detection of edges.

The goal of creating an efficient, yet powerful, computer vision system has been pursued for many years. In most cases, however, the designers were limited by practical engineering constraints to methods that were fast and used little memory. This led to the development of specialized techniques using coarsely sampled images, very localized operators, and clever techniques for reducing the resulting problems of noise and quantization. Unfortunately, these methods often work poorly outside of their limited domains.

When we study the human visual system, though, we find that it samples the world at a high resolution and applies operators that use information from a large number of photoreceptor cells. This great processing power can overcome the limitations of specialized techniques and is one of the keys to the richness of human sight. By dedicating specialized hardware processors to the vision task we can begin to realize this level of performance in a computer vision system—provided that we have an *effective theory* that tells us how to deal with this wealth of information.

Studying the biological evidence on how the eye and brain process visual data, we can develop such a theory. The theory is significant in several respects. Not only does it suggest how a computer vision system can most effectively employ intensity information to find edges, but it leads to further understanding of the early stages of *human* visual processing as well. This article discusses the theory of edge detection developed by David Marr, myself, and other colleagues at MIT. First, though, let's look at some of the previous results—both in computer and biological vision systems—that influenced our work.

Past Approaches to Edge Detection

Mathematically, we can regard an image as a function of two variables. For every location, *(x,y)*, in the image, there corresponds an intensity, *I(x,y)*. Although a *digitized* image must necessarily be divided into discrete picture elements (pixels), normally stored in a two-dimensional array, let's consider *I*, to be a continuous function defined in a plane. Most computer vision systems detect edges by a *differential* analysis of the image, which I can explain in terms of this intensity function.

Taking the first derivative of the image—in effect, differentiating *I(x,y)*—yields a new function that shows how rapidly intensity is changing at each point in the image. Since edges correspond to the largest local intensity changes, we can locate edges by finding the high points *(peaks)* in the differentiated image. Figures 1*a* and 1*b* illustrate this principle in a one-dimensional cross section (in two dimensions, the *gradient* operator, ∇^2, is used).

Now suppose we differentiated the raw image twice. Points that appear as peaks in a once-differentiated image now appear as points where the twice-differentiated intensity function crosses zero. These *zero-crossings* correspond to edges in the raw image. Figure 1*c* demonstrates this effect.

This might be all there is to edge detection—*if* all edges occurred at the same scale. Unfortunately, they do not. If we look at the digitized image, the intensity changes from pixel to pixel. The boundaries of most "real-world" objects produce sharp edges—the intensity function changes steeply over a small number of pixels. In many cases, however, the edges are fuzzy; their corresponding intensity function increases gradually in smaller steps over a larger number of pixels. Edges occurring at different scales often overlap. We often find, for example, a slow intensity change due to a shading effect superimposed on a sharp, high contrast edge caused by a discontinuity in the surface.

We would like to be able to look for edges occurring at different scales. That way we can find local changes, taking place over short distances, as well as fuzzy edges in the image. We want to look at the intensity function first at a resolution near the initial image resolution and detect edges at this scale. Then, we want to *smooth* the intensity function over larger numbers of pixels and detect the edges that occur at larger and larger scales. Figure 2 shows the results of smoothing an image at different scales.

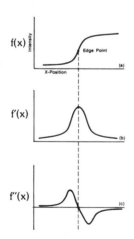

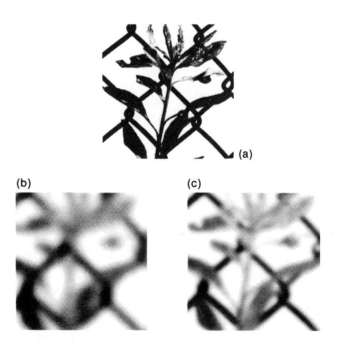

Figure 1. The principle of edge detection. Part (a) shows, for simplicity, a one-dimensional intensity function. The edge point is found where the intensity function has its maximum slope. In (b), we see the first derivative of the intensity function. The edge point can now be found at the peak of the derivative. Finally, in (c), we see the second derivative of the intensity function. Now we can find the edge point where the second derivative crosses zero—the zero-crossing.

Figure 2. An image smoothed by a convolution with a Gaussian filter. The original 320 × 320 pixel image appears in (a). In (b) and (c) we see the original image convolved with Gaussians having scale factors of σ = 4 and σ = 8 picture elements, respectively.

Another problem is that since the real image is a discrete function, not the continuous one we assumed, the derivative operations are not mathematically defined. The solution to this problem, as well as to how to accomplish the smoothing at various scales, is realized in the process of image *convolution*. (See sidebar "Image Filtering and Convolution".) Conventional edge detection techniques commonly use an operator with a Gaussian or step-pulse profile for smoothing, combined with either first or second order differentiation to detect intensity changes, and followed by various thresholding or thinning operations. I recommend several reviews for a more thorough discussion of these methods. [2, 4, 11, 13]

We shall see now how many of these same operations are embodied in the early processing stages of the human visual system.

Insights from the Human Vision System

Human vision first registers light intensity with the array of photoreceptors in the retina. We sense images at quite a high resolution. *One square inch,* viewed from a distance of three feet, covers an array of about 200×200 = 40,000 photoreceptors in the central fovea of the eye. Several layers of cells in the retina process the detected light intensity, culminating with the output of the *retinal ganglion cells*—whose axons form the optic nerve fibers that carry information to the visual cortex.

Neurophysiological recordings in the retinas of cats and monkeys have told us that the receptive fields of retinal ganglion cells are organized as shown in Figure 3. [6, 12] Light striking the center of the cell's receptive field excites the activity of the cell, while light striking the surrounding area inhibits it. We also know how the cell's sensitivity to light varies across its area. The shape of the sensitivity distribution can be described mathematically as the difference of two concentric Gaussian distributions:

$$G_1(x,y) - G_2(x,y) = \frac{1}{2\pi\sigma_1} e^{-\frac{r^2}{2\sigma_1^2}} - \frac{1}{2\pi\sigma_2} e^{-\frac{r^2}{2\sigma_2^2}}$$

where r is the radius from the center, and σ_1 and σ_2 are the spatial scale factors of the exitatory and inhibitory distributions, respectively. The significance of the scale factors will become apparent shortly. It is the *shape* of this distribution that is significant, not its absolute magnitude.

Extensive psychophysical experimentation has given us evidence about how the visual cortex processes information from the retina. Cortical cells have differing response patterns, some sensitive to orientation, others to

motion, etc. [5] Of particular interest to us are the "simple" cells, which have "bar" and "edge" shaped receptive fields, selective to the orientation of the stimulus. Our theory contends that the processing done in the retina is non-oriented, with the simple cells processing the results to find edges. [8]

There are also different *channels* of information processing, tuned to different scales of filtering. [15] The evidence suggests that there are at least four such channels, whose initial operators are also shaped like the difference of two Gaussians. The sizes of the operators are separated by an octave—if the size of the smallest operator covered w retinal cells, then the other channels would have sizes of 2w, 4w, and 8w. The sizes of the operators at a particular point in the visual field increase linearly with its distance from the central fovea. The shape, size, and frequency bandwidth of these operators had considerable influence on our work.

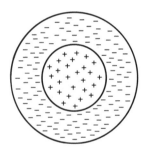

Figure 3. Receptive fields in the retina's ganglion cells have an antagonistic center-surround organization. Light falling on the central region excites the cell, while light falling on the surrounding area inhibits its output.

Marr-Hildreth Edge Detection

Our approach to edge detection came about by asking how the human visual system worked. Yet we can apply the results of our biological study to computer vision systems as well. The theory can be supported and discussed—as I am about to do— solely on computational grounds.

Recall that edge computation proceeds in three steps: The image is smoothed, differentiated, and scanned for peaks (if the first derivative is used) or zero-crossings (if the second derivative is used).

Suppose that we first smooth the image by convolving it with a Gaussian filter, as shown in Figure 2. We can describe the smoothed image symbolically as $G * I$. Our next step is to differentiate the smoothed image. For this we can use the Laplacian operator,

$$\nabla^2 = \frac{\partial^2}{\partial x^2} + \frac{\partial^2}{\partial y^2}$$

The Laplacian gives a non-oriented second derivative. We can write the new image as $\nabla^2(G * I)$.

At this point, a mathematical property of the convolution comes into play:

$$\nabla^2 (G * I) = (\nabla^2 G) * I.$$

This means that we can combine the ∇^2 and the Gaussian, G into a single operator, whose shape is

$$\nabla^2 G = (2 - \frac{r^2}{\sigma^2}) e^{-\frac{r^2}{2\sigma^2}}$$

Then $\nabla^2 G$ is convolved with the image. This is the same process— $\nabla^2 G * I$—that we believe occurs in the retina!

The $\nabla^2 G$ operator is intimately related to the Difference-of-Gaussians (DoG) profile measured in biological experiments. The shape of the DoG pattern becomes identical to that of $\nabla^2 G$ when the spatial scales of the two Gaussian profiles are close, as they are in the biological case. The $\nabla^2 G$ operator is illustrated in Figure 4. We have an edge-detector justified mathematically that is also consistent with biological experience.

The meaning of the $\nabla^2 G$ operator should now be clear. The G part smooths the image to an appropriate scale, then the ∇^2 part performs a second order derivative. The next step is to find the zero-crossings, which may signify edges in the image. Figure 5 shows each step in our edge detection process.

Given a collection of zero-crossings, we still have to extract and represent edge evidence in a useful way. Our representation, suggested by David Marr, is called the *raw primal sketch*. [7] To create it, we chop our collection of zero-crossings into a series of short line segments. For each segment, we compute its location in the image, its two-dimensional orientation, its length, and the rate at which the $\nabla^2 G$ output changes as it crosses zero—the latter measurement is related to the contrast or sharpness of the edge.

The important properties of the $\nabla^2 G$ operator are that it is (1) localized in space and in frequency, (2) of adequate size to reduce the effects of noise, and (3) non-directional. Its spatial scale factors determine both the region within which evidence may make a contribution and the resolution or "response" of its smoothing filter. The size of the operator is important in determining its sensitivity to noise as well. The smallest operator we use in our system, having a spatial scale of $\sigma = 8$ pixels, is able to respond to a step change in intensity even with the signal to noise ratio of the image almost one!

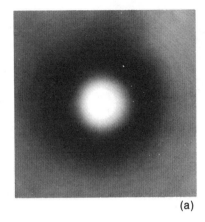

(a)

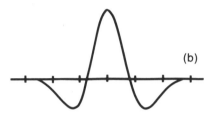

(b)

Figure 4. The $\nabla^2 G$ operator. Part (a) shows a top view of the distribution obtained by plotting its amplitude as brightness with 0 = gray. Part (b) gives a side view cross-section of the operator.

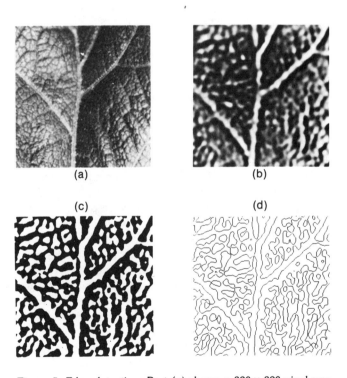

Figure 5. Edge detection. Part (a) shows a 320 × 320 pixel raw image. Part (b) depicts the image after it has been transformed by a $\nabla^2 G$ operator with a size w = 9 pixels. In this image, 0 is represented by medium gray. In (c), the positive values in the $\nabla^2 G$ output are shown as white, the negative values as black. Finally, in (d), we see the zero-crossings, the edges of the image in (a).

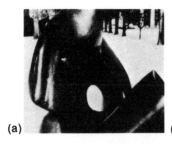   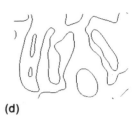

(a)   (b)   (c)   (d)

Figure 6. Zero-crossings from different sized operators. Part (a) shows the raw image. Parts (b), (c), and (d) show the zero-crossings when the image has been transformed by $\nabla^2 Gs$ with sizes w=6, 12, and 24 pixels, respectively.

Multiple Channels of Visual Information

Suppose that we have created a number of primal sketches, each corresponding to zero-crossings detected at a different spatial scale. (Figure 6 shows edges at several different scales.) We are now faced with one of the difficult and elusive problems in edge detection: How do we integrate information from primal sketches made by different sized operators?

One possibility is to use zero-crossings at various scales to confirm the existence of edges. Zero-crossings do not necessarily correspond to a physical boundary. However, a significant edge in a scene will tend to produce zero-crossings in adjacent channels which coincide spatially. This spatial coincidence, shown in Figure 7, might serve as a criterion for establishing the existence of an edge, but there are still a number of unsolved problems. [4]

Later processes which use the primal sketch—stereopsis, motion analysis, texture computation, and so on—may also integrate primal sketches from various channels. In stereopsis, for example, the major computational problem is to match elements in the images provided by the right and left eyes (or cameras), and measure the disparity in their positions. This disparity gives us the depths of elements in the scene.

According to a recent theory of human stereo vision [9], the elements we match are the zero-crossings from different sized operators. Because zero-crossings are such simple and primitive elements, there is a significant problem with *false targets*. That is, for a particular zero-crossing in the left image, there are several candidate zero-crossings in the right image to which it could correspond. To solve this problem, the algorithm first matches zero-crossings from large operators, where the zero-crossings are less dense, and therefore contain fewer false targets over a given distance in the image. This provides a rough idea of where things are in depth. The zero-crossings from successively smaller operators refines this information. This approach has been embodied in a computer system for stereo range measurement. [3]

Studying these later processing stages may also tell us whether the information contained in the raw primal sketch is enough. Are other description primitives needed, such as bars, blobs, or terminations? For stereopsis, the zero-crossings were sufficient, but we need to explore the later operations to gain a further understanding of edge detection.

Figure 7. The spatial coincidence of zero-crossings from two different scales. Edges that represent real object boundaries tend to be found in several adjacent scales.

Conclusion

Our theory of edge detection has prompted the building of special purpose edge detection hardware [10]. Members of our laboratory have already built a device that, in one second, transforms a 1000 × 1000 pixel image with difference-of-Gaussians approximation to $\Delta^2 G$. We are now building a linear array camera and a zero-crossing detector of comparable speed. With our special-purpose hardware, the performance of machine edge detection will begin to approach edge detection by the human visual system.

Though the complete "vision problem" has yet to be solved, studies of the human vision system have greatly contributed to our understanding of edge detection. These studies have shown us the importance of sampling the image at high resolution, applying a range of differential operators whose shape is $\nabla^2 G$, and detecting the zero-crossings in the operator's output. The result is an edge detection scheme which is simple, efficient, and can be applied to computer vision—using present technology. The computational study of human vision, which began to blossom with the work of David Marr, will continue to provide insights for the development of general computer vision systems. ▣

Acknowledgement: This article describes research done at the Artificial Intelligence Laboratory of the Massachusetts Institute of Technology. Support for the laboratory's artificial intelligence research is provided in part by the Advanced Research Project Agency of the Department of Defense under the Office of Naval Research contract N00014-75-C-0643 and in part by National Science Foundation Grant MCS77-07569.

IMAGE FILTERING AND CONVOLUTION

One of the basic operations of image processing is to filter an intensity array to reduce the effects of noise and quantization. By adjusting the "response" of the filter, the same technique will smooth the image at various scales to reduce the effect of edges more narrow than the scale of the filter. One simple way to smooth an image is to average the intensity of the pixels in the neighborhood of a given image location and use the result as the intensity at that point in the smoothed image.

Figure (a) shows a 3 × 3 image filter "window", centered at location (i,j) in the image, where the intensity is $I(i,j)$. In the smoothed image, the new intensity of (i,j) will be

$$S(i,j) = \frac{1}{9} \sum_{u=-1}^{1} \sum_{v=-1}^{1} I(i-u, j-v).$$

Now, let's rewrite this equation in a slightly different form. First, let $f(i,j) = 1/9$ if both i and j lie between –1 and 1 (inclusive) and 0 otherwise. Then, $S(i,j)$ can be rewritten:

$$S(i,j) = \sum_{u} \sum_{v} f(u,v) \, I(i-u, j-v).$$

Mathematically, this double sum is a discrete convolution, where the image—$I(i,j)$—is convolved with the "operator" $f(i,j)$. Mathematical readers will recognize this equation as the discrete form of the two-dimensional convolution integral

$$S(x,y) = \int_{-\infty}^{\infty} \int_{-\infty}^{\infty} f(u,v) \, I(x-u, y-v) \, du \, dv.$$

In mathematical shorthand, "*" is used as the symbol for convolution. In the previous equation, $S = f * I$.

By writing $S(i,j)$ in the form of a convolution, we can explicitly see the characteristics of the operator that smooths the image. The operator's *shape* and *size* are two of these characteristics. Shape simply refers to the kind of function (say a sine function) that describes the operator. An operator's size is the number of points that contribute significantly to the output of the convolution. The *shape* of our $f(i,j)$ operator is a step pulse with a *width* of 3 pixels (in the discrete space). (See (b).)

Operators of certain other shapes can also be convolved with an image to smooth it. One such operator is the 2-D Gaussian,

$$G(x,y) = 2\pi\sigma \, e^{-(x^2 + y^2)/2\sigma^2}$$

where σ is a scale factor that determines the operator's width. Figure (c) shows a one-dimensional version of the Gaussian. Figure 2 in the text shows how images can be smoothed by convolving them with Gaussian operators of different sizes.

Most methods of edge detection in gray scale pictures are based on some form of convolution. They filter the image to smooth it, apply differential operators to the smoothed image, then detect edges as peaks or zero-crossings. Convolution has an important property that greatly simplifies this process. It allows an image differentiation to be combined with the smoothing process and performed in one "pass" of the image.

To understand why this is so, let's look at the first derivative of a one-dimensional convolution:

$$(f * g)' = \frac{d}{dx} \int_{-\infty}^{\infty} f(t) \, g(x - t) \, dt.$$

Since only the function g depends on x, the derivative operation passes through the integral and acts only on g, so that

$$(f * g)' = \int_{-\infty}^{\infty} f(t) \, g'(s-t) \, dt = f * g'.$$

If now we make the change of variable $s = x-t$, we can rewrite the undifferentiated convolution integral as

$$f * g = \int_{-\infty}^{\infty} f(x-s) \, g(s) \, ds,$$

where $x-t$ replaces t as the parameter of f. Taking the derivative of the convolution as before, we see that, since now only the function f has a dependence on x, $(f * g)' = f' * g$ as well. This result is important, for it means that to smooth the image and differentiate it at the same time, we can simply convolve it with the *derivative of the smoothing operator*. This gives us a way to define differentiation on the discrete digitized image array—and it is the key to using the $\nabla^2 G$ operator discussed in the text.

(i–1,j–1)	(i–1,j)	(i–1,j+1)
(i,j–1)	(i,j)	(i,j+1)
(i+1,j–1)	(i+1,j)	(i+1,j+1)

(a)

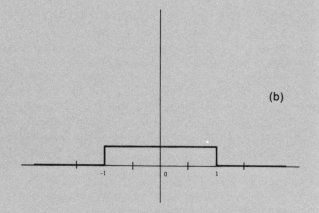

(b)

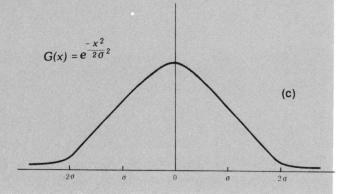

$G(x) = e^{\frac{-x^2}{2\sigma^2}}$

(c)

References

[1] Binford, T. "Inferring Surfaces from Images." *Artificial Intelligence Journal*, Vol. 16, Special Issue on Computer Vision, J. M. Brady (ed.). *Interesting work on the interpretation of edges in a single, static image, and their constraint on surface structure.*

[2] Davis, L. (1957) "A Survey of Edge Detection Techniques." *Computer Graphics and Image Processing 4*, 248-270.

[3] Grimson, W. E. L. (1981) *From Images to Surfaces: A Computational Study of the Human Early Visual System.* MIT Press, Cambridge, Ma. *An in-depth computational study of the Marr-Poggio stereo algorithm, and the interpolation of full surface descriptions from sparse data. It is an excellent example of the computational approach to studying human vision.*

[4] Hildreth, E. C. (1980) "Implementation of a Theory of Edge Detection." MIT TR-579. *Contains further discussion of the work described in this article, as well as an extensive review of other edge detection schemes.*

[5] Hubel, D. H. & Wiesel, T. N. (1968) "Receptive Fields and Functional Architecture of Monkey Striate Cortex." *J. Physiol. (Lond.) 195* 215-243. *Presents some of the pioneering studies of the behavior of cells in the visual cortex.*

[6] *Kuffler, S. W. (1953) "Discharge Patterns and Functional Organization of the Mammalian Retina." J. Neurophysiol. 16, 37-68. A pioneering study of the behavior of cells in the retina.*

[7] Marr, D. (1982) *VISION*. W. H. Freeman Co. San Francisco. *An excellent exposition of the computational approach to understanding human vision. It integrates* studies of computation, neurophysiology, and psychophysics in a way that spans the depth and breadth of our present knowledge of processing of information by the human visual system.

[8] Marr, D. & Hildreth, E. C. (1980) "Theory of Edge Detection." *Proc. R. Soc. Lond. B. 207,* 187-217.

[9] Marr, D. & Poggio, T. (1979) "A Computational Theory of Human Stereo Vision." *Proc. R. Soc. Lond. B. 204,* 301-328.

[10] Nishihara, K. & Larson, N. (1981) "Towards a Real Time Implementation of the Marr and Poggio Stereo Matcher," Proceedings of the ARPA Image Understanding Workshop, pp. 141-120.

[11] Pratt, W. (1978) *Digital Image Processing.* John Wiley & Sons, N.Y.

[12] Rodieck, R. W. & Stone, J. (1965) "Analysis of Receptive Fields of Cat Retinal Ganglion Cells." *J. Neurophysiol. 28* 833-849. *One of the earliest studies to describe the shape of the receptive fields of retinal cells.*

[13] Rosenfeld, A. & Kak, A. (1976) *Digital Picture Processing,* Academic Press, New York.

[14] Tennenbaum, M. & Barrow, H. (1978) " Recovering Intrinsic Scene Characteristics from Images." in: *Computer Vision Systems,* A. R. Hanson & E. M. Riseman (eds.). Academic Press, N. Y.

[15] Wilson, H. R. & Bergen, (1979) "A Four Mechanism Model for Threshold Spatial Vision." *Vision Res. 19,* 19-32. *Presents a quantitative psychophysical study of the spatial frequency channels in human vision, as well as references to other work in this area.*

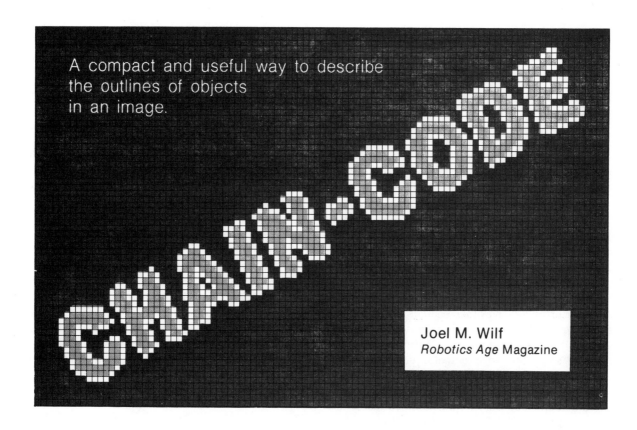

A compact and useful way to describe
the outlines of objects
in an image.

Joel M. Wilf
Robotics Age Magazine

Introduction

Anyone who works with images knows how they can burden a small system. An image represented as a 512 × 512 array of picture elements (*pixels*), with 8 bits of brightness (*gray-scale*) information, takes more than a quarter megabyte of storage. In robot vision systems, image understanding often needs multiple images. The system can spend most of its time just shuffling pictures between main memory and secondary storage devices. Clearly, a vision system needs to use either a very large memory or an image with less data.

Even if main memory were large and inexpensive enough to solve the storage problem, raw vision data would still offer a serious drawback: it is not a very convenient form for image understanding. Most image understanding strategies—including those of the human brain—work in a series of successive stages. Each stage performs progressively intricate operations on a progressively smaller amount of more highly structured data. The problem, then, is not only to compress image data. The data must be represented in a form useful to techniques of image understanding.

Can a raw image be represented by only a portion of its data? To answer this, suppose we partitioned an image into regions, each one having just a single brightness level. We could then represent each region by its outline. To reconstruct the whole image we simply fill each outline with the appropriate brightness. Because we can usually describe outlines with much less data than regions, outlines are useful for data compression. Since we can reconstruct the whole region from its outline, outlines can be a good form for data representation.

Chain-code, the subject of this article, is a data structure for storing outlines in computer memory. Chain-code is a list. Its first entries are the *x, y* coordinates of the point from which to begin tracing the outline. Subsequent entries are numbers giving the direction between each point in the outline and its neighbor. There are eight possible directions between a point and its neighbor. As figure 1 shows, these eight directions are numbered "0" through "7," counter-clockwise. Figure 2 shows an outline and its chain-code.

Edge Detection

Edge detection, for this discussion, refers to the process that takes a standard gray-level image array as input and

Reprinted from Robotics Age, *Volume 3, Number 2, March/April 1981*

outputs an array which, ideally, contains only the outlines of objects in the image.

Though edge detection is still a vital research problem, vision system designers can choose from quite a few techniques currently in use. Rather than discuss these techniques in general, let's follow one particular edge detection process, from start to finish. For those readers who want a more thorough and general discussion of edge detection, I recommend Davis's survey (*see references*).

Smoothing is the first step in our edge detection process. Smoothing eliminates "salt-and-pepper" type noise. That is, it removes isolated pixels whose gray-level is radically different from those of its neighbors. Noise of this kind can come from defective photosensors or analog-to-digital roundoff error, among other sources. If not eliminated, salt-and-pepper noise can cause spurious edges to appear later, when we compute edge information.

In the next step, the edge detector calculates the brightness difference between each point and its neighborhood. The size of the neighborhood or *window* examined by the edge detector determines how sensitive the edge detection is to noise and to detecting real edges in the image. There is always a tradeoff between the reliability of an edge detector and its speed. For example, a Sobel-type

operator, which examines a 3 × 3 window, is shown in figure 3. Though relatively unsophisticated, a Sobel operator has the advantage of being easy to implement in hardware for high-speed edge detection.

When the edge detector has finished determining the brightness differences at every pixel, the image is in the form of a *gradient edge map*. Each pixel in the gradient edge map represents how sharply the gray-level changes at that point.

The next step in our process separates edge points from the rest of the image. The brute force approach to this problem would be to set a single threshold value: Every point with an edge intensity above this threshold would be set to "1," and the others set to "0." But his approach ignores variations in contrast over the image. In regions of high contrast, it would detect spurious edges. In regions of low contrast, it would miss edges. Instead, then, of setting just one threshold for the image, we use a function that sets a *local threshold* for each neighborhood. When the

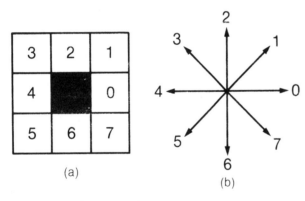

(a)

(b)

Figure 1. *The dark square at the center of (a) represents the pixel we are currently examining. As shown above, the pixel has eight neighbors, numbered "0" through "7" counter-clockwise. As (b) indicates, these numbers can also represent the directions we can travel from the center pixel to its neighbors. Notice that our choice of eight pixels is somewhat arbitrary. We could construct a four pixel neighborhood—by considering just the four pixels which share a side with the center (numbered "0", "2", "4", and "6"). This would give* four-direction *chain-code, as opposed to the* eight-direction *chain-code discussed in this article.*

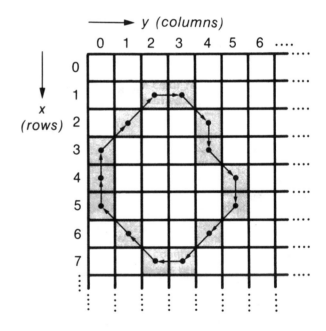

Figure 2. *If we use pixel (1, 2) as the first point in the chain, and move clockwise along the curve, we obtain "0, 7, 6, 7, 6, 5, 5, 4, 3, 3, 2, 2, 1, 1" as the chain-code for this outline. We can traverse the chain in the opposite direction by reversing the order of the chain and adding "4" (modulo 8) to each link. "5, 5, 6, 6, 7, 7, 0, 1, 1, 2, 3, 2, 3, 4" is the counter-clockwise chain-code for this curve. Notice that (1, 2) still represents the starting point.*

local thresholding has been applied to every pixel, the output image is a *binary edge map*. Every edge point has a value of "1"; every non-edge point has a value of "0". Figure 4 depicts the entire transition of an image—from gray-level array to binary edge map.

Ideally, every edge in the binary edge map would be a simple, closed curve, corresponding to the outline of an object in the image. Unfortunately, edge detectors usually pick up edges that are not on the boundary, and they miss edges that are. To remove spurious edges or complete broken contours, edge detection needs feedback from a process that "knows" how complete outlines in the image should look.

Chain-Encoding

Assume, now, that the image is a binary edge map, as shown in figure 4C. Each pixel in the image array has a value of "1" if it lies on an edge, and "0" otherwise. Notice that the majority of pixels have a value of "0". Therefore, we waste memory when we store the whole array. More importantly, this arrangement wastes processing time because the entire image must be scanned in operations

A	B	C
D	E	F
G	H	I

(a)

$(i-1, j-1)$	$(i-1, j)$	$(i-1, j+1)$
$(i, j-1)$	(i, j)	$(i, j+1)$
$(i+1, j-1)$	$(i+1, j)$	$(i+1, j+1)$

(b)

Figure 3. For convenience, we can label a pixel's 3=3 window as shown in (a) or (b). "E" is the pixel under examination, and its neighbors are labelled A, B, C, D, F, G, H, and I, using the labelling shown in (a). Then the Gradient Edge Value, $g(E) = |A + 2B + C - G 2H - I| + |A + 2D + G - C - 2F - I|$. The same 3 + 3 window can be used to obtain a local threshold value. Using the labelling shown in (b), let

$$S(i, j) = \sum_{k=-1}^{1} \sum_{L=-1}^{1} W_{kL} * g(i+k, j+L)$$

where the W_{kL} are programmable integer "weights", and W_{00} is minus the sum of the other weights. Each point in the binary edge map is set to one (edge) if $S(i, j)$ is positive, and to zero (no edge) otherwise.

(a)

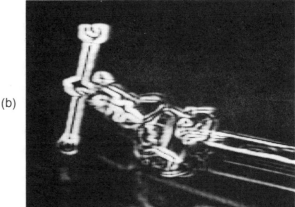

(b)

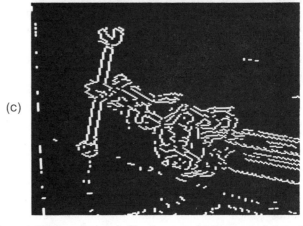

(c)

Figure 4. The unprocessed video image in (a) has a resolution of 244 + 188 pixels, with a gray-scale resolution of 256. Image (b) shows the gradient edge map made from image (a). Notice that bright areas in the raw image, have become areas with bright fringes and dark insides in the gradient edge map. Image (c) shows the binary edge map made from image (b). Notice that most edges form broken edge segments rather than simple, closed curves.

which involve only edges. It's time we converted the image to chain-code.

A binary edge map can be chain-coded in three-steps. In the first step, we scan the image for points from which to start a chain. In the second step, we chain-encode all edges that do not form simple, closed loops. The third step then picks up the simple, closed boundaries. Let's consider each step in detail.

In the first step, do the following: Raster-scan the image. At every edge point, count how many of its eight neighbors lie on an edge. A point with no neighbors on an edge is *isolated*. Store it as complete chain-code, since that point is the only one on the edge. A point with one neighbor on an edge terminates that edge. Store it in a list of *end points*. If a point has two neighbors on an edge, it lies in the middle of the edge. Ignore it. Finally, with three or more neighbors on an edge, the point lies at the fork between two edge branches. Store it in a list of *node points*. Figure 5 illustrates this process.

In the second step, we construct the actual chain-code. Do the following: Start each chain with an end-point or a node-point. Examine this point's eight neighbors in clockwise fashion, starting with the immediate righthand neighbor (direction "0"). When a neighbor is found which lies on the edge, add it to the chain. The new edge-point's neighbors are then examined, counter-clockwise, starting in the previous edge-point's direction. In this way, "walk-around" the outline, adding points to the chain. As each point is added, delete it from the image so it will not be picked up twice. Do not delete node-points, though, since more than one edge emanates from them. Continue scanning the neighbors of a node-point, tracing all the edges, until the node-point has no neighbors left in the image. *Then* delete it.

Elegant mathematical relationships exist between an outline and the region it bounds. These enable us to gather certain statistics about a region while walking around its boundary. At every point, we can calculate partial sums for the length of the chain, its "chain-height" and "chain-width", and a set of shape statistics known as *moments*.

For any region, we can generate an infinite series of moments. But for most purposes, the first six terms of the series suffice. The first term of the moment series is the region's area; and the second and third terms are the moments of inertia (familiar from college physics) around the x-axis and y-axis, respectively. The next section discusses how to use moments to recognize objects in the scene. For now, mathematically inclined readers can find out how to calculate moments from outlines in the "Mathematician's Corner" (at the end of this article).

By the third step, the image will be empty of everything but simple closed curves (since the first step ignored points in the middle of an edge). Raster-scan the image until reaching an edge point. Beginning with this point, chain-encode the curve, deleting points as they are added to the chain. By the time we have completely scanned the image, it will be empty, and chain-code for all its edges will reside in memory.

Image Understanding

By now, the vision system has converted the image to a collection of chain-coded edges and some statistics about them. The question remains: How can this information be used to recognize objects?

One commonly used recognition technique is *template-matching*. As the name implies, template-matching compares the outline of an object in the image with stored outlines ("templates") of known objects.

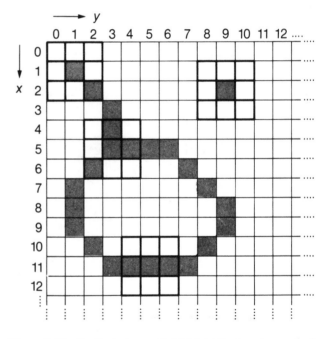

Figure 5. The 3 × 3 window grids highlight the four types of edge points in an image. The point with coordinates (2,9) is an isolated point. None of its neighbors lie on an edge. It can be considered a complete edge in itself. (1, 1) is an end point. With only one neighbor "turned on", it terminates the edge. (5, 3) is a node-point, since more than two of its neighbors are "on". Notice how the node-point forms a fork between two edge segments. (11, 5) has two "turned on" neighbors. It therefore lies in the middle of an edge.

ROBOTICS AGE: IN THE BEGINNING

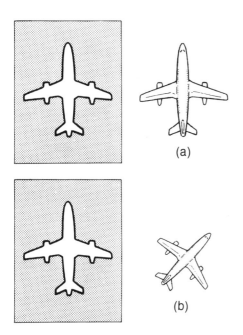

Figure 6. The airplane on the right side of (a) can be recognized by matching it with its template (on the left side), stored in computer memory. The airplane in (b), however, must be rotated and rescaled if it is to be compared with the template lying to the left of it.

Figure 6, which illustrates this process, also reveals one of its weaknesses: Before an object can be compared with a template, it has to be moved, rotated, and rescaled to fit the template's frame of reference. To prepare an outline for template-matching, we need to define coordinates which depend only on the outline itself; not on the outline's location in the image. Making the outline's center-of-mass the origin of its coordinate system is the first step. The next is to orient the outline along an axis intrinsic to the outline, not its orientation in the image. The *axis of minimum moment of inertia* is one such axis, which we can calculate from moments. (Remember, we calculated the first six moments as we chain-encoded the outline). Finally, we can scale the outline by adjusting its "chain-height" to the template's height. The reader can find formulas for "normalizing" chain-coded outlines in the "Mathematician's Corner".

Discrimination is another, more sophisticated method of object recognition. It also requires an internal model of the object being sought. However, this model is a list of "features". Average brightness, center-of-mass, region area, and any other measurable quantity associated with an object may be included in the feature list (often called a *feature vector*). The most useful features, though, are those which remain constant no matter how an object is moved around or rotated in an image.

Moment invariants—certain linear combinations of moments—are features which have this trait. No matter how an object is translated, rotated, or rescaled in an image, moment invariants remain constant (though digitization and approximation errors do creep in). Like the moment sequence itself, an infinite number of moment invariants can be generated. We may need only a few, however, for a given recognition task. The curious reader can turn to the "Mathematician's Corner" to see what the first few moment invariants look like.

Moment invariants can characterize objects in two-dimensional scenes—where we can ignore depth information. Wong and Hall (*see references*) reported that moment invariants helped classify printed characters with an accuracy of 95%. They also mention that moment invariants have been used to characterize X-rays and identify aircraft. In their own work, Wong and Hall used moment invariants to match aerial radar images with optical images of the same scene.

Limitations

Chain-code has many advantages: It is compact, easily constructed, easy to understand, convenient for some kinds of mathematical manipulation, and useful for some types of statistical object recognition. It has, though, several important limitations.

One problem is that we are unable to construct chain-code in a single raster-scan of the binary edge map. This is a drawback in a robot vision system which must analyze changing scenes in real-time. A number of researchers are trying to speed up the chain-encoding process.

Cederburg (*see references*), however, has proposed discarding chain-code in favor of a different, chain-code-like structure called RC-Code (for Raster Chain-Code). RC-Code uses only the four chain-code directions that lie in the "forward scan" direction (the chain-code directions 0, 7, 6, and 5). RC-Code has the advantage of being constructable in a single raster-scan of the edge map.

Certain basic operations, such as arbitrary rotations and scale changes, are very awkward to perform on chain-code. Rotations which are multiples of 90 degrees are simple enough: just add (in module 8) multiples of 2 to each link in the chain. For all other rotations, however, we must transform the chain-code back into coordinates, rotate the coordinates, and re-encode the chain. Most scale changes also force us to convert back to coordinates.

Chain-code stores outlines; but outlines are not the only way to represent an image. Sometimes, we might prefer to work with two-dimensional regions in an image. When we want to store a region in computer memory, we can use a data structure that records the sizes and number of two-

The Mathematician's Corner

Notation

Let x and y be variables, and let (x, y) be the coordinates of an arbitrary point (pixel) in the image. Let x_L and y_L represent the value of x and y, respectively, at link L along the chain-coded curve. (x_0, y_0) is taken to be the starting point for the curve. For any $L \geq 1$, let

$$\Delta x_L = x_L - x_{L-1} \qquad \Delta y_L = y_L - y_{L-1}$$

If a curve is approximated by an arbitrary polygon, Δx_L and Δy_L can take on any value. By adopting the notation given here, we make most of the formulas given below apply easily to any polygonal approximation of a curve. Chain-code can be viewed as a special case of polygonal approximation, where Δx_L and Δy_L can take only the values –1, 0, or 1.

The Length of a Chain

An even link is defined as a link whose chain-code is 0, 2, 4, or 6. An odd link, conversely, is one whose chain-code is 1, 3, 5, or 7.

Let n_e be the number of even links in a chain, and n_o be the number of odd links. Then, the length of the chain is

$$\text{LENGTH} = n_e + \sqrt{2} * n_o$$

Chain-Height and Chain-Width

Let

$$X_i = \sum_{L=1}^{i} \Delta x_L + x_0 \qquad Y_i = \sum_{L=1}^{i} \Delta y_L + y_0$$

Then,

$$\text{CHAIN-WIDTH} = \underset{i}{\text{MAX}} \, X_i - \underset{i}{\text{MIN}} \, X_i$$

$$\text{CHAIN-HEIGHT} = \underset{i}{\text{MAX}} \, Y_i - \underset{i}{\text{MIN}} \, Y_i$$

If a box were drawn around the curve so that the curve touched all four sides, then the width of the box would be the chain-width, and its height would be the chain-height.

Moments Defined for a Region

In general, a $(p + q)$th order moment, m_{pq}, is defined by the equation

$$m_{pq} = \iint\limits_{-\infty}^{\infty} f(x, y) \, x^p \, y^q \, dx\,dy,$$

where $f(x, y)$ is a density distribution function. In the case of a uniformly bright region in an image, we can take $f(x, y)$ to be 1, if (x, y) lies in the region and 0, otherwise. For a uniformly bright region, **R**, the definition of its $(p + q)$th order moment becomes:

$$m_{pq} = m_{pq}(\mathbf{R}) = \iint\limits_{\mathbf{R}} x^p y^q dx\,dy.$$

Center of Mass and Angle of Minimum Moment of Inertia

We can use moments to compute the centroid, $(\overline{x}, \overline{y})$ of a region by

$$\overline{x} = \frac{m_{10}}{m_{00}} , \quad \overline{y} = \frac{m_{01}}{m_{00}}$$

We can also use moments to compute the angle of a region's axis of minimum moment of inertia, Θ. This quantity defines a region's orientation within a two-fold degeneracy. It is given by

$$\Theta = \frac{1}{2} \tan^{-1} \left[\frac{2 \, (m_{00} m_{11} - m_{10} m_{01})}{(m_{00} m_{20} - m_{10}^2) - (m_{00} m_{02} - m_{01}^2)} \right]$$

The quantities x, y, and Θ are useful in object recognition because they specify the position and orientation of regions. This, in turn, allows us to define a transformation between image and template coordinates.

If a region is symmetrical, its Θ has further degeneracies. In the extreme case, for a perfectly circular region, every axis that passes through its centroid is an axis of minimum moment of inertia. Because of digitization errors (etc.), though, regions are rarely symmetrical.

Calculating Moments

The moments for a region can be calculated while "walking around" its boundary. Let n be the number of links in a chain-coded (or polygon-approximated) curve.

Let

$$A_L = x_L \, \Delta y_L - y_L \, \Delta x_L$$

Then, the first six moments are given by

$$m_{00} = \frac{1}{2} \sum_{L=1}^{n} A_L$$

$$m_{10} = \frac{1}{3}\sum_{L=1}^{n} A_L \left(y_L - \frac{1}{2}\Delta y_L\right)$$

$$m_{01} = \frac{1}{3}\sum_{L=1}^{n} A_L \left(x_L - \frac{1}{2}\Delta x_L\right)$$

$$m_{20} = \frac{1}{4}\sum_{L=1}^{n} A_L \left(x_L^2 - x_L\Delta x_L + \frac{1}{3}\Delta x_L^2\right)$$

$$m_{11} = \frac{1}{4}\sum_{L=1}^{n} A_L \left(x_L y_L - \frac{1}{2}x_L\Delta y_L - \frac{1}{2}y_L\Delta x_L + \frac{1}{3}\Delta x_L\Delta y_L\right)$$

$$m_{02} = \frac{1}{4}\sum_{L=1}^{n} A_L \left(y_L^2 - y_L\Delta y_L + \frac{1}{3}\Delta y_L^2\right)$$

Moment Invariants

Moment invarients are best calculated in several stages. In the first stage, we find quantities called central moments—the region's moments in its center-of-mass coordinate frame. For a uniformly colored region, **R**, its $(p + q)$th order central moment, μ_{pq} is defined by

$$\mu_{pq} = \iint_{\mathbf{R}} (x-\bar{x})^p (y-\bar{y})^q\, dxdy$$

The first few central moments are

$$\mu_{00} = m_{00}$$
$$\mu_{10} = 0$$
$$\mu_{01} = 0$$
$$\mu_{20} = m_{20} - \bar{x}m_{10}$$
$$\mu_{02} = m_{02} - \bar{y}m_{01}$$

After we find the low order central moments, we calculate the normalized central moment, η_{pq}, by the equation.

$$\eta_{pq} = \frac{\mu_{pq}}{\mu_{00}^s} \quad \text{where } s = (p + q)/2$$

Finally, we can calculate some low order moment invariants. Let φ_n be the nth moment invariant. Then

$$\varphi_1 = \eta_{20} - \eta_{02}$$
$$\varphi_2 = (\eta_{20} - \eta_{02})^2 + 4\eta_{11}^2$$
$$\varphi_3 = (\eta_{30} - 3\eta_{12})^2 + (3\eta_{21} + \eta_{03})^2$$
$$\varphi_4 = (\eta_{30} + \eta_{12})^2 + (\eta_{21} + \eta_{03})^2$$

I leave it as an exercise for the reader to prove that these quantities are invariant with respect to translation, rotation, an scale change. Ambitious readers should turn to Hu and Wong and Hall (*see references*) to further explore the mathematics of moments.

dimensional blocks which cover that region.

The most serious limitation of chain-code, though, is that it is not a good structure for representing three-dimensional scenes. It works well enough in two dimensional situations—when aerial scenes are analyzed, for example, or when parts on a conveyor belt are inspected from directly overhead. However, we need other methods when we need to model the three-dimensional world. And the problem of modelling the three-dimensional world—in all its glory—has not yet been conclusively solved. Ⓡ

References

[1]	Cederberg, R. "Chain-Link Coding and Segmentation for Rater Scan Devices," *Computer Graphics and Image Processing*, Vol. 10, 1979, pp. 224-234.

[2]	Davis, L. S. "A Survey of Edge Detection Techniques," *Computer Graphics and Image Processing*, Vol. 4, 1975.

[3]	Eskenazi, R. and Wilf, J. Low Level Processing for Real-Time Image Analysis. Publication 79-79, Jet Propulsion Laboratory, Pasadena, Calif., Sept. 1979. This describes in detail a hardware implementation of real-time edge detection.

[4]	Freeman, H. "Computer Processing of Line-Drawing Images," Computing Surveys, Vol. 6, No. 1, March 1974, pp. 57-97. This is the standard reference on chain-code. I have heard it called "the chain-code bible."

[5]	Hu, M. "Visual Pattern Recognition by Moment Invariants," IRE Transactions on Information Theory, IT-8, 1962, pp. 179-187. This is a highly mathematical exposition on the theory of moment invariants.

[6]	Wilf, J. and Cunningham, R. Computing Arbitrary Moments from Chain-Coded Boundaries. Publication 79-49, Jet Propulsion Laboratory, Pasadena, Calif., Nov. 1, 1979.

[7]	Wong, R. and Hall, E. "Scene Matching with Invariant Moments," *Computer Graphics and Image Processing*, Vol. 8, 1978, pp. 16-24.

[8]	Wong, V. Computational Structures for Extracting Edge Features from Digital Images for Real-Time Control Applications. PhD. Thesis, Division of Engineering, California Institute of Technology, Pasadena, California, Feb. 1979.

BUILD A LOW-COST IMAGE DIGITIZER

Don McAllister
4709 Rockbluff Drive
Rolling Hills Estates, CA

Would you believe that you can build a complete image digitizer for only about five dollars in parts? Well, you can—provided that you have Radio Shack's new TRS-80 Color Computer or some other micro with the same useful features. This machine makes it easy to get started in robotics experimentation. It comes with two 6 bit digital-to-analog converters (DACs) and four analog-to-digital converter (ADC) channels. Also included are a serial communications port, a 40 pin connector that brings out all the bus signals of the 6809 CPU, built-in cassette I/O, and, of course, its color graphics capability: 12 different colors with resolution up to 192×256 picture elements (pixels). All this, with an extended BASIC interpreter, in a package small enough to mount as a single board computer inside a robot. I'll be writing about that latter application in a future article. But for now, I'll show you how I used the Color Computer to get started with image processing for robot vision.

The Image Sensor

The key to this low-cost digitizer lies in using one of the Color Computer's joysticks as a pointing device. On the end of the joystick's handle, I mounted a Radio Shack solar cigarette lighter, which is a small parabolic reflector with a fixture at its focal point. To this fixture I bonded a CdS photocell (Radio Shack part #276-116) with Superglue, so that the light collected by the mirror is focused on the photocell's surface. The digitizer operates by reading the photocell output and the joystick location as you manually move the joystick in a scanning pattern. For use in a robot, of course, you would want to replace the manually positioned joystick by some form of automatic pointer, but this simple trick lets you work on robot vision software—using real images—with the barest minimum of hardware. Figure 1 shows the sensor assembly mounted on the joystick.

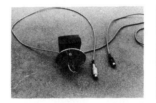

Figure 1. The reflector and photocell are mounted on the end of one of the Color Computer's joysticks.

The cadmium sulphide photocell is sensitive to light through the entire visible spectrum, with a response curve as shown in Figure 2. It is a resistive type photocell—varying from 500K ohms in the dark, to 1.7K ohms in one foot-candle (ftc.) illumination, down to 100 ohms at 100 ftc. The size of this particular photocell and the geometry of the mirror resulted in a field of view approximately 1.5 degrees wide. A smaller photocell could give a higher resolution. But since the computer can only resolve the joystick position to within a degree or so, this arrangement was satisfactory for my experiments.

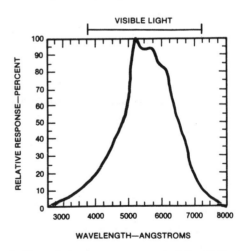

Figure 2. The relative response curve of the CdS photocell.

Reprinted from Robotics Age, Volume 3, Number 5, September/October 1981

Interfacing the Sensor to the ADC

Each joystick on the TRS-80 CC contains two potentiometers that read its position in two dimensions (degrees of freedom). Figure 3 shows the schematic of a joystick and its connector. Pin 5 provides 5V to power the divider, and pins 1 and 2 lead to a pair of ADC channels in the computer. As you move the joystick, the voltages on pins 1 and 2 vary in the range from 0 to 5V. Pin 4 is grounded by the push button "fire" switch on the joystick. I used the right joystick to point the sensor.

The Color Computer performs A/D conversion by a built-in BASIC intrinsic subroutine, called JOYSTK, that uses the two DACs. It accesses each DAC by writing a number in the six most significant bits of its memory location. The 6 bit value (0-63) is latched and converted to an analog voltage between .25V and 4.75V (by a resistor ladder type of converter). To perform an A/D conversion, one of the analog channels is selected by an analog switch and compared with the DAC output by an analog comparator. Each DAC has two such channels that can be selected for conversion. The intrinsic software varies the DAC output until the voltages match, then returns the corresponding 6 bit number. A call to JOYSTK requires a little over 4ms.

To sense the light on the photocell, I used a simple resistor divider to convert the cell's change in resistance to a voltage in the 0-5V range used by the joysticks. This let me use the port for the unused left joystick as the input port for the image sensor. The joystick port also supplies the 5V, as shown in Figure 4. I used coaxial cable to bring the signal back from the photocell, minimizing noise due to RF interference, etc. The 10K pot, used to adjust the sensitivity of the sensor, is bonded directly to the DIN plug at the joystick connector. The sensor output is connected to pin 1 of the plug—JOYSTK(2) reads the voltage on this pin.

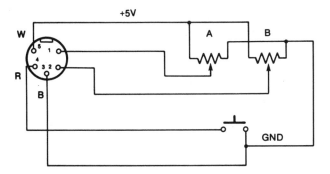

Figure 3. Joystick schematic.

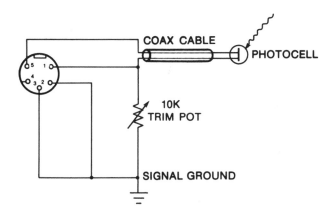

Figure 4. Photocell interface schematic.

Using the Sensor

There are a number of ways you can use the output of this simple sensor. Since the computer can read the pointing angles of the joystick, it doesn't depend on the accuracy of your manual scan. It can use the two joystick position values as an array index and store the digitized value at the right "coordinates" in the image. As a demonstration program, I used the sensor readings to form a "false color" image which I displayed using the machine's color graphics.

Figure 5 gives the listing of this program. The range of values returned by the sensor is partitioned into four bands, each displayed as a different color. the resolution of the display varies depending upon the color mode you select. In the four-color mode I used in the program the display has 192x256 pixels. I scaled up the joystick readings to fill the TV screen. Scale factors of 4 and 3 for the horizontal and vertical display coordinates, respectively, gave a reasonably undistorted "mapping" from joystick to screen coordinates. the PAINT statement fills in the boundary created by the CIRCLE statement in a particular color.

The program will continue to refresh the image as long as you keep scanning the sensor, with new colors painting over old if the image varies. Figure 6 shows a "false color" image of a light bulb.

By storing image values in an array, you can begin to do true image processing. You can write programs that scan the stored image to find light sources or to perform other vision operations. As you try to make effective use of this rich source of information, you will quickly find out that sight is an incredibly complex sense. This simple sensor

will let you start exploring robot vision—*without* expensive
TV cameras and complex video interfacing. Ⓐ

Figure 6. Image generated by scanning a light fixture.

```
20 ':    LOW COST IMAGE DIGITIZER PROGRAM
30 '                 BY
40 '             DON MCALLISTER
50 '
80 ' SET COLOR MODE FOR (192×256) 4 COLOR MODE
90 PMODE 3,1
100 ' CLEAR HIGH RESOLUTION SCREEN
110 PCLS
120 SCREEN 1,1
130 ' READ RIGHT JOYSTICK INPUTS
140 J0=JOYSTK(0)
150 J1=JOYSTK (1)
160 ' READ PHOTOCELL OUTPUT (J2)
170 J2=JOYSTK (2)
180 J3=JOYSTK (3)
190 ' IF PHOTOCELL OUTPUT > 50 SET COLOR = ORANGE
200 IF J2>50 THEN C=0
210 'IF PHOTOCELL OUTPUT<50 AND>30 SET COLOR = BLUE
220 IF J2<50 AND J2>30 THEN C=3
230 'IF PHOTOCELL OUTPUT<30 AND >20 SET COLOR = GREEN
240 IF J2<30 AND J2>20 THEN C=2
250 ' IF PHOTOCELL OUTPUT <20 SET COLOR = BACKGROUND (WHITE)
260 IF J2<20 THEN C=1
270 ' SCALE HORIZONTAL AND VERTICAL AXES
280 ' FOR RIGHT JOYSTICK OUTPUT
290 X=J0*4
300 Y=J1*3
310 'DRAW CIRCLE AROUND RIGHT JOYSTICK COORDINATES
320 'WITH FALSE COLOR  = C, AND RADIUS 4 PIXELS
330 CIRCLE (X,Y),4,C
340 ' COLOR INTERIOR OF CIRCLE WITH COLOR = C
350 PAINT (X,Y), C
360 ' RETURN TO SAMPLE PHOTOCELL OUTPUT
370     GOTO 140
380 END
```

Figure 5. False color image display program.

"Many people make the assumption that
robot experimentation is beyond the
resources of the average hobbyist. The simple
interfaces described here provide the means,
both in hardware and software, for
interesting experimentation at
minimum cost."

Don McAllister
4709 Rockbluff Drive
Rolling Hills Estates, CA

MULTIPLE SENSORS FOR A LOW-COST ROBOT

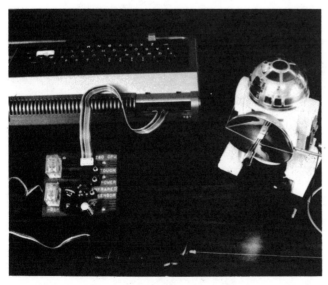

*Figure 1. Test setup for a multiple-sensor robot. Both the ADC
and the drive output hardware are contained on a single circuit
card. (Radio Shack #276-155)*

Robot behavior involving multiple sensory functions is a
fascinating field of research, for it is in the complex
interactions between sensor, control functions, and the
environment that behavior similar to that of animal life
forms can be produced. Until recently such investigations
were well beyond the financial resources of most hobbyists, but with the advent of inexpensive home microcomputers, this is no longer the case.

This article will describe a method of constructing an
extremely inexpensive multi-channel analog to digital
converter (ADC) that can be used with the Radio Shack
TRS-80 computer, allowing programs written in Level II
BASIC to acquire input from several sensors. Behavior
modes using the sensors are demonstrated through direct
computer operation of a radio-controlled toy, in this case a
Kenner "R2D2." Although the techniques presented here
are for the TRS-80/Kenner-R2D2 combination, they are
sufficiently general to be adapted to any microcomputer or
test vehicle.

The components of the experimental setup are shown in
Figure 1. The interface board plugs directly into the

Reprinted from Robotics Age, Volume 2, Number 1, Spring 1980 expansion port of the TRS-80. Output lines from the interface connect to the drive switches in the Kenne: R/C transmitter. However, since there is no return radio link, signals from the sensors mounted on the robot must be carried by wires. The use of lightweight coax minimizes the introduction of electrical noise. The sensors shown include a phototransistor infrared detector (used with a spherical collector mirror) and a microswitch touch sensor. Other sensors may also be used, such as a microphone or temperature or pressure transducers.

Interfacing Multiple Sensors to the TRS-80

A key to the simplicity of the ADC interface is that part of the conversion process is performed by the computer software. The interface hardware converts the signal from an analog level to a pulse whose duration depends upon the level. A program loop in the computer measures the length of the pulse by counting the number of loop cycles that the interface signal is high, exiting with the measured value when the signal returns to low. The interface circuit, shown in Figure 2, may be expanded up to six sensors. By using only one 74367 bus driver chip, data from multiple sensors can be input to the computer using only one 555 timer chip per sensory channel.

Referring to the figure, let's consider the functioning of a typical sensory channel. Each 555 timer is configured as a monostable pulse generator. [1] When triggered on pin 2 by a negative-going OUT' pulse from the CPU, a positive output pulse is generated by on pin 3. The duration of the pulse is determined by the sensor input voltage applied to threshold/discharge terminals of the 555 (pins 6 & 7). The pulse duration should range from 10 microseconds to 30 milliseconds as a function of the input level from the sensor.

This configuration, although easy to build, does impose some constraints on the sensor output. Before the timer is triggered, pin 7 is held near ground by a normally on transistor inside the IC. When the device is triggered, the discharge transistor is turned off, and current from the sensor is allowed to charge the sensor capacitor. The timer output will remain high until the voltage on pins 6-7 matches the reference voltage on pin 5 (2/3 of the supply voltage, in this case 3.33V). Thus, the pulse duration will depend upon both the output level and the output impedance of the sensor used. A low sensor output causes a long output pulse and vice versa. A resistor network or a one-transistor amplifier may be needed to provide suitable output characteristics for a particular sensor.

To allow the CPU to measure the pulse duration, the output of each timer is connected to one bit of the 74367 tri-state bus driver. When the driver receives a negative-going IN' pulse from the TRS-80, the values of the timer outputs (either logic 1 or 0) are strobed onto the corresponding bits of the computer's data bus. When IN' is high (IN=0) the driver chip presents an open circuit to the bus to allow other computations to be performed regardless of the timer output values.

In this circuit no I/O address or port decoding is used. Whenever IN' goes low, the data on the inputs of the 74367 are placed on the CPU bus. Obviously, no other devices may use the same data bits on the expansion port for input unless suitable device selection logic is added. In this interface, up to six 555 timers may be used, each with its own unique timing pulse. The timers are assigned to data bit D_7 through D_2, and their outputs may be read by the computer at any time using the Z-80's "IN" instruction.

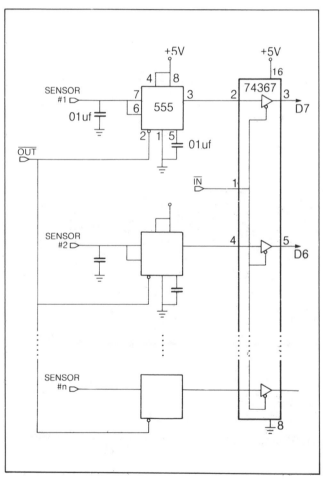

Figure 2. The Analog to Digital Converter (ADC) interface, which plugs directly into the TRS-80 expansion port.

TABLE 1

Multiple Sensor Analog/Digital Conversion Routines in Z80 Machine Language

```
D3 10      A$:      OUT (10H), A      ; trigger A to D converter
CD 7F 0A            CALL 0A7FH        ; initialize converter
44                  LD B,H            ; with most significant byte
4D                  LD C,L            ; least significant byte
03         LOOP1:   INC BC            ; increment counter
CB 58               BIT 3,B           ; limit max converter value
C0                  RET NZ            ; return if value exceeded
DB 10               IN A, (10H)       ; input sensor data D7
CB 7F               BIT 7,A           ; check if D7=1 [see note below*]
ED 43 01 70         LD (7001H), BC    ; store count in 7001H [see note 2*]
20 F2               JR NZ, LOOP1      ; keep counting if D7=1
C9                  RET               ; stop count when D7=0
                                      ; and return to basic
```

*Note: Op-codes for BIT n,A with n=0, 1, 2.., 7 are:

n=7: CB 7F	n=4: CB 67	n=1: CB 4F
n=6: CB 77	n=3: CB 5F	n=0: CB 47
n=5: CB 6F	n=2: CB 57	

*Note 2: To deposit the results at different addresses, change the last two bytes of the LD instruction to the desired address, low-order byte first, so that "ED 43 03 70" will store the result at 7003-4, etc.

TABLE 2

Basic Program for Analog/Digital Conversion and Display

```
10    A$=" SENSOR #1 A/D CONVERSION"
20    B$=" SENSOR #2 A/D CONVERSION"
30    C$=" SENSOR #3 A/D CONVERSION"
40    X=VARPTR(A$)
50    POKE 16526, PEEK(X+1): POKE 16527, PEEK(X+2)
60    L=USR(0)
70    X=VARPTR(B$)
80    POKE 16526, PEEK(X+1): POKE 16527, PEEK(X+2)
90    L=USR(0)
100   X=VARPTR(C$)
110   POKE 16526, PEEK(X+1): POKE 16527, PEEK(X+2)
120   L=USR(0)
130   S1=PEEK(28673) + 256 · PEEK(28674)
140   S2=PEEK(28675) + 256 · PEEK(28676)
150   S3=PEEK(28677) + 256 · PEEK(28678)
160   PRINT S1, S2, S3
170   GOTO 40
200   END
```

Sensors such as switches that have only two states (logic 0 or 1) may be connected directly to the inputs of the bus driver and have their state gated directly onto the bus. Of course, the software that reads such a sensor should only test the sensor state once instead of attempting to measure the duration of the signal.

The actual analog to digital conversion is performed by subroutines written in Z-80 assembly language and called by the "USR" command in Level II BASIC. [2] The subroutines, one of which is shown in Table 1, are each embedded in a single line string definition statement in the Level II BASIC program. The machine language code is deposited into the same locations formerly occupied by the characters in the original string statement. This procedure eliminates the need for loading system tapes into protected memory and makes the machine code an integral part of the BASIC program.

To understand how this is done and how the ADC subroutines are used, refer to the BASIC program in Table 2, which reads values from sensors 1, 2, and 3 and prints them. Source statements 10, 20 and 30 are used to store the machine code for the conversion routines for the respective sensors. The address of the subroutine is the byte immediately following the first quote mark of the string definition.

Machine code can either be POKE'd into the string definition locations, or, more conveniently, it can be directly entered into the string location using a monitor program such as RSM or TBUG, residing in high memory (well out of the way of BASIC programs). RSM has the additional advantage that it disassembles the machine code, thus providing a check on the Z-80 opcode mnemonics. The BASIC program should first be entered using arbitrary keyboard characters to fill out the strings (remember to use no fewer characters than the number of bytes in the subroutine plus one zero byte at the end). Then, the machine code in Table 1 can be inserted into the proper string, using the BIT instruction that tests the data bit for the desired sensor. The subroutine shown deposits the result of the conversion into location 7001H. Subroutines for the other sensors may use different locations, such as 7003, 7005, etc. Refer to the footnotes in Table 1 for the code modifications necessary to test other data bits or use different result locations.

Once the machine code is entered, the BASIC program may be stored on tape. The next time it is loaded, the machine code will still be there and the program may be used immediately. The two restrictions on using string definitions to store machine code subroutines are that the subroutine must be no more than 255 bytes and must contain no byte with a value of zero. A zero byte will cause the BASIC interpreter to terminate the string definition prematurely. Careful coding can usually get around both these restrictions. In the ADC subroutine, for example, zeroes in the code are avoided by using 10H as the (arbitrary) port address in the OUT instruction and 7001H, etc., as the result address.

Line 40 in the BASIC program sets the variable X with the address of the ADC subroutine for sensor #1. Line 50 deposits this address in the location used by the USR command, and Line 60 invokes the subroutine. After the conversion is completed, the two byte result is stored in high memory locations 7001 and 7002 (hexadecimal,

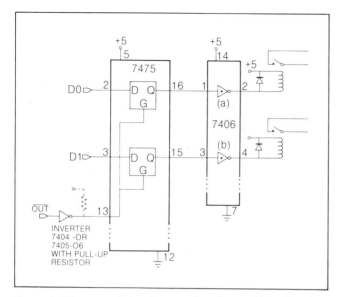

Figure 3. The control signal output interface, which also plugs directly into the TRS-80 expansion port.

corresponding to 28673-4 decimal). Lines 70 through 120 repeat this process for Sensors 2 and 3, whose ADC subroutines deposit their results in consecutively higher locations. Lines 130-150 convert these data to a usable form stored in variables S1, S2, and S3, respectively.

There are other options for the design of the ADC subroutines. Since no output port address decoding is used, the "OUT (10H),A" command in Table 1 will cause all the 555 timer chips to output pulses of duration proportional to their respective sensor voltages. In the subroutine shown, only the desired sensor bit is tested. It would also be possible to configure a routine that looped until the pulse lengths of all timer outputs had been counted. Another alternative would be to write a param-eterized routine that could read the value of any one sensor. The number of the desired sensor could either be POKE'd into a parameter location or passed via the USR command linkage. The BASIC program could even POKE the proper instructions directly into the code to access the desired sensor.

In the ADC subroutine given in Table 1, the USR parameter linkage is used to provide an initial value for the loop counter. The "CALL 0A7FH" statement loads the (two byte) value that was given as the argument to the USR statement into registers H and L of the Z-80 CPU. These are then moved into the 16 bit register combination BC which is then incremented each loop cycle. Thus, L=USR(0) initializes the ADC to zero whereas L=USR(100) begins the count at 100. Also, the code shown limits the maximum converter output values to 2047 by exiting when bit 3 of the result becomes a "1." A different limit could be set if desired.

We can now see the great flexibility and ease of implementation achieved by performing part of the A to D function in software. Even more dramatic is the cost savings obtained by this simple method, which requires

only a few dollars in parts and for many applications performs as well as multi-channel ADC boards costing hundreds of dollars.

Control of the Robot's Motion

The Kenner R/C robot used in these experiments is controlled by two switches that govern the toy's trans-lation and rotation. For use as a computer-controlled robot, these switches are interfaced to the TRS-80 using the circuit shown in Figure 3. Latched data bits set by the CPU under program control operate relays that in turn key the R/C transmitter.

Upon receiving a positive-going enable pulse, the 7475 4-bit latch chip reads the values on data lines D_0 and D_1 and holds them on the corresponding outputs until the next enable pulse occurs. In this case, the enable pulse is produced as shown by converting the OUT' pulse pro-duced by the CPU. Each relay is driven by a 7406 open-collector inverter, which grounds one side of the coil when the output of the latch is high (one). When the data bit is low (zero) the coil floats at 5V, leaving the relay inactive. Radio Shack relays #275-004 were chosen because they can be driven directly by the inverter/drivers, needing only 10 milliamps to engage.

The "OUT" command in Level II BASIC has the same function as the Z-80's OUT machine command, placing the desired data on the output bus and generating the OUT' pulse. In this case an "OUT 0,1" command in a BASIC program will cause the robot to move forward while a "OUT 0,3" command is used to set the direction of the movement. "OUT 0,0" will stop all motion. The amount of the motion is proportional to the duration of the movement—once a motion is initiated by the CPU, it will continue until the CPU generates a new movement command. By using the OUT command and variable time delays it is very easy to program any desired motion sequence in Level II BASIC. [3]

As in the sensor input interface, no I/O port selection logic is used. An OUT command to any port will cause the relays to operate. Thus, the OUT machine code statement used by the ADC subroutines to start the timers could also affect the robot's motion. Since the USR statement used to call the ADCs always sets Z-80 register A to zero, the OUT statement in the ADC code will turn off the relays. If the conversion interval is brief, however, the robot may not have time to respond to this stop command, provided that the controlling program sends out the proper move-ment command after reading a sensor. Other alternatives would be to read sensor inputs only when the robot is

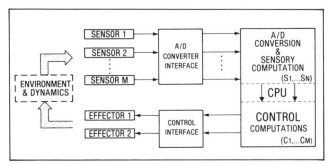

Figure 4. Functional block diagram for the TRS-80 controlled robot.

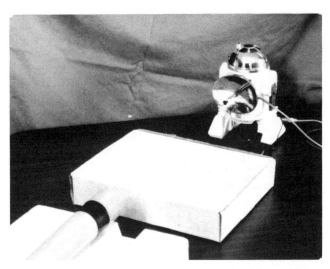

Figure 5. Experimental arrangement for obstacle avoidance behavior.

stopped or to let the same OUT command that controls the robot serve also to start the timers. However, leaving the robot turned on while writing to the cassette or any other peripheral may have unpredictable results!

Investigating Robot Behavior Modes

For a robot with N sensors, the state of all the sensor outputs at a given time can be regarded as an N-dimensional vector:

$$\mathbf{S} = (S_1, S_2, \ldots, S_N)$$

Similarly, if there are M control or actuator states, we can define:

$$\mathbf{C} = (C_1, C_2, \ldots, C_M)$$

to be the M-dimensional control vector.

Complex robot behavior can be defined as the process of computing the feedback control vector $\mathbf{C}$ as a function of the sensor state vector $\mathbf{S}$. The control function can consist of logical tests on the sensor values, as used for the task described below, or else it may involve arithmetic operations on the sensor values or be dependent on the past history of the process. Even with the simple sensory and control capabilities of the inexpensive system described here, complex behavior may be investigated simply by changing and improving the robot control computer program.

With the robot sensory and control interfaces implemented, we can write BASIC programs that will determine a specified robot behavior pattern in response to entered commands and sensory measurements. A functional block diagram for the TRS-80 controlled robot system is shown in Figure 4. The sensors mounted on the robot make measurements on various analog voltage levels. These voltages are converted by the sensor interface, which sends numbers to the control program that are a known function of the sensor measurements.

These data are then processed by the CPU and used as input to a decision-making procedure, which then produces control commands that are sent to the robot by the output interface. In this application, only the two output

bits D_0 and D_1 are needed, but this interface can be easily extended (by connecting additional output latches and relays, etc.) to provide up to eight control bits, allowing the program to select one of up to 256 distinct control states by proper choice of the output byte "n" in the BASIC statement "OUT 0,n."

Complex behavior patterns may be implemented in software by coding the desired control responses as a function of the sensor inputs. These responses may just rely on the sensor inputs at a given moment (no sensory memory), or may be based on sensory observations made at earlier moments during the robot's operation.

The process of producing a complex behavior pattern may be illustrated by the following experiment in which the TRS-80 controlled robot detects and tracks an infrared source. In this example, only two sensors were used: the infrared phototransistor scanner and the microswitch touch sensor, both shown earlier in Figure 1. Both the scanner and the touch switch were mounted to the front of the Kenner R2D2, and their signals were returned to the ADC interface by means of lightweight coaxial cable. The R/C transmitter was used for the control signals from the computer output port.

The experiment involves placing the robot at an arbitrary position and orientation relative to the light source, as shown in Figure 5. An obstacle is placed so that it blocks the straight line path between the robot and the light. The robot's task is to detect the light source and go to it, avoiding obstacles as it moves. Breaking this task into primitive operations results in the following procedure:

1. Perform a sequence of IR readings separated by incremental rotations, so that the IR sensor scans the robot's surroundings.
2. Detect the light source, and begin to move toward it.
3. If the touch sensor detects an obstacle, initiate an avoidance maneuver, then repeat from step 1.

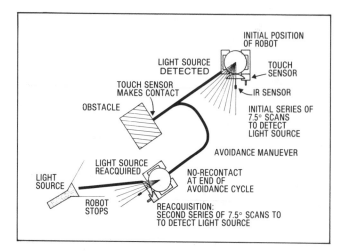

Figure 6. Robot obstacle avoidance behavior. The computer-controlled robot will move toward the light, automatically avoiding an obstruction in its path and stopping when it reaches the light.

4. Continue moving toward light source until the IR reading drops below a threshold, then stop.

In step 1 the robot is rotated in azimuth by 7.5 degree increments. At the end of each increment the IR sensor output is converted and checked to determine if the light has been seen. This is accomplished by testing the IR sensor output against a selectable threshold, using the single BASIC statement:

IF S1>IR THEN GOTO (movement procedure)

The value in S1 may be obtained from a single call to the ADC subroutine, or it may be an average value derived from, say, 10 conversions. Averaging can be used to reduce the likelihood of a false detection if the sensor readings vary greatly due to the presence of electrical or IR noise. The threshold value in variable "IR" is set low enough to give an acceptable probability of detection and high enough to give an acceptably low false alarm rate.

During the forward motion toward the light source, the touch and IR sensor ADC subroutines must be entered periodically by means of the USR command and the resulting values checked. If the touch sensor is found to be on, an appropriate obstacle avoidance routine is called, and repeated until the touch sensor is no longer on at the end of an avoidance cycle. When the robot is very near the light, the collector mirror becomes unfocused, and the IR sensor reading drops below the threshold. A typical ground trace of the robot's motion during this experiment is shown in Figure 6.

Conclusion

Many people make the assumption that robot experimentation is beyond the resources of the average hobbyist. The simple interfaces described here provide the means, both in hardware and software, for interesting experimentation at minimum cost, given that a microcomputer such as the TRS-80, with access to machine code subroutines, is available. The interfaces described here may easily be elaborated by the addition of port selection logic or other hardware, but just as they are they will allow you to begin your own robot experiments with minimum expense and effort. ▣

References

[1] Don Lancaster, "The TTL Cookbook," Howard Sams & Co., 1978.

[2] William Barden, Jr., "TRS-80 Assembly Language Programming," Radio Shack, 1979.

[3] Don McAllister, "Low Cost Robot," *Byte Magazine*, June 1980.

The radio-controlled R2D2 (price appx. $30) can be ordered from: Kenner Products, 2940 Highland Ave., Cincinnati, OH 45212.

GLOSSARY

Complement or logical negation:
   Frequently used in digital logic to indicate a signal having the opposite logical value of another named signal. The complement is indicated by an apostrophe following the signal name or a bar over the name. For example, either Q' or $\bar{Q}$ may be used to represent the complement of signal Q. Thus, when signal Q goes low, for example, Q' must go high, and vice versa.

ADC:
   Analog to Digital Converter—a device for measuring the voltage level of an analog signal by converting it to a corresponding binary number which can be used by a digital computer.

About the Author

The author, Don McAllister, has built several robot interfaces for his microcomputer. He has worked in the areas of guidance and control as applied to the Apollo and Minuteman missile programs and is currently involved in sensor data processing at the Aerospace Corporation. He obtained his B.S. in Physics at the California Institute of Technology and his M.S. in Astronautics at the Massachusetts Institute of Technology.

OPTO "WHISKERS"

for a Mobile Robot

Suggestions for improvement on a basic optoelectronic proximity sensor

Martin Bradley Winston

In my research of obstacle detection schemes, the most obvious common requirement that emerged is the need for an inexpensive circuit that could simply say "there's something here." Precise descriptions of the object being detected, its absolute or relative position and other specifics concerning it are of interest, of course, but the first priority is to detect its presence.

Some of the primitive mechanisms that have appeared in published articles rely on actual mechanical contact between the mechanism and the obstacle. Bumpers that actuate microswitches, pressure-sensitive ribbon switches and stiff wires attached to microswitch actuator levers are among the several forms of mechanical contact-sensing switches that have been tried, all with some success.

But this is analogous to wending our ways about with blindfolds on, requiring that we touch a wall or tree or bit of furniture in order to make any progress. This is because the inherent limitation of contacting obstacle detectors (and of our finger, when we're blindfolded) is that they can only tell us where things *are*; it would be much to our advantage if we could design them to also tell us where things *aren't*.

So consider, if you will, a theoretical system of precisely positioned "wires" of no mass and no substance, but with the ability of signalling whether or not they are being intercepted by any object. If a biological analogy will help, you might consider the whiskers of a cat. With enough of these "whiskers," our mechanism could be perfectly aware of the space surrounding it—at least in terms of free paths and obstacles within the sensor's range.

There are ways to achieve this type of device in practice, including radar and SODAR* systems and reflective optical systems. The high cost of the two former approaches tends to obviate them in favor of the latter where a large number of such sensors are anticipated.

Some Problems

If the entire world were dark and all obstacles were vertical cylindrical mirrors, all we would need to do is shine a single, omnidirectional light and rotationally scan for

---

*SOound Direction And Ranging, using a directional ultrasonic range measurement circuit.

Reprinted from Robotics Age, Volume 3, Number 1, January/February 1981

159

reflections. But there are other lights to cope with in the real world, requiring that we somehow "label" our light by giving it some special characteristic not likely to be duplicated in the environment the mechanism must negotiate.

Another difficulty is that most optical detectors can be swamped by ambient light within their sensitivity range, spectrally speaking. Still another is that the spreading of the outgoing beam and the off-axis sensitivity of the detector combine to both waste power and limit the resolution of the "whisker."

And another problem—indeed, one we've already touched upon—is that we must try to solve all these problems inexpensively.

Fortunately, we're not the first to face these problems. In industry, for example, both beam-break and reflective systems (including both reflective beam-break and reflective beam-make designs) have been used in the positioning of machinery and the counting of parts.

Enter the PLL

In a search for a better answer to problems associated with using light emitters and detectors in positioning applications, Richard Oliver of Centralab Electronics, West Lafayette, Indiana, developed a simple circuit based on the NE567 phase-locked loop tone decoder IC. [1] His circuit, with only minor modifications, appears in Figure 1.

In this circuit, the pin 6 output of the current-controlled oscillator (a $1V_{P-P}$ exponential triangle waveform with an average DC level of half the supply voltage) drives IC2, an ordinary 741 op amp. The 741 squares up this waveform and drives Q1, an inverting power driver which turns on D1 (an infrared-emitting diode) in implicit sync with the PLL. When Q2, a phototransistor, sees the output of D1 and provides its signal to the input of IC1 (the NE567), the loop locks and turns on its output, pin 8, which lights indicator LED D2.

A breadboard of this circuit with the components indicated is capable of detecting a human hand at about 2½ inches, a piece of white paper at about 4 inches.

It should be noted that this performance is without the benefit of optics external to the lens-packaged infrared-emitting diode and lens-packaged phototransistor, and with all power provided by a standard 216-type 9 Volt "transistor" battery.

Obviously, this level of performance is unsuitable for the kind of "whisker" function we're considering. Both the output power of the emitter and the sensitivity of the

detector are going to have to be greatly improved.

While far from ideal, the circuit shown in Figure 2 represents an impressive first effort. As shown (and again, without external optics and using only a common 9 Volt battery), this circuit can detect a human hand at about *2 feet*, a piece of white paper at about 3 feet. For some applications, this may prove adequate; for others, simple lenses may adequately improve performance. In any case, further experimentation (both by readers and the author) is strongly suggested.

Improving Radiated Power

The XC881 emitter used in Figure 1 is rated for about 3 mW at drive currents near 1 Ampere; unfortunately, junction heating would quickly destroy the device if it were run continuously at this level, or even at the 50% duty cycle of this circuit. The manufacturer recommends limiting pulse duration to 100 microseconds at a 10 pps rate.

Of course, the battery is another limiting factor here, since these batteries cannot sustain a current requirement beyond a few hundred milliAmps for more than a short period.

Another problem is that the switching transistor must itself be capable of delivering this level of current drive (and pulse response) without unduly loading the driving circuit.

Fortunately, we are at a very good time in the developing state of the art of manufacturing infrared emitters. The Xciton SC-88-PC "Super High Output Infrared Emitter" used here, for example, is rated at 10.5 mW total radiant output power at 100 mA drive, or a peak pulse current of 3.0 Amps (10μsec, 100 Hz) can deliver over 200 milliWatts. [2] Furthermore, the lens on its TO-46 style package is designed to provide a 15° angle between half power intensity points. This Xciton device is not the only one available with these capabilities, although Xciton prices bear watching because of the company's commitment to producing products specifically for high volume markets (and thus tends to quickly feel the price-reducing efficiencies of high-volume production).

AEG-Telefunken, for example, offers type CQX-19, rated 500 mW at 10 Amps drive (20μsec, 1500 pps) and a 40° half-intensity angle. [3] Similar devices are also available from RCA [4]; Spectronics [5]; and Texas Instruments [6].

The pulsed operation helps make battery operation more reasonable, too, by greatly reducing the average current requirement on the battery. The 100μF capacitor across the supply provides enough charge storage to handle this requirement while also providing ample decoupling ballast to help keep pulse current requirements from appearing in the receiver; in addition, 0.01µF bypass capacitors are sprinkled generously throughout the circuit.

The op amp squaring buffer of Figure 1 is duplicated, with a slight increase in gain. This drives a CD4047, configured as a monostable multivibrator with a very narrow pulse width, well under 1µsec. The output of the 4047 drives a 2N2222 and TIP3055 pair of NPN transistors in a Darlington configuration, which in turn drive the emitter.

An effort was made to use exclusively off-the-shelf, easily-available components, which are both inexpensive to obtain and not terribly critical in terms of current, voltage or thermal operation.

Improving Detector Sensitivity

The BPX25 phototransistor [7], a Ferranti part, is not remarkable for any unusual sensitivity (typically 55µA/ℓ/ft² minimum sensitivity rating), but does offer an advantageously narrow 3.5° half-sensitivity/response acceptance angle. The package is a metal TO-18 with a glass lens, nearly identical to that of the emitter.

The first area where improvements were ventured is the emitter bias resistor. A substantial range improvement resulted from raising its value to 220K (from 68K); further increases in this resistance with this particular device resulted in insufficient signal and no output from the 567. Also, the coupling capacitor was raised in value to 0.022 µF to help lengthen the time constant of the input circuit and hopefully let a little additional energy through while still rejecting slow changes in detected energy. This input circuit, when breadboarded, exhibited no tendency to swamp (saturate) with high ambient light levels.

The only other modification—again, a significant one— was to add a 470K resistor to ground at pin 1 of the 567; this came straight out of the manufacturer's device data as a suggested means of improving the input sensitivity of the 567.

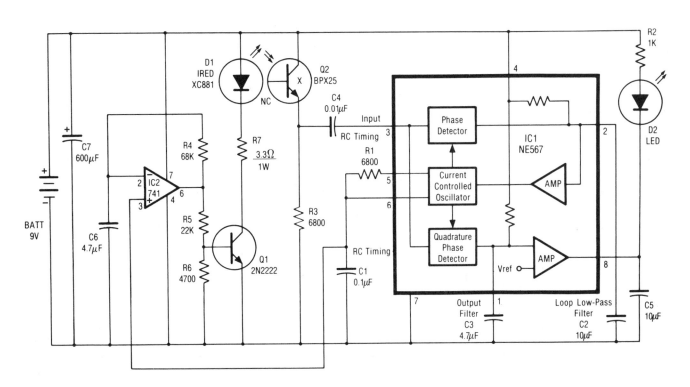

Figure 1. Modulated optoelectronic proximity detector, adapted with only minor revisions from 1975 EDN article [1] (see text). As shown, circuit responds to human hand at 2½ inches, white paper at 4 inches.
Signal from IC tone decoder PLL oscillator drives emitter through op amp and transistor; detected reflection triggers response from IC. This scheme offers very high immunity to ambient light.

Isolating the Output. And Notes on Further Improvements

Since the absolute maximum supply voltage that the 567 is rated for is 10 Volts and the specific system to which the "whisker" will be added may operate at some other voltage, a standard 4N36 optoisolator was added to the output of the 567 to provide a more general interface. A red LED indicator is included here for ease of observation during experimentation.

Obviously, from the notes here, a great deal of improvement is still possible. Photodiodes, photosensitive FETs and photodarlingtons all offer promise, but time has so far prevented further experimentation with them; furthermore, there are a number of alternative phototransistors very much more capable than the BPX25 (which is itself a huge improvement over the FPT100, for example). Improvements in the detector sensitivity offer the greatest promise for increasing the range of this circuit.

Optics are another area where range improvements can readily be made. Simple lenses are all that's required. A lens in front of the detector can combine with its built-in lens to gather more light and to further narrow its acceptance angle; similarly, a lens properly positioned in front of the emitter can further narrow its beam width.

Emitter and detector characteristic wavelengths also require some attention. As Xciton points out in its data sheets on the emitters used here, for example, their 880 nm wavelength significantly improves coupling to silicon phototransistors, offering *twice* the coupling efficiency of more common 940 nm infrared emitters.

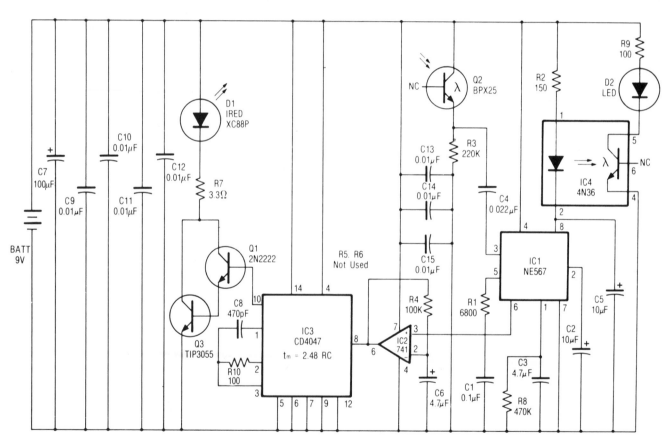

Figure 2. Improved circuit incorporates monostable narrow-pulse high-power driver, higher-power narrower-beam emitter, increased detector and PLL sensitivity and isolated output.

As shown and without external optics, a breadboard of this circuit was capable of responding to a human hand at 24 inches, white paper at 37 inches.

Component callouts have been normalized to Figure 1. C4, C7, R2, R3, R4 and D1 values have changed; C8-15, R8-10, IC3, IC4 and Q3 have been added; R5 and R6 have been deleted.

Detectors that exhibit inconvenient tendencies toward swamping in the presence of high levels of ambient light can be desensitized by the simple expedient addition of an infrared filter in their optical path. (Obviously, there is no benefit in doing the same for the emitter!)

During experimentation, it was found that the intensity of the indicator LED varied to some degree (inversely) with distance; this might provide some facility for determining a reflectivity and/or range value for the detected object—bonus data for this very simple circuit.

Readers are encouraged to further develop this and similar circuit ideas and to communicate with the author or the editors via *Robotics Age*. Indeed, the capabilities of an improved inexpensive "whisker" of this sort extend beyond robotics to aids for the visually handicapped, alarm systems, automotive collision avoidance systems, automatic door controls and many other application areas.▣

References

[1] Richard Oliver, "Improve Photo Sensors with a Phase-Locked Loop IC," *Electronic Design News*, April 5, 1976, p. 112.

[2] XC-88-PC "Super High Output Infrared Emitter, Xciton Corp., Shaker Park, 5 Hemlock St., Latham, NY 12110.

[3] CQX-19 IR Emitter, AEG-Telefunken Semiconductors, Rte. 22, Orr Drive, Sommerville, NJ 08876.

[4] RCA Optical Communications Products, Solid State Division, Electro Optics and Devices, Lancaster, PA 17604.

[5] Spectronics, a division of Honeywell, Inc., 830 E. Arapaho Rd., Richardson, TX 75081.

[6] Texas Instruments, Inc., PO Box 5012, Dallas, TX 75222.

[7] BPX25 Phototransistor, Ferranti Electric Inc., E. Bethpage Rd., Plainview, NY 11803.

Martin B. Winston is the author of Android Design, *(Hayden Books, Rochelle Park, NJ) and is the author of numerous technical articles on home robot experimentation.*

TEACH YOUR ROBOT TO SPEAK

Tim Gargaliano and
Kathryn Fons
Votrax Division
Federal Screw Works
Troy, MI

Machines have communicated with machines for a long time. But as robots enter our lives, how will they speak to *us*? Unlike machines, humans can't be connected to a standard RS232C or IEEE488 interface.

Traditionally, machines have given us information through display devices, such as the CRT. But displays have disadvantages. They require visual attention. And, in their present state of development, full-screen displays come in two flavors: fragile and cumbersome—or extremely expensive.

Why not have machines use the *spoken* word? We can listen to speech while doing almost anything else. In addition, current speech technology offers devices that are small, rugged, and inexpensive.

This article is about speech technology. It briefly covers the basics of synthesizing speech. It describes inexpensive hardware that can give an unlimited vocabulary to your robot. And it explains the programming techniques you need to teach your robot how to speak.

Synthesizing Speech

There are two fundamentally different ways to synthesize speech. In the first method, we reconstruct a *time domain* signal from digital samples taken from a human source. As Figure 1 shows, a time domain signal records how the sound's loudness varies with time. The amount of data depends on the rate at which we sample the signal and the number of quantized levels used in the analog-to-digital conversion. Data compression techniques can reduce the overall bit rate. But time domain synthesis still needs between 16,000 and 64,000 bits/second.

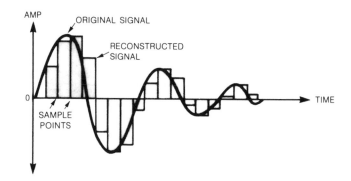

Figure 1. Reconstructing a signal in the time domain.

In the second approach,' we synthesize the frequency content of speech as it varies with time. Frequency information for speech is contained in bands called *formants*, shown in Figure 2. During dynamic speech, these formants change in their center frequency and bandwidth. Since *frequency domain* information changes much more slowly than time domain information, this technique needs much less data. Typically, a formant synthesizer has a data rate of between 1000 and 2000 bits/second.

A formant synthesizer, shown in Figure 3, consists of a filter network, plus pitch and noise sources that excite the filters. In this system, filter parameters and excitation amplitudes are under external control.

You could control the formant synthesizer by individually setting each of its parameters. But suppose these parameters naturally occurred in certain combinations. Then, we could encode these combinations. We could feed the code to a parameter generator/interpolator

Reprinted from Robotics Age, *Volume 3, Number 6, November/December 1981*

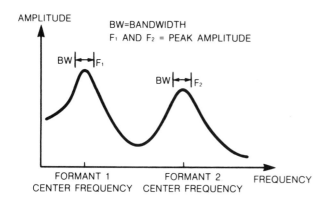

Figure 2. A formant plot shows the frequency content of speech at a given instant.

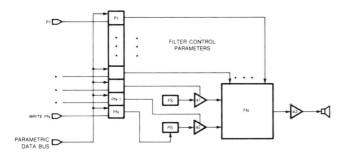

Figure 3. Block diagram of a speech formant synthesizer.

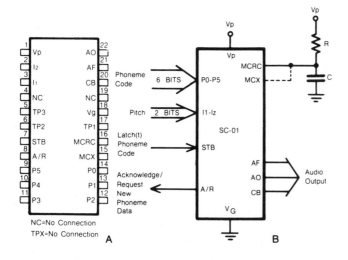

Figure 4. The SC01 speech synthesizer chip: (a) pin out, (b) functional diagram.

which, in turn, would supply a continuous flow of parametric data to the formant synthesizer.

Certain combinations of formant parameters do occur in human speech. These are *phonemes*, the basic speech sounds we use to create words. A *phoneme synthesizer* uses phoneme codes to control a formant synthesizer. The main advantage of this approach is its low data rate—typically under 100 bits/second. At this rate, a 2716 EPROM could store about 200 words built from phonemes.

The SC01 Phoneme Synthesizer Chip

The Votrax SC01 is the first single chip phoneme synthesizer. The SC01 contains an analog model of the human vocal tract, using switched capacitors to emulate resistors for a small chip size. During continuous speech, the SC01 needs only about 70 data bits per second—though this figure varies with the master clock frequency. As shown in Figure 4, the chip comes in a 22 pin DIP. Since it uses CMOS logic, its power consumption is low enough for battery-powered systems.

As you can see in Figure 5, the block diagram of the SC01, the chip consists of a latch (for bits P0-P5), a phoneme controller, pitch and noise sources, a filter network (simulating the vocal tract), an audio preamplifier, handshake logic, and an onboard oscillator (master clock). The master clock determines all internal timing functions. Its frequency should be set to about 720 KHz by connecting an external resistor and capacitor to the MCRC (master clock resistor/capacitor) pin. As an alternative, you can supply an external master clock to the MCX pin. By varying this clock's frequency, you can create different voices and sound effects.

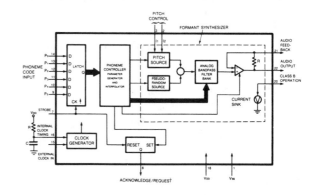

Figure 5. Internal block diagram of SC01.

The input latch on pins P0-P5 hold the phoneme code, which is fed to the phoneme controller. Here, the chip generates parameters to dynamically control the "vocal tract" and the pitch and noise sources. The output of the vocal tract is preamplified before it exits the audio ouput pins. Two additional pins—audio feedback (AF) and class B (CB)—provide for external amplification.

At the heart of the SC01, a phoneme synthesizer produces 64 speech sounds. In English, at least 45 phonemes are needed to produce every word in the language. The additional 19 sounds, called *allophones*, are variant pronunciations of certain phonemes. Table 1 lists the 64 sounds produced by the SC01, along with key words that contain each sound. By selecting and sequencing these phonemes—writing phonetic programs—you can teach the speech synthesizer to speak.

How to Program Phoneme Sequences

As a phonetic programmer, you'll create a sequence of phonemes, which the synthesizer will ouput as speech. The better you program the sequence, the clearer, more accurate, and more human the synthesizer sounds.

How can you translate a word into sounds? You could begin by looking up the written word in the dictionary. But most written characters have more than one possible pronunciation. Therefore, we need an alphabet—different from the written system—that represents phonemes.

Each dictionary has such an alphabet, defined by its "pronunciation key." Also, linguists have created a standard International Phonetic Alphabet (IPA). Votrax uses the phoneme symbols given in Table 1. For this article, we'll stay with Votrax's phoneme symbols. Whenever we use phoneme symbol that might be confused with a written character, we'll enclose the phoneme in slash marks. For instance, /A/ means the vowel sound in the word "tame."

Now, with grammar rules and a dictionary in hand, you can begin phonetic programming. The programming procedure typically consists of the following eight steps:
1. Select a word to synthesize.
2. Pronounce the word aloud.
3. Count the number of sounds in the word.
4. Translate the word into a phoneme sequence. For each sound counted in step 3, find its appropriate phoneme symbol, using Table 1.
5. Enter the phoneme sequence into the synthesizer, and activate the speech output.
6. Listen to the synthesizer's output. Check how accurately it pronounces the word you programmed.

TABLE 1

VOTRAX PHONEME TABLE AND PRONUNCIATION KEY			
CONSONANTS		VOWELS	
PHONEME SYMBOL	KEY WORD	PHONEME SYMBOL	KEY WORD
B	Bat - ruB	A	tAme - pAI1 - mAke
CH	CHeese - maTCH	A1, A2	
D	DaD - raiD	AE	sAd - plAId
DT	buTTer	AE1	
F	Fake - cuFF	AH	tOp - fAther
	PHone - lauGH	AH1, AH2	cAll - Office
G	Get - loG	AW	
H	Have - Help	AW1, AW2	
J	Jazz - JuDGe	AY	tAme - pAll - stAY
K	Car - KiCK	E	bEEf - Even
L	Low - caLL	E1	
M	Mat - diM	EH	lEg - rEAdy - sAId
N	No - soN	EH1, EH2,	
NG	riNG - driNk	EH3	
P	Pack - haPPy	ER	bIRd - hEARd - ovER
R	Race - haRd	I	pIt - swIm
S	Soup - lightS -	I1, I2, I3	
	City - aSk	IU	YOU - mUsic
SH	SHeep - acTion	O	fOr - tOrn - bOld
T	Tip - askeD	O1, O2	
TH	THing - maTH	OO	tOOk - pUt - cOUld
THV	THe - moTHer	OO1	
V	Van - paVe	U	mOve - schOOl - JUne
W	Wake - WHen -	U1	
	qUit	UH	cUp - Around - sOn
Y	Yard - berrY	UH1, UH2	
Z	Zap - haZe -	UH3	randOm - statIOn
	panS - glasseS	Y1	You - mUsic
ZH	pleaSure - aZure		
PA1	(long pause)		
PA0	(short pause)		
STP	(no sound)		

7. Adjust the phoneme sequence wherever the synthesizer pronounces inaccurately.
8. Repeat steps 6 and 7 until you have the speech output you want.

Step 7 is where you'll spend most of your programming time. This is because step 4 only gives the primary sound sequence of a word. In order to match human pronunciation—which injects rhythm and meaning into a word—you have to adjust the sequence to add stress.

In grammatical rules, stress is often shown as an accent placed on a portion of the word. When programming a phoneme sequence, you can add stress by increasing the duration, pitch, or amplitude of the vowel sound in the accented part of the word. Table 2 shows how the words "family," "sensory," and "visual" advance from an initial program, through trial sequences, to the final refined program. The major difference is stress.

Notice that we start by placing equal stress on all syllables in the initial program. To begin with, we use the

TABLE 2

WORDS	INITIAL PROGRAM	TRIALS	REFINED PROGRAM
FAMILY	F-AE-M-I-L-E	F-AE-M-I3-L-E1 F-AE1-M-I3-L-E1 F-AE1-EH3-M-I3-L-E1	F-AE1-EH3-M-L-E1
SENSORY	S-EH-N-S-O-R-E	S-EH1-N-S-O2-R-E1 S-EH1-N-S-ER-E1	S-EH2-EH3-N-S-ER-E1
VISUAL	V-I-ZH-U-UH-L	V-I1-ZH-U1-UH3-L	V-I1-ZH-W-UH3-L

TABLE 3

VOWEL PLACE-OF-PRODUCTION CHART			
FRONT VOWELS	MEDIAL VOWELS	BACK VOWELS	MOUTH
E		U	CLOSED
E1		U1	
I		OO	
I1, I2, I3	ER	OO1	
	UH	O	
A	UH1, UH2, UH3	O1, O2	
A1, A2		AW	
EH		AW1, AW2	
EH1, EH2, EH3		AH	
AE		AH1, AH2	
AE1			OPEN

longest lasting vowel sound that translates the vowel letter. During the trial sequencing, we substitute shorter lasting vowel sounds in the unaccented syllables.

The key to achieving a refined program is selecting the appropriate vowel durations for each syllable. This must be done by listening closely to the speech output. The durational relationship of one vowel to another must *sound* correct. In the word "family," for example, the first syllable lasts the longest, while the last syllable is of medium duration. The middle syllable has the shortest duration. It is virtually absent in the best pronunciation of "family."

Advanced Techniques

Sometimes you can select phonemes more accurately if you know how the human vocal tract produces them. The study of how humans produce sound is called *articulatory phonetics*. When you make a sound—say the vowel sound /AH/—notice how you do it. Notice that your mouth is wide open and that the sound resonates from the back. In terms of articulatory phonetics, this means that /AH/ is an *open back vowel*. Table 3 shows some articulatory features of vowel phonemes and allophones.

This information is helpful when you want to program *diphthongs*—two or more vowel sounds in sequence, spoken as a single unit. In a diphthong, the tongue moves rapidly from one position to another. By noting how you make each vowel sound in the diphthong and referring to charts like table 3, you can select the right phonetic representation.

Take the sound of the long "i" in the word "time," for example. When you say the long "i" diphthong, notice how your tongue starts in the back of your mouth, then travels quickly to the front as your lips close slightly. This means that the first sound in the diphthong is an open back vowel, and the second sound is a closed front vowel. Using table 3, we combine /AH/ and /E/ to create the sound.

This gives us a first approximation to the sound of "i" in "time." But after you listen to the synthesizer say the sequence, you'll realize that we need shorter lasting sounds in order to perceive the sequence as a single unit. This leads us to select the allophones /AH1/ and /E1/ to produce the sound.

Once again, you'll feed the combination to the synthesizer. When you hear the output, you'll notice that in a one syllable word, like "time," the "i" sound should be slightly longer in duration. Using table 3, we can select a vowel that falls between our original two sounds and insert it into the sequence. The /EH3/ qualifies, as does the /I3/. The former sound favors the more open production of /AH/, while the latter favors a more closed production, like /E/. Either selection could work. Choose the one that sounds best to you.

It's easier to program consonants than vowels. Still, there are cases that articulatory phonetics can help. Take, for example, *stop-plosive* sounds. These are the sounds produced when we block the vocal tract (with our lips, for instance), then suddenly release the sound. /B/, /D/, /P/, and /T/ are examples of stop-plosives.

Stop-plosives are difficult to perceive, even when spoken by a human speaker. When synthesizing these sounds, it is helpful to double the consonant sound, or combine it with its *voiced* (uses the vocal chords) or *voiceless* (doesn't use the vocal chords) counterpart.

The programs for "eighteen" and "ninety" are examples of this. "Eighteen" has the sequence A1-AY-Y1-T-T-E1-N. This stop-plosive /T/ is doubled. "Ninety" is programmed as N-AH1-I3-Y-N-D-Y. In this case, the stop-plosive /T/ is replaced by its unvoiced counterpart, /D/. Table 4 shows how the vocal tract produces these and other consonants.

Given the subtle problems just discussed, you'll find it helpful to use established phoneme sequences as patterns for creating new words. Table 5 lists some words that apply to robotics, along with their phonetic programs.

There are no steadfast rules for phonetic programming, only pronunciation guidelines. You're free to experi-

ment with different phoneme combinations to create a special effect. For example, you may want to hear a certain dialect of English or a word from a foreign language. In either case, you need only follow the eight-step procedure that we outlined earlier.

Special Effects

You can produce different voices and sound effects simply by varying the master clock frequency to the SC01. With slow minor variations, you can create various voices. Fast drastic changes in frequency produce sound effects.

In either case, you might want to control these changes with software from a computer. There are several ways to do this. One is to vary the frequency of the SC01 master clock oscillator by switching in different resistor/capacitor combinations. A circuit that does this is shown in figure 6. An 8 bit binary weighted resistor ladder can be selectively connected to a fixed capacitor and the SC01. A CD4016 analog switch is used to select each resistor in the ladder into or out of the circuit. As more resistors are switched into the circuit, their effective resistance becomes lower and the master clock frequency becomes higher.

TABLE 4

CONSONANT MANNER and PLACE -OF-PRODUCTION CHART		
MANNER-OF-PRODUCTION		PLACE-OF
VOICED	VOICELESS	PRODUCTION
B,D,G	P,T,K,DT	STOP PLOSIVES
Z,ZH,V,THV	S,SH,F,TH,H	FRICATIVES
J	CH	AFFRICATES
M,N,NG		NASALS
R,L,W,Y		SEMIVOWELS/GLIDES

TABLE 5

SAMPLE WORD LIST	
PHONETIC PROGRAM	
arm	AH1-UH3-R-M
backward	B-AE1-EH3-K-W-ER-D
come	K-UH1-UH3-M
computer	K-UH1-M-P-Y1-IU-U1-T-ER
down	D-AH1-UH3-U1-N
emergency	I2-M-R-R-D-J-I2-N-S-Y
find	F-AH1-EH3-Y-N-D
forward	F-O2-O2-R-W-ER-D
go	G-OO1-O1-U1
hello	H-EH1-UH3-L-UH3-O1-U1
help	H-EH1-UH3-L-P
human	H-Y1-IU-U1-M-EH3-N
left	L-EH1-EH3-F-T
leg	L-EH1-I3-G
me	M-E1-Y
no	N-OO1-O2-U1
noise	N-O1-UH3-I3-AY-Z
off	AW2-AW2-AW2-F
on	AH1-UH3-N
please	P-L-E1-Y-Z
right	R-UH3-AH2-Y-T
robot	R-O2-O2-B-AH1-UH3-T
stop	S-T-AH1-UH3-P
up	UH1-UH2-P
yes	Y1-EH3-EH1-S

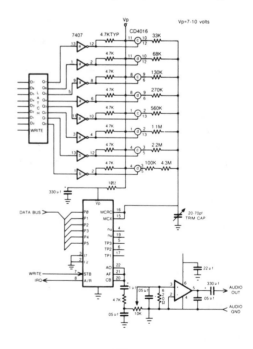

Figure 6. An SC01 application circuit. The switched resistor network can vary the clock frequency to generate sound effects.

ROBOTICS AGE: IN THE BEGINNING

```
          EXAMPLE RAMP SOUND EFFECT ROUTINE

 ;  Generate a ramp number sequence
 ;
 ;
 ;  B = repeats   C = increment size
 ;  D = start     E = stop
 ;  HL = timer counts between increment

 REPEAT:  LD    A,D          ;  A = start number
 OUTPUT:  OUT   (LATCH),A     ;  output current number
          PUSH  HL           ;  save timer value
          CALL  TIMER        ;  wait for selected interval
          POP   HL           ;  recover timer value
          ADD   C            ;  add increment
          CP    E            ;  at end?
          JR    NC, OUTPUT   ;  brif not
          DJNZ  REPEAT       ;  repeat ramp
          RET

 LATCH: EQU 10      ;  output port location
 TIMER: EQU 1000H   ;  delay subroutine location
```

Figure 7. Subroutine to produce a ramp sound effect. (Timer subroutine and output latch locations may vary.) The ramp modulates the phonemes output by the SC01.

```
 1   ;  use the data setting directives found on
 2   ;  an assembler to pack the phoneme lexicon
 3
 4   ARM:        DC  AH1, UH3, R, M+FLAG
 5   BACKWARD:   DC  B, AE1, EH3, K, W, ER, D+FLAG
 6   COME:       DC  K, UH1, UH3, M+FLAG
 7   COMPUTER    DC  K, UH1, M, P, Y1, IU, U1, T, ER+FLAG
 8   DOWN:       DC  D, AH1, UH3, U1, N+FLAG
 9   EMERGNCY:   DC  I2, M, R, R, D, J, I2, N, S, Y+FLAG
10   FIND        DC  F, AH1, EH3, Y, N, D+FLAG
11   FORWARD:    DC  F, O2, O2, R, W, ER, D+FLAG
12   GO:         DC  G, OO1, O1, U1+FLAG
13   HELLO:      DC  H, EH1, UH3, L, UH3, O1, U1+FLAG
14   HELP:       DC  H, Eh1, UH3, L, P+FLAG
15   HUMAN:      DC  H, Y1 IU, U1, M, EH3, N+FLAG
16   LEFT:       DC  L, EH1, EH3, F, T+FLAG
17   LEG:        DC  L, EH1, I3, G+FLAG
18   ME:         DC  M, E1, Y+FLAG
19   NO:         DC  N, OO1, O2, U1+FLAG
20   NOISE:      DC  N, O1, UH3, I3, AY, Z+FLAG
21   OFF:        DC  AW2, AW2, AW2, F+FLAG
22   ON:         DC  AH1, UH3, N+FLAG
23   PLEASE:     DC  P, L, E1, Y Z+FLAG
24   RIGHT:      DC  R, UH3, AH2, Y, T+FLAG
25   ROBOT:      DC  R, O2, O2, B, AH1, UH3, T+FLAG
26   STOP:       DC  S, T, AH1, UH3, P+FLAG
27   UP:         DC  UH1, UH2, P+FLAG
28   YES:        DC  Y1, EH3, EH1, S+FLAG
29   FLAG:       EQU 1000000B   ;  last phoneme flag bit
30
```

Figure 8. Creating a lexicon. This code assumes that the EQUates for the SC01 phoneme symbols have already been done.

Figure 7 shows a software routine that generates a "ramp" sound effect using this resistor ladder. By selecting different phonemes—holding the phoneme steady during the whole effect—you can get a panoply of different sounds.

As you generate a vocabulary or accumulate phoneme programs, you may want to create a lexicon (word table) to be stored on disk or in ROM. Figure 8 gives a Z-80 assembly language program that packs words into a lexicon. Figures 9 and 10 give SC01 calling and driving routines. (The driving routine is intended to be interrupt driven.)

```
 MSG1:   LD    DE,PLEASE    ;  DE points to please
         CALL  SPEAK        ;  speak
         LD    DE,GO        ;  DE points to go
         CALL  SPEAK        ;  speak
         LD    DE, LEFT     ;  DE points to left
         CALL  SPEAK        ;  speak
         RET

 SPEAK:  LD    A,(DONE)     ;  speech busy?
         CP    0            ;
         JR    Z, SPEAK     ;  brif yes
         LD    (NXTPH) DE   ;  save pointer
         LD    A,0          ;
         LD    (DONE),A     ;  set busy flag
         RET
```

Figure 9. An example of a speech calling procedure. A call to SPEAK sets the pointer NEXTPH to the next phoneme string and sets the flag DONE to false (0). On the next interrupt from the SC01, phoneme output will begin.

```
 SRV-SC01:  LD    A, (DONE)    ;  speech pending?
            CP    0            ;
            JR    NZ, .SRVC2   ;  brif not
            LD    DE, (NXTPH)  ;  load pointer
            LD    A, (DE)      ;  get a phoneme
            INC   DE
            LD    (NXTPH), DE
            BIT   7, A         ;  test for flag bit
            JR    Z, .SRVC1    ;  brif not
            LD    (DONE), A    ;  set flag
 .SRVC1     OUT   (SC01), A    ;  output phoneme
            RETI
 .SRVC2     LD    A, PA1       ;  A= pause code
            OUT   (SC01), A    ;  output pause
            RETI
 DONE       DC    0
```

Figure 10. SC01 interrupt routine. It outputs the next phoneme (pointed to by NEXTPH) and increments the pointer. It sets DONE to true if it's the last phoneme.

Conclusion

Robots that talk are now a reality. Technical advances in the last five years have shrunk speech synthesizers to the size of an LSI chip. Complete lines of speech-based products have been put on the market. (See the sidebar on "Speech Products.")

We are entering an era of talking machines. Someday soon, our cars will tell us when they need gas, oil, or repairs. Our ovens will tell us how to make our favorite recipes. Our hobby kits will tell us how to build them. Our games will tell us how to play them. For the average consumer, talking machines may still be a few years in the future. But for those with a computer and a speech synthesizer, that future is now. ℝ

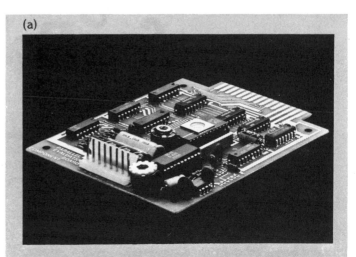

(a)

Speech Products

Phonetic programming—though easy once you learn the routine—remains a very time consuming task. There are several speech-based products designed to make life easier for you. The examples we give here are all currently offered by Votrax.

The Votrax Speech PAC board, shown in figure (a), is built around the SC01 speech chip. In addition, it contains a parallel interface, a 1 watt audio amplifier, and a prestored vocabulary of 250 words. If you need to modify any of these words, the PAC board allows you to override any of the stored phonemes.

For more advanced applications, there is the Votrax VSM/1 (Versatile Speech Module), shown in figure (b). This board contains an SC01 speech chip, an M6800 processor with support circuitry, an RS232C interface, a parallel interface, a 1 watt audio amplifier, sound effect circuitry, 1300 words of prestored vocabulary, 1K byte of RAM, and a terminal operating system. The VSM/1 can serve as a board for experimenting with speech and sound effects. Or it can work as a stand alone controller (as in industrial applications).

Those who want to avoid phoneme programming can use a text-to-speech translator. The Votrax Type 'N Talk, shown in figure (c), automatically translates a string of ASCII characters into phoneme sequences. Using its own text-to-speech algorithm, it connects to any RS232C compatable source of ASCII codes. You can send the output phoneme strings to either the onboard SC01 speech synthesizer or back to the host computer. (This device is described in "New Products," *Robotics Age*, May/June 1981.)

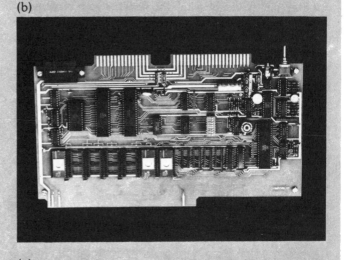

(b)

(c)

SECTION 3

APPLICATIONS AND DEVELOPMENT

What good is a robot unless it can be applied? And how do we develop new applications? To answer the latter question, we engage in experimentation and a quest for approaches. And in search of applications, every use of the technology at present is in some sense experimental. In this final section of the collection, we find several articles on approaches to developing new applications.

We lead off with a short piece that emphasizes an important theme: today's toy becomes tomorrow's necessity. A short staff article "Inside 'Big Trak™'" is an account of one of the first mobile autonomous robot toys. Continuing on the theme of making such toys practical, Martin B. Winston ponders design of a "Robot 'Digestive Tract,'" an important detail of a practical autonomous robot for the household, office, or factory. In "A Robot Arm Without a Budget," Timothy Nagle and V. Michael Powers discuss an experimental approach to engineering an inexpensive manipulator.

One of the most fertile areas of experimentation is that of mobile autonomous robots—the real engineered versions of the Big Trak toy. Starting out on this experimental theme, Don McAllister provides a two-part article "SUPERKIM Meets ET-2" on the integration of an inexpensive single-board computer with a mobile platform, then adding the necessary sensors. Another experiment in mobile platform robot design is provided by John Blankenship in his two-part account of "TIMEL: A Homebuilt Robot." In both cases, we see examples of the leading area of electronics/computer experimentation in the 1980s.

Turning to an application to machine tools, Jan Rowland supplies an article on "A Homebuilt Computer-Controlled Lathe." Jan's lathe is an experiment, suitable for exploring some of the problems of controlling machine tools. Robots in industrial applications are the first big practical use of the technology. Vern Mangold discusses applications of "The Industrial Robot as Transfer Device." Continuing on a production theme, John Meacham discusses the application of industrial arms in "Tungsten Inert Gas Welding with Robots."

Larry Leifer presents "Rehabilitative Robots," an account of strategies aimed at implementation of one of the most satisfying uses of robotic technology: replacing functions of broken or worn out body parts. The idea of using a microprocessor-controlled interactive real-time system to help augment speech or mobility is a major robotics research theme in our view. We conclude this section and the book with a short bibliography of the literature of robotics from the point of view of the creative experimenter. David Smith and Mike Schoonmaker III have assembled "The Robot Builder's Bookshelf" as a collection of pointers into other sources of literature on the subject.

INSIDE "BIG TRAK"*

New computerized toys
may provide a source
for cheap robot hardware.

When we heard about Milton Bradley's new "Big Trak™" programmable toy tank, we wondered how they were able to include a microprocessor controller and still retail the toy for so little. (Our local toy shops have it for about $40.00.) To find out, we contacted Milton Bradley and spoke with Mel Taft, Milton Bradley's Senior Vice

President of R & D, who kindly offered us a Big Trak, and gave us a few facts about it and information on his company's plans for future computer-controlled toys.

The tank is programmed by a paper-thin membrane keyboard (see photos), allowing the user to enter a sequence of up to 16 different commands. Most commands take a two-digit argument that is interpreted differently depending upon the command, specifying the amount of time, distance, rotation, etc. Apart from the motor drives, two program-controlled outputs are available, one that activates an accessory port and another that drives a lightbulb, coincident with sound effects for the tank's "phasers." A dump-truck trailer is available that uses the accessory port, and other accessories are planned.

The toy's controller is TI's TMS 1000 4-bit micro-computer-on-a-chip. The 1024 instruction device includes an ALU, ROM, and I/O in a 28-pin package. The TMS 1000 scans the keypad directly, composes the sound effects that accompany the tank's operation, and, using the only other IC in the unit, a 75494 output driver, sends its orders to the tank.

High volume production, of course, is the key to the

*"Big Trak" is a registered trademark of Milton Bradley, Inc.

toy's relatively low cost. Milton Bradley expects to market over one million of the tanks this year, according to Taft.

The major toy manufacturers are convinced that computerized toys will sustain their already great popularity, and continued development of innovative products is vital. Mr. Taft told Robotics Age that Milton Bradley is currently spending over $600,000 per year for research on robot- and computer-related products. *(This is roughly comparable to the entire NASA robotics research budget—ed.)*

One of the products of this effort is a proprietary voice synthesis chip which MB expects to begin using later this year. The chip can store in ROM about 22 different words and will cost the manufacturer a little over $3.00. Additional 22 word increments may be added at a cost of about $1.65 per ROM chip. Using these, future models of Big Trak will talk back to their owners.

But this is just the beginning, according to Taft. Eventually, MB toys will feature sonic and contact sensing, and even a robot manipulator is in the works.

All this should be important news to Robotics Age readers. As the toy industry, already past masters at producing inexpensive mechanisms, discovers the vast potentials of robotics, we can expect to see an increasing array of cheap, practical devices suitable for computer

interfacing, providing an attractive alternative to time-consuming hand-built items and expensive commercial equipment.

One final note—Mr. Taft tells us that Milton Bradley licenses the designs of about half of the devices it uses in its toys from outside sources, so you may want to send them that unique new sensor you've built! ▣

ROBOT "DIGESTIVE TRACT"

An obedient, semi-intelligent charger and
monitor for gelled electrolyte batteries.

Martin Bradley Winston

Those of us with an interest in experimental mobile
robots and similar mechanisms have by and large settled
on batteries as the only practical means of providing power
for the beasties. My own preference is for gelled electrolyte
batteries, and the circuit described here is intended
specifically for use with a Globe-Union type U-128
Gel/Cell®, a 12 Volt 28 Amp-hour battery; nevertheless, it
should prove satisfactory for use with many other batteries, including both automotive lead-acid types (as well as
those of other chemistry) and gelled electrolyte batteries of
different capacities. In any case, the basic method is
appropriate, even if component values must change.

I describe this circuit as "obedient" or semi-intelligent
because it contains no decision-making circuitry and is
designed to both supply information to and receive
instructions from a controlling microprocessor. All input
and output lines to the micro are 5-Volt logic compatible.

Operational Overview

The AC input to the charger is both bypassed and
surge-protected, as well as protected by a self-resetting
circuit breaker. The only element that is permanently
connected to the AC input is a General Instrument
Optoelectronics type MID-400 AC line monitor, a special-
ly-configured optoisolator that maintains a low (open
collector) output as long as suitable AC voltage is
maintained with no dropout (low voltage) exceeding a
couple of cycles. A pullup to a battery-derived regulated 5
VDC supply is provided as an ACOK output at pin 11 of
the DIP connector.

Once the processor is satisfied that the charger has
been plugged in to a "valid" socket, it can provide an
active-high ACON signal to pin 9 of the DIP connector,
which turns on a small 5 Volt DPDT relay (an ITT device,

Reprinted from Robotics Age, *Volume 3, Number 2, March/April 1981*
© *1981 Martin Bradley Winston*

available as part number 275-215 from Radio Shack), connecting the AC line to the primary of the power transformer.

The U-128 battery requires a maximum of 14.4 VDC at 6 Amps for proper recharging. The transformer you select will involve a number of trade-offs, including size, weight and cost. One easy choice is the commercially-available Triad F-257U, which provides 20 VCT at 6 Amps, which weighs in at 5.7 pounds and measures 3¾ (H) x 3⅛ (W) x 3⅛ (D). [1]

The geometry of my beastie required something shorter, so I decided to connect two 3 Amp transformers in parallel. This approach is acceptable only if the transformers are of identical manufacture, otherwise one of them may end up driving the other as a load, perhaps with unpleasant consequences. With this caveat in mind, two good choices are the Radio Shack 273-1514, 18 VCT at 4 Amps, which measures 4x2x2½; and the Digi-Key 308C, 22 VCT at 3 Amps, 2.6 x 3.1 x 2.7 [2], which is shown here. An additional word of warning: *be sure you wire the transformer secondaries in phase.* After tying one side of the two secondaries together, check to be sure there is zero AC voltage between the other two ends before connecting them, so that you don't short a series-aiding circuit!

The rectifier is a type PB-10 25 Amp 100 PRV Minibridge® by EDI [3], which is brought to a regulated 16.5 VDC by a National Semiconductor LM338 regulator—a nominal 5 Amp device that seems quite capable of delivering the 6 Amps peak current the charger requires. So far, pretty normal.

The next step allows processor intervention: a second connect relay controls the path between the output of the regulator and the balance of the circuitry. This relay is switched by a 2N2222 transistor, controlled by an active-high DCON signal at pin 10 of the DIP connector. A number of good choices for this relay are available inexpensively from Electronic Surplus, Inc. [4] The Sigma relay shown cost a whole dollar!

Now we get tricky again. Three series 3 Amp diodes are connected to four parallel 10 Ohm 10 Watt resistors (Radio Shack 271-132), and a similar arrangement parallel to the first. The result is a ballast resistor that both limits the maximum battery current to 6 Amps and provides a voltage proportional to the current through the resistors *added to* the 3 forward voltage drops of the diodes. Here's why.

The manufacturer's data on the U-128 indicates that charging is complete when current to the battery drops to somewhere between 180 and 460 mA, corresponding to a voltage across the ballast resistors of from 0.23 to 0.58

VDC. Added to the three diode drops, this is from 2.2 to 2.5 VDC. We can monitor this voltage handily with an optoisolator (the 4N33 in the schematic) to provide a recharge-in-progress RIP output at pin 14 of the DIP connector. This signal is low only when the charger is operating properly and the charging cycle has not yet completed.

DIP Connector Off-Board Interface Pinout

PIN 1:   Ground Reference
PIN 2:   "Hunger" monitor, *extreme emergency* level.
PIN 3:   "Hunger" monitor, *emergency* level.
PIN 4:   "Hunger" monitor, *extreme urgency* level.
PIN 5:   "Hunger" monitor, *urgent* level.
PIN 6:   "Hunger" monitor, *alert* level.
PIN 7-8: Not used.
PIN 9:   CONNECT AC input.
PIN 10: CONNECT DC input.
PIN 11: ACOK output.
PIN 12: Not used.
PIN 13: DCON output.
PIN 14: RIP (recharge in progress) output.
PIN 15: Auxiliary +5 VDC output or reference.
PIN 16: Auxiliary battery +V output.

The last bit of circuitry adapts a Texas Instruments type TL489 bargraph display driver as an independent state-of-discharge "hunger" monitor for the battery. With the 1N4735 Zener (6.2 Volts, Radio Shack 276-561) and resistor values shown, monitor points are provided for charge levels of 10.8, 9.7, 8.1, 7.0 and 6.0 VDC; a battery charge remaining of at least these voltages is indicated by a logic low at pins 6, 5, 4, 3 and 2 of the DIP connector, respectively.

Finally, ground and +5 VDC references are brought out at pins 1 and 16; pins 7, 8, 12, 13 and 15 are uncommitted.

You might have noticed the author's penchant for providing availability information and parts call-outs more often than you might expect in articles of this sort. This is to absolutely assure the reader's ability to reproduce the circuit shown with a minimum of frustration—and to circumvent Mr. Murphy's laws—*not* as a commercial forum for the products, manufacturers or distributors mentioned.

The AC Section

A GE surge suppresser and line de-glitcher is soldered directly across the connector terminals of the AC power-line (a standard tevision "cheater cord"), providing full-time service no matter how much of the remaining circuit is connected. A small 1 Amp self-resetting circuit breaker is wrapped with vinyl electrical tape and held in a vinyl cable clamp—a good practice with any glass-encapsulated device. A pair of $0.01\,\mu F$ 200 Volt capacitors bypass RF and noise, not so much because any part of this circuit would prove sensitive to it but because general design practice supports using these cheap and lightweight ounces of protection.

At this point, the AC line is connected to both a double-pole DIP relay set of contacts and to a special optoisolator (U1), a General Instrument Optoelectronics (formerly Monsanto) MID-400 power-to-logic interface, used as a simple AC voltage monitor. With just three passive components added (R1, R2 and C3), this provides a 5 Volt logic compatible signal to pin 11 of the off-board DIP connector [see pinout listing].

The AC-connect relay is controlled by a signal at off-board DIP pin 9, which turns it on through Q1, a 2N2222,

The Robot Digestive Tract

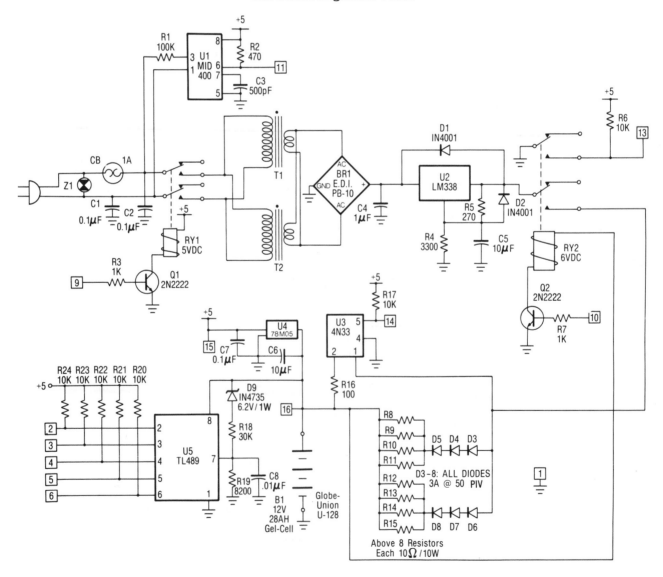

with current limiting by R3. While 5 Volt logic is assumed throughout, even 12 Volt CMOS should drive it comfortably—providing a dedicated buffer or latch output is used. When energized, this relay connects AC power to the transformer primary (or primaries, as shown).

The Regulator

National Semiconductor has somewhat eased the task of designing a regulator capable of handling the battery's peak current demand: the LM338K, an adjustable 5 Amp (nominal, 6 Amp max) regulator does the job with a minimum of external components.

Globe-Union suggests the use of a voltage regulated, current limited recharger with their Gel/Cell batteries. The data sheet for the U-128 12 Volt 28 Amp-hour battery used here suggests a 14.4 VDC maximum charge at up to 6 Amps. Let's take a closer look at this requirement to see how that translates to particular parts values for the regulator's peripheral circuitry.

The schematic shows ballast resistors R8 through R15 and diode strings D3/4/5 and D6/7/8 in series between the regulator and the battery, and we need to sum the *end-of-charge* voltage drops across these elements with the 14.4 VDC the battery requires to determine the output

Parts List:					
B1	Battery, 12 VDC 28 Amp-Hour Globe-union type U-128 Gel/Cell®	RY1	5V DPDT relay (Radio Shack 275-215)		either Digi-Key or Electronic Surplus)
C1	.01µF 200 Volt	RY2	6-9 V DPDT relay (check with Electronic Surplus, ref. [4])	U5	Texas Instruments type TL489 5-step analog bar graph driver
C2	.01µF 200 Volt	ST	Twin screw terminal strip [to connect battery] Radio Shack 274-663	Z1	General Electric V150LA20A transient absorber (available from Electronic Surplus)
C3	500 pF				
C4	1µF tantalum	SK1	16-pin low profile DIP socket (for RY1)		
C5	10µF tantalum	SK2	16-pin long-lead wire-wrap-type DIP socket (Radio Shack 276-1994), used to bring off-board signals to chassis level from circuit board		**References for Parts:**
C6	10µF tantalum				[1] Triad F-257U, Triad-Utrad Distributor Services, Div. of Litton Industries, 305 N. Briant St., Huntington, IN 46750.
C7	0.1µF				
C8	.01µF				
CB1	1 Amp 125 VAC self-resetting circuit breaker (available from Electronic Surplus, ref. [4])	T1, T2	Power transformer (see text), 22 Volts at 3 Amps		[2] Digi-Key Corporation, PO Box 677, Thief River Falls, Minn. 56701
D1	1N4001	U1	General Instrument Optoelectronic type MID-400 AC Line Monitor, power to logic optoisolated interface IC (available through Hamilton-Avnet)		[3] Electronic Devices Incorporated, 21 Gray Oaks Ave., Yonkers, NY 10710.
D2	1N4001				
D3-8	3 Amp/50 PIV diode, 1N5400 or similar (available from Electronic Surplus, ref. [4], or as Radio Shack part 276-1141)				[4] Electronics Surplus Incorporated, 1224 Prospect Ave., Cleveland, OH 44115.
		U2	National Semiconductor type LM338K 5 Amp [nominal, 6 Amp rated maximum] adjustable regulator (also available through Hamilton-Avnet)		
D9	6.2 Volt 1 Watt Zener, type 1N4735 (available from Radio Shack, part 276-561)				**Other Items:**
					• Terminal lug strips
BR1	Bridge rectifier, 25 Amp 50 PIV unit used is EDI part PB-10 (see ref. [3]—actually, this is rated 100 PIV), or use Radio Shack 276-1185	U3	4N33 Darlington output optoisolator (Radio Shack 276-133, which is closer to a 4N31, is fine, or check with Electronic Surplus)		• Printed circuit breadboard (Radio Shack Archer 276-170 in prototype)
					• 1½ x 4½ x 8 inch chassis box (Bud AC-1407 or equivalent)
R1-R7	¼ Watt, values as shown	U4	78M05 5 Volt 3-terminal regulator (or use a 7805 or LM340T-5.0 or Radio Shack 276-1770 or order through		• Appropriate heat sinks (Digi-Key part 690-3B or HS110-3 with Aavid 5791, or equivalent)
R8-R15	10 Ohm 10 Watt wirewound (Radio Shack 271-132)				
R16-R24	¼ Watt, values as shown				• Chassis-mounted AC "cheater" connector
Q1, Q2	2N2222				• AC "cheater" line cord

requirement of the regulator. The manufacturer specifies an end-of-charge current for the U-128 of 180-460 mA (substitute data for your battery). This current through the ballast resistors (composite parallel resistance of eight 10 Ohm 10 Watt resistors is 1.25 Ohms at 80 Watts) yields a drop of 0.225-0.575 Volts, which is added to the forward drops of the series diodes.

The bottom line is a requirement for something close to 17 VDC out of the regulator, accomplished according to the LM338 data sheet by 3300 Ohms at R4 and 270 Ohms at R5. The only capacitors needed are the $1\mu F$ at C4 and $10\mu F$ at C5, assuming you use tantalums. Multiply these values by 5 or 10 if you have to use standard electrolytics, and get ahold of the LM338 spec sheet for more information.

Diodes D1 and D2 provide reverse surge protection—an unlikely eventuality with this circuit, but again, a cheap ounce of protection.

The output of the regulator is connected *only* to the DC-connect relay, and is connected to the balance of the charger circuitry only when this relay is energized. The relay driver is a duplicate of that at the AC-connect relay, and is driven by an appropriate signal at pin 10 of the off-board DIP connector.

All About Ballast

Let's look at a couple of figures: what happens when the 6 Amp maximum current we talked about is flowing through the 1.25 Ohm ballast resistor? Obviously, there's a 7.5 volt drop. Now, turn this around. If there's 14½ Volts (more or less) on the supply side of the ballast, only a battery depleted to 7 Volts charge remaining would cause a 7½ Volt drop, and only then would the full 6 Amps be drawn through the ballast.

Practically speaking, we will want to have recharged the battery long before this. A 5 Volt regulator, for example, usually wants about 8 Volts at its input in order to function properly. The manufacturer doesn't talk about any life being left (although there's always *some*, practically speaking, with a gelled electrolyte battery, even after severe discharge) below 7.2 VDC.

If the battery is ever allowed to drop to 6 Volts remaining charge, this ballast resistance would mean a 7 Amp draw, which is beyond the rated capabilities of the transformers (depending on your selection) and the regulator, and some damage to the battery may have occurred. You take your chances recharging at this point.

So, recognizing a practical lowest discharge limit of 7 Volts at the extreme, the 1.25 Ohm ballast resistor

provides practical current limiting within the manufacturer's suggested limits for the battery. With eight 10 Watt resistors in parallel (for 80 Watts capacity), this ballast would be comfortable with 8 Amps going through it, so it's very comfortable with the 6 or less it will encounter here. In any case, the resistors were mounted close to the chassis on terminal lugs, and placed in contact with the chassis with a layer of heat sink compound, helping radiate any heat that's produced.

The Ohm's Law performance of the ballast keeps current proportional to the voltage difference between the supply and the battery, with constant tapering as the battery charges up.

The Monitors

There are three functions being monitored by charger circuitry here and reported on appropriate pins of the off-board DIP connector.

The first monitors the status of the DC-connect relay through the contacts of a second pole. Pin 13 provides a 5 Volt logic compatible low when this relay is energized, a high otherwise.

Optoisolator U3 (a 4N33 here, but a 4N31 or something else could work as well—the Darlington ouput isn't an absolute requirement) monitors the voltage drop across the ballast resistors and associated diodes (in other terms, between the regulator through the relay and the battery). In fact, the only reason for the series diode strings is to increase this voltage drop to a point where it's more easily monitored by the optoisolator. The forward drop across the LED in U3 is 1.5 volts. With R16 (100 Ohms) to limit current through the diode, U3 is on as long as there's a charging current through the ballast resistors and diodes *greater than* the end-of-charge current, at which point the LED is quenched off.

So a logic low at pin 14 means specifically that a recharge cycle is in progress; a logic high means that either the charger isn't powered (which can be double-checked at pins 11 and 13) or that the end of charge has been reached.

U5 provides a battery monitor function that is incorporated both out of convenience and space availability—it could be implemented independently of the charger. It provides five logic outputs that correspond to specific levels of remaining battery charge.

An inexpensive bar graph driver is used, the Texas Instruments TL489, which is fundamentally five comparators in an 8-pin DIP. An expanded scale is provided (normally, the driver switches at 200 mV intervals from 0-1

Volt) by knocking the top 6.2 Volts off the battery voltage with Zener D9, then dividing what's left with R18 and R19.

Now let's see how the battery monitor corresponds to "real life."

The Battery Monitor

At what point does a battery need recharging? This depends a great deal on the type of application. If the powered device is a power-failure safety light, the number of choices is reduced. If it's involved in a safety task (say paramedical support equipment, or a rescue robot), its ability to complete the task on the charge remaining is an important consideration.

You'll have to answer this question for yourself, or provide software to determine it. The battery monitor here is designed to aid that decision.

Five specific "state-of-discharge" trip points have been established, above which the output on the corresponding off-board DIP connector pin is at a 5 Volt logic compatible low, below which it's high.

The highest of these, an "alert" level, corresponds to a 10.8 Volt trip level. This is a very comfortable discharge level, and the device may request recharging at its convenience once this level has been reached. The *need* for recharging is less than critical, and may be approached leisurely *if* the device is not engaging in a high-current-demand task. If a high-current-demand task is anticipated and this level has been reached, it would be preferable (but not mandatory) to recharge before engaging in the task. This signal appears on pin 6.

The next, at pin 5, trips at 9.7 Volts and corresponds to an "urgent" level. Unless the device is engaged at an urgent task, recharging should be considered mandatory at this point. A battery (similar to the U-128) delivering 15 Amps for an hour or 25 Amps for thirty minutes discharges to this level rapidly toward the latter moments.

At pin 4, a signal corresponding to an 8.1 Volt charge provides an "extreme urgency" warning. At this point, 5 Volt regulators are within a few tenths of a Volt of losing their grip, motor speed and power available are reduced, and almost all reserve capacity from the battery is gone.

Pin 3, a 7.0 Volt alert, signals an "emergency" condition. At this point, the recharger going full blast can still "save" the battery, though many on-board systems may have long since begun behaving undependably or erratically.

Pin 2 triggers at 6.0 Volts, signalling an "extreme emergency" condition. At this point, its best to shut all systems down, inspect them scrupulously and baby the battery back to health. At least temporarily, virtually everything has "had it."

It's helpful, as some of the popular books on hobby robotics suggest, to think of these levels of discharge as degrees of "hunger." The analogy is complete if we consider the battery to be the robot's "stomach" and this circuit its "digestive tract."

Plus Some Extras

For convenience, 3-terminal regulator U4 provides a regulated 5 Volts for on-board functions and off-board 5 Volt logic compatibility. This 5 Volts is also brought to pin 15 of the DIP connector for facilitating testing or powering (modest) connected controls. Ground is available at pin 1, and the full battery voltage at pin 16.

Connections to the battery are made through a two-screw terminal on the top of the chassis. Remember, this particular circuit needs to carry up to 6 Amps (normally, possibly a bit more), so be generous with the wire gauge you select for connecting the battery.

Conclusion

It doesn't take much to add a little sensitivity to the brute force approach of most battery chargers. Indeed, 6 Amp automotive-style chargers have been on sale for under twenty dollars recently, and these would easily accomplish the basic recharging job. But a dumb charger isn't the most intelligent choice, especially when it's going to be used with intelligent systems.

I would suggest that whatever controlling intelligence governs the device's behavior under critical conditions be CMOS, some other low-power technology, or independently or failsafe powered. The Motorola MC146805E2L is a CMOS version of the 6805, compatible with a large number of 6800-family instructions, and an excellent choice.

Checkout of the circuit is very straightforward. I used a Hickok MX-333 DMM with logic probe and audio monitor functions in conjunction with a Heathkit IP-18 variable supply and got the whole thing done while watching an episode of *Buck Rogers* (how about those arms on the Crichton robot!). The values indicated are the result of actual measurements on my prototype, and may vary with other component choices.

If you have refinements or modifications to suggest, please send them directly to **Robotics Age**. And more power to you!

A ROBOT ARM WITHOUT A BUDGET

Timothy Nagle and
V. Michael Powers
Electrical & Computer Engineering
Department
Oregon State University
Corvallis, OR

Introduction

A robot's not a robot in any of the fantasy or practical senses unless it can manipulate its environment. To us, a robot is little fun unless it can hand out mail, flip a switch, scare a friend, etc.

This article describes a portion of a robot project dedicated to building a low-cost manipulator; an arm, if you will, which can pick up and move light objects. The "Fanny Pincher" (FP) has fingers for grasping and a wrist and arm for positioning or carrying objects such as a glass of water or a soldering iron.

FP was built as part of a collection of projects underway by the Robot Group, a continuing group of Oregon State University students sponsored (but rarely funded) by the student IEEE chapter and the Electrical and Computer Engineering Department. Most of the current projects are spinoffs resulting from ideas which arose while contemplating the Micro-Mouse problem posed by the IEEE *Spectrum* magazine. Some students simultaneously complete the requirements of a credit-carrying project in EE, while others squeeze their participation into their schedules as an extracurricular activity.

The one project currently involving the most students is a three-wheeled cart named "Sparko." Sparko has one powered, steerable wheel and is capable of carrying a full-grown person. A single-chip microcomputer (Intel 8748) has been programmed to read light sensors and keep the tricycle on track while it follows a trail consisting of an adding machine tape laid on the floor. While plans continue to add a single-board Z-80 computer for higher levels of coordination, feelers, and acoustic and optical sensing, the FP project was undertaken to provide an arm mounted on top of the vehicle that could swing around and up and down and do whatever seemed useful, amusing, or most importantly, challenging. (Figures 1 & 2)

No budget is given, not because cost was unimportant but because it was paramount. Many of the materials and items used were scrap; other choices and other mechanisms might have been implemented from a different scrap pile. Nonetheless, the methods reported here work, and others may find them or variations thereon useful in their own work.

The Arm's Degrees of Freedom

An arm is a movable appendage. In the world of manipulators, an arm is categorized first in terms of the *degrees of freedom*, or the number of different ways it can move, and then, further, in terms of which kinds and how far each different movement is. In that sense, the FP arm has four degrees of freedom: limited rotation on top of the tricycle (side to side), extension of 30 cm., elevation of +15 to –22 degrees and wrist rotation of three turns around the longitudinal axis of the arm. Respectively, these motions

Figure 1. Sparko, a three-wheeled cart built by students of the Robot Group at Oregon State University, carries FP, the arm, up to a bicycle which proves too heavy to lift.

Figure 2. FP was built as an arm to fit on top of Sparko. It can rotate on its "Lazy Susan" base, elevate by rocking around the center support, extend and retract its rotatable wrist and a "hand" capable of grasping small objects such as a flower.

roughly correspond to human arm motions of swinging from side to side across the body (mostly shoulder movement), reaching out (in the human, a combination of shoulder and elbow action but in FP a linear motion), raising and lowering the hand (familiar to any student), and twisting the wrist (try it with your arm extended, and see how many motions are really involved).

The degrees of freedom are the number of different kinds of motion available to the controller of the arm; the number of basic strategies available for putting together a sequence of controls that will move the arm to its desired destination (Figure 3). To the builder or to the programmer of these actions, which must be carefully controlled and synchronized, the amount of work necessary for successful arm operation may suggest that the term "degrees of slavery" may seem more appropriate than "degrees of freedom!"

In industrial applications, some people feel that six degrees of freedom are necessary to do assembly work, while others feel that only three degrees of freedom are necessary for the majority of assembly tasks [1, 2]. We felt that when combined with the mobility of the tricycle, the four degrees of freedom listed above were sufficient for many tasks, and included a sufficient number of interesting types of motion.

Control

Any robot must, we feel, be able to move to a goal. Sometimes this procedure is very complicated, involving some of the yet unsolved problems of Artificial Intelligence. We are concerned first, however, with a simpler problem. As discussed in the following sections, each of the movements (degrees of freedom) possible has a mechanism and a motor to allow movement. The general control problem which first appears when trying to put these mechanisms to work is: how to move it from one position

precisely to another, desired position. Once this problem is solved satisfactorily, the robot builder can proceed to sequences of movements, simultaneous movements, and choosing which of the possible sequences of motion is appropriate to reach a goal. (The latter problem can be so complex as to seem unsolvable).

"Open-Loop" Direct Control

The direct approach to control of a single movement is to turn on the motor for a certain calculated time and then turn it off. At the end of that time, the motor, coupled through the drive train, has moved the object. Ideally, one can calculate just what that time should be for a given

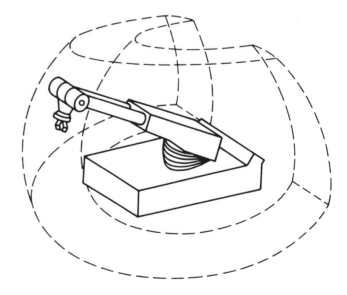

Figure 3. With any robot arm, such as the one sketched here, the work envelope consists of all the space which can be usefully reached by the hand as a result of combinations of the arm's movements in each of its degrees of freedom. A spherical coordinate system is shown.

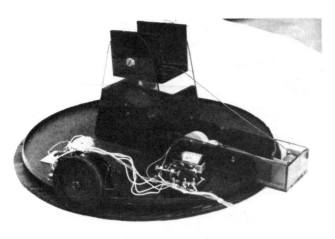

Figure 4. Rotation is provided by a wheel made from a pulley and some foam rubber, and equipped with a brake. The wheel causes the "Lazy Susan" to rotate on its base, and swings the arm from side to side.

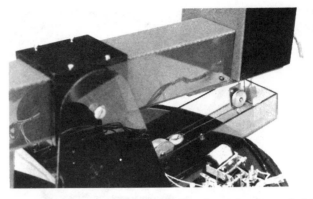

Figure 5. Elevation comes from pulling down the rear end of the arm with a combination of a motorized winch and a couple of pulleys.

motion as in navigation by "dead reckoning." But gears and wheels slip, arms have inertia, real materials bend and shake, and moving the same distance in two different directions takes different amounts of power. We often don't even know the physical parameters of the system precisely enough to make possible the complicated calculation [3, 4], even if we wanted to perform it.

Accurate motion control requires that the actual movement resulting from a command signal is somehow compared with the desired one in a "closed-loop" system that can automatically correct for errors in performance. To illustrate the impracticality of the open-loop control approach for a useful robot, consider trying to get in your car and drive down to the corner store—with your eyes closed.

There are practical applications of open-loop control, of course. One example is the ordinary electromagnetic relay. The coil is energized full on, and the contact moves. When the coil is turned off, the contact returns. In each case, the power supplied is more than adequate to reach the goal, and the moving element has a built-in limit, or stop, at the end of travel. This sort of limited travel between fixed stops has many applications in industrial automation for parts transfer, etc.

Feedback Control

In closed-loop or "feedback" control, the movement is monitored, and a sensor will "feed back" a measure of progress to the controller, which can then determine how to move in order to get closer to the goal. There is an information-flow loop, then, between the controller, the motor, the moving object, the sensor, and back to the controller, hence the term "closed-loop" control.

One simple type of feedback control system uses detectors at the far extremes of the permissible movements. At the end of the allowable travel of some part is a

limit switch—a microswitch or optical detector. Once the arm reaches maximum extension or rotation, the limit switch trips, causing the controller to stop.

Another more sophisticated approach commonly used is a servo system. The input to the servo system is set to the desired point or path. The controller attempts to match it, driving the load object (the arm, for example) appropriately. As the load lags behind or overshoots the target position, the difference or error signal is used to advance or retard the load's progress.

Feedback loops can be implemented either with digital hardware or in software. The latter approach is the most flexible, and is commonly used in robots. Here the robot controller senses or observes the movement of its own part in relation to the goal location. Control of the motors, particularly when close to the goal, can be computed from the difference between present position and goal position, etc. This feeding back the measurements of actual motions to the controller is the sort of function that a servo system performs automatically, but the software control method can be used where hardware servos are not available or not practical.

One example of this form of feedback control is employed in our experimental robot. The powered wheel on Sparko has reflective foil spots around its rim, and a photosensor detecting them can report the robot's progress. If Sparko is commanded to go, say, twenty feet down the hall, a subfunction can count turns of the wheel. When the counting nears the equivalent of twenty feet, this subfunction could slow down or begin to stop the wheel motion.

Difficulties in using the software form of feedback control include the difficulty of finding and interfacing a means for monitoring motions and relating the measurements to a meaningful frame of reference, the sophisticated mathematics sometimes necessary for attaining ideal behavior, and obtaining the proper timing of measurements and feedback control computations.

Nonetheless, future developments of Sparko and his arm are planned to include more such feedback schemes, using optical sensing (vision) and acoustics (sonar).

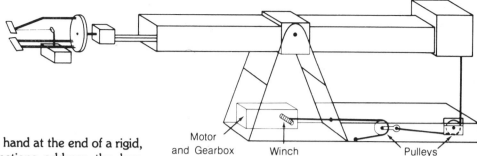

Figure 6. A sketch of FP showing the elevation mechanism, with the wrist and hand mounted ready to slide in and out.

Construction

We begin with the notion of a hand at the end of a rigid, straight arm. The following sections address the low-budget construction of mechanisms to give it motion in a spherical coordinate system.

Shoulder Rotation

One of the handiest items in our parts bin was a turntable. Because it is large, all kinds of things can be mounted on it, including the arm itself, and the drive motor and gear box for turning. There is enough room for future addition of microphones, sonar transducers and a digital TV camera. These items roughly correspond to hearing, depth perception and vision. In people, these senses are mounted in a turnable head. Even though in humans the arm doesn't rotate with the head (unless you have a stiff neck from a bad night's sleep), it would make sense for a robot to have these mounted on the rotation platform along with the arm. This saves having to build another platform just for the sensors.

The base of the platform is a 15″ plywood circle. On top of the wood base is the rotation platform. This is a simple turntable (a Rubbermaid "Lazy-Susan"). This turntable consists of two surfaces separated by ball bearings in a grooved channel. (This arrangement is sometimes called a *thrust bearing.*) The entire platform is driven by a friction wheel turned by a small motor and gear box. The motor and gear box (like most of the others in this project) came out of a broken toy (Figure 4).

The motor and gear box are mounted near the edge of the platform, with the drive wheel extending over the edge. The radius of the wheel is just a bit more than the distance between the center of the wheel and the base. This helps insure good contact between the edge of the wheel and the base.

The drive wheel is made from a pulley. On the outside edge there is a strip of ¼″ foam rubber for traction. In addition to glue to hold the foam strip on, there is a tightly wound wire around it, forcing it somewhat into the edge of the pulley.

The platform rotates well, but one problem to be kept in mind is momentum. This could be taken care of either with position feedback and correction, or a system for stopping, or both. Stopping can be done by reversing the motor momentarily, or by a brake. Since there was a brake in the parts box, that's what was used. There are a couple of problems in stopping a system. One is that attempting to quickly stop a moving mass puts a lot of strain on the

structure. Second, it is difficult to instantly stop any mass. Remember that this gets worse when the arm is extended, or when weight is added to the end of the arm as when it is holding a payload.

While the platform could rotate 360°, its movement is currently more limited. The cables from the motors and feedback sensors are rather bulky and hang off the back end. This limits its rotation to ±45°, which for our uses is entirely adequate.

Elevation

At this point, our robot can move its rigid arm from side to side (shaking hands with this robot would be impossible—unless you were lying on your side). It would be nice to have it be able to pick things up or move things from one plane to a higher one. Performing tasks such as washing dishes or picking things up from an assembly line requires an arm which can move in a vertical direction.

Moving in the vertical plane can be done by raising the entire arm—as in an X-Y-Z coordinate system. Devices such as elevator mechanisms or screw drives could be used to do this. We designed our arm to be rotated in a vertical plane rather than be raised and lowered in it.

This rotation could be done by attaching the back end to the platform and raising and lowering the front end. This creates a few problems. One is that the front end can't be lowered below the base. But the most important is the weight problem. The drive system must support all the weight. In hopes of avoiding some of these problems a fulcrum method was used. The arm is mounted with the center of balance just in front of the pivot point. This way gravity would provide the downward force. For the raising action we used a winch. The winch was housed under the fulcrum and the cable ran along the platform to a point directly under the "back" end of the arm. The cable then ran around a pulley and up to the arm.

When the winch is turned on, the winch begins winding the cable and the back end of the arm is pulled down to raise the front end. Some of the workload on the winch can be reduced by sharing the load with a pulley system. (Figures 5 & 6)

The vertical rotation of the arm is limited by the support structures. It can be lowered until it reaches the edge of the platform, about –22°. It can be raised until the back end reaches the pulley box, about +15°.

No Elbow

Now that we can get our hand to any position on a portion of the surface of a sphere, it would be nice if we could also move along the radius. With this, we could access anywhere in a space bounded by its rotational angles, outer radius, and inner radius. Some of the commercial robots use a joint in the middle of the arm which resembles an elbow. As humans we use a combination of shoulder and elbow movements to position our hand between our shoulder and the end of our reach along a linear track. However, in keeping with the spherical coordinate system we have chosen, our arm telescopes in and out in the radial direction instead of using an elbow.

We mounted the wrist and hand at the end of a 40 cm push rod. This rod was set into a 45 cm U-channel so it could track easily. Some roller bearings were strategically placed to cut down the friction as the push rod moves along the track. Attached to the U-channel, at the end opposite the hand, are the motor and gear box. Viewed from above, the output shaft of the gearbox is in line with the center line of the channel and rod. The output shaft is connected to a threaded drive rod. This drive rod runs up to the front where it terminates in a bearing assembly. This

should be carefully checked for proper alignment. The back end of the push rod is attached to the drive rod by a nut, which is prevented from rotating by mounting it in a block which is fixed to the push rod. (Figure 7)

If everything is properly aligned, the whole mechanism should operate without binding. As the motor turns the threaded drive shaft, the nested nut is driven longitudinally. This is turn moves the push rod and hand. The whole action is guided by the U-channel.

The entire mechanism should now be able to slide in and out, giving the extension effect which a human achieves with shoulder and elbow.

The Hand and Wrist

The hand is the business end, the part that adds usefulness to the arm. Usually this is designed to grasp (or pinch) something. Often this grabber is made to be detachable so that some special-purpose mechanism be added. Some of the commercial machines have paint sprayers or welders which attach to the end of the arm. But for us and most other experimenters, a general purpose grabber was the most useful.

Our grabber has two foam pads about an inch and a half apart. Imbedded in one pad is a small light and in the other is a photocell. This way we can tell when the grabber is positioned right, and it keeps us from getting a handful of air.

The pads are brought together by a lever action

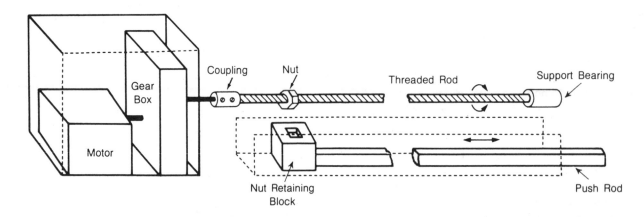

Figure 7. For extension, a push rod slides in and out. The back end of the push rod is fixed to a nut, which moves along the length of a threaded rod as the threaded rod turns.

(powered by a geared motor) and squeeze the object. With this arrangement we've been able to pick up and hold pens, pencils, soldering irons, coffee cups (very light ones), chess pieces; and to pull light switches and to flip switches on a computer. (Figure 8)

The grabber can be a simple structure with no degrees of freedom, or it may have several degrees. For Fanny Pincher we decided that one degree (the wrist) would be enough to do most of the tasks we had in mind. With this much, once it gets hold of something it can turn it (like a knob or screwdriver) or tilt it (like a test tube or cup of coffee). This allows enough activity to keep several programmers busy for a long time.

The wrist is a rather simple device. It is simply a motor and gearbox attached to the end of the push rod. The output shaft of the gearbox is attached to the hand. The wrist can turn several revolutions. The only limiting factor is the cable from the hand; as the wrist rotates it winds up the cable. Our configuration permits a ±540° range of motion.

Summary

We have presented here our design choices in building a "no-budget" hand and arm for experimental robotics study. Providing controls through a computer interface has proved relatively easy. Very important in making this control accurate and useful will be the implementation of feedback sensors on the manipulator joints. If we can measure progress of motion in a given degree of freedom, we can better control the "reach" to a desired position, and perhaps pour from the pitcher into the glass rather than onto the floor!

A more involved long-range project is concerned with the higher levels of organization in the general robotics problem of achieving goals in terms of sequences of actions. Meanwhile, we will continue to enjoy the implementation problems and their solutions.   ▣

References

[1] R. Nevins, D. Whitney, "Computer-Controlled Assembly," *Scientific American*, 238, 2 February 1978.

[2] J. F. Engelberger, "Robot Arms for Assembly," 78-WA/DSC-37, Dynamic Systems and Control Division, ASME, December 1978.

[3] R. H. Taylor, "Planning and Execution of Straight Line Manipulator Trajectories," *IBM Journal of Research and Development 23*, 4, July 1979.

[4] R. Paul, "Robots, Models and Automation," *Computer*, July 1979.

Figure 8. The business end. Near the front end of the arm, one motor rotates the wrist to bring the hand to any desired angle. Another pulls tight the short cable or sinew to close the padded fingers.

SUPERKIM Meets ET-2

Don McAllister
4709 Rockbluff Drive
Rolling Hills Estates, CA

This article presents some of my experiences in interfacing and programming a SUPER-KIM single board computer (SBC) for the control of a Lour Control ET-2 robot shell (Figure 1). The ET-2 (*E*xperimental *T*ransmobile with *2* drive motors) consists of a three level frame powered by two separately driven wheels and balanced by a free caster. The lower level contains the drive motors and gearbox, a 32 amp-hour 12V motocycle battery, and two driver electronics boards. The upper levels are available for the installation of user equipment. In this case, the SBC is mounted on top.

The ET-2 may be operated under computer control using only four TTL command lines. Each motor has two control bits, one to turn it on and another to set its direction (by a reversing relay). The driver boards provide the amplification necessary to convert from TTL logic levels to the 12 volt power for the motors and relays. Control of motor speed is obtained by varying the duty cycle (the percentage of time the bit is on) of a low frequency (10-20Hz) square wave signal applied to the motor's drive bit. The inertia of the motor and robot effectively average the signal to a proportionally lower DC level at the motor.

The drive motors are Ford permanent magnet windshield-wiper motors, which, besides having built-in gear reduction that produces a good deal of torque, are also less expensive than PM motors with comparable performance and are readily obtainable. Each motor can be independently driven in the forward or reverse direction. Lour states that to turn the shell, the preferred method is to drive one motor forward and the other in reverse so that the robot spins on its vertical axis. Turns with only one motor driving are not recommended, due to the increased loading of the motor. Reversing a motor while it is in operation can put a tremendous strain on the motors and drive system. Thus, both motors should be programmed to stop briefly between commands.

The SUPERKIM, by Microproducts, Inc. is a complete, powerful microcomputer control system based on the 6502 microprocessor, contained on a single 11.5 x 11.5 inch PC board. The board is fully socketed for easy servicing and expansion to 4Kbyte RAM and 16K EPROM on board. It comes with 1K RAM, and the address space is fully decoded so that with additional boards up to 64K of memory or I/O may be used. For this purpose, the CPU bus lines are brought out on wire-wrap pins that may also be used with standard in-line ribbon cable connectors to expand the bus.

The SUPERKIM has eight priority interrupts which are individually vectored and resettable under software control—a feature useful for real-time robot control systems. Four SYNERTEK 6522 Versatile Interface Adapter (VIA) sockets are provided on the board; one 6522 comes with the board. This IC is indeed a very flexible I/O device, containing two bidirectional 8-bit parallel ports with handshaking (with each bit separately programmable for input or output), an 8-bit, bidirectional serial to parallel shift register, and two 16-bit programmable counter/timers. The board comes with a 6530 interface chip as well.

The ports on the 6522's could also be used for implementing analog to digital converters (ADC's). A full complement of 6522's would permit up to eight 8-bit ADC's for interfacing to robot sensors, etc.

Interfacing the SUPERKIM to the ET-2

The SUPERKIM is mounted to the topmost PVC platform on the ET-2 with machine screws and .75″ spacers. 12V from the battery is supplied to the SUPERKIM's on-board 5V regulator through a SPDT switch.

Figure 2 shows the location of the pins on the 6522 that are used as output ports to the ET-2. The four control lines of the ET-2, D1, D2, E1, E2, are connected to the control bits in the SUPERKIM's J5 VIA parallel output port as shown in Table 1. There are, of course, many alternate possibilities for configuring the interface. For convenience, the motor drive signals were assigned to bits 0 to 1 of port A (address 1302H) and the reversing relays to the corresponding bits of port B (address 1303H). Note that since the drivers on the ET-2 invert the logical sense of their inputs, a logical 0 (low) on an output will turn the corresponding motor or relay ON, and a logical 1 (high) will turn it OFF. Thus, writing to addressed 1302 and 1303 controls the motors and relays directly, with sixteen possible control states.

Due to the action of the power on reset, the I/O ports of the 6522 are initialized to be output ports, and zeroed. Therefore, as soon as the SUPERKIM is turned on, the ET-2 will lurch forward if the motor drivers are connected to the interface. To eliminate this problem, a 2-pole switch is used between the 6522 outputs and the motor drive inputs, which should be open when the computer is switched on. After location 1302H is set to 03, the motors may be engaged. The switch also comes in handy as a panic switch if your program causes the ET-2 to run amok!

Figures 3 and 4 show examples of ET-2 turning maneuvers. In Figure 3 the left motor is driven in reverse while the right motor runs forward, resulting in the preferred spin turn. In Figure 4 the right motor is driven forward with the left motor turned off, so that the left wheel is the axis of the turn, and the turn is more gradual. As mentioned above, the spin turn should be used for best results.

We will now describe how to reproduce these and more interesting movements using the SUPERKIM, both directly from the keyboard and then under program control.

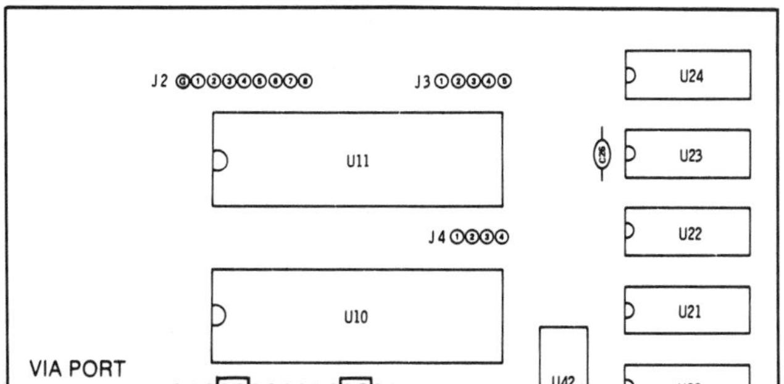

Direct Command Mode

With the SUPERKIM interfaced to the ET-2 as previously described, constant motion modes can be commanded

Figure 2. SuperKim/ET-2 robot control interface connections.

CONTROL LINE	FUNCTION (WHEN LOW)	J5 PIN	
	TABLE 1		
D1	RIGHT MOTOR ON	PIN 3	(A0)
D2	LEFT MOTOR ON	PIN 4	(A1)
E1	REVERSE RIGHT	PIN 11	(B0)
E2	REVERSE LEFT	PIN 12	(B1)

Figure 3. An on-axis turn. With one motor reversed, the ET-2 can turn in place.

Figure 4. A "one-motor" turn. With one motor off, ET-2 turns with the stopped wheel as an axis.

directly from the keyboard as follows:

Step 1: Make sure that the motor switch is turned off (motor drivers disconnected from the computer) and then turn on the computer power switch. The display should light up.

Step 2: As described in the SUPERKIM manual, initialize the keyboard interrupt vectors as shown in Table 2. These values make the single step (SST) and stop (ST) keys work correctly.

Step 3: The ET-2 can now be commanded manually by entering the desired control states into address locations 1302H and 1303H. Table 3 shows the results of various output settings. *Note that the ET-2 should not be driven with both motors reversed, as the caster turns inwards and makes the unit unstable.*

Step 4: After the desired state is entered, turn the motor switch ON. *WARNING: In this mode the unit can only be stopped by turning the motor switch off, disconnecting the driver inputs from the computer!)*

Movement Under Program Control

While the direct command mode will allow you to check out your wiring, more complex sequences of control states

must be commanded by machine language programming. Programs can be entered and debugged directly from the hexadecimal keypad on the SUPERKIM and then saved using the board's build-in cassette tape interface.

A highly desirable alternative to machine language programming is the use of a 6502 development system (APPLE, etc.). Instead of keying your program into memory in hex code, programs can be prepared on the development system using an assembler and then downloaded to the SUPERKIM through its serial interface. The advantages of using an automatic assembler to translate opcodes and compute the addresses for a new code file will become obvious the first time you have to add an instruction into the middle of an existing machine language procedure.

Table 4 is a listing of a 6502 machine language program for moving the ET-2 in a roughly octagonal pattern. It makes use of two nested time delay subroutines, LDELAY (long delay) at 0300H and SDELAY (short delay) at 0310H. SDELAY itself consists of two nested delay loops, each counting down from FFH to 0 (256 cycles) resulting in a delay of about 0.25 sec.

The byte at 0301H sets the loop count of the LDELAY subroutine, and is originally set to 2 as shown, for an aggregate delay of about half a second. Different delays may be obtained by using that byte as a subroutine parameter,

TABLE 2

ADDRESS	DATA
17FA	00
17FB	1C
17FE	00
17FF	1C

TABLE 3

ADDRESS	CONTENTS	CONTROL STATE
1302	00	BOTH MOTORS ON
	01	RIGHT MOTOR ON
	02	LEFT MOTOR ON
	03	BOTH MOTORS OFF
1303	00	BOTH RELAYS ON
	01	RIGHT RELAY ON
	02	LEFT RELAY ON
	03	BOTH RELAYS OFF

TABLE 4

ADDRESS	CONTENTS	LABEL	OPERATION	COMMENTS
0200	A9 03		LDA #$03	;POLYGON PROGRAM
0202	8D 03 13		STA $1303	;TURN RELAYS OFF
0205	A9 00	LOOP:	LDA #$00	
0207	8D 02 13		STA $1302	;BOTH MOTORS ON
020A	20 00 03		JSR LDELAY	;WAIT
020D	A9 03		LDA #$03	
020F	8D 02 13		STA $1302	;BOTH MOTORS OFF
0212	20 00 03		JSR LDELAY	;WAIT
0215	A9 01		LDA #$01	
0217	8D 02 13		STA $1302	;RIGHT MOTOR ON
021A	20 00 03		JSR LDELAY	;WAIT
021D	A9 03		LDA #$03	
021F	8D 02 13		STA $1302	;BOTH MOTORS OFF
0222	20 00 03		JSR LDELAY	;WAIT
0225	4C 00 02		JMP LOOP	;KEEP ON GOING
		;		
0300	A0 02	LDELAY:	LDY #$02	;SET DEFAULT COUNT
0302	8C 20 03	LOOP1:	STY COUNT	;SAVE IT
0305	20 10 03		JSR SDELAY	;CALL SHORT DELAY
0308	AC 20 03		LDY COUNT	;GET COUNT
030B	88		DEY	;COUNT DOWN 1
030C	D0 F4		BNE LOOP1	;CONTINUE TIL ZERO
030E	60		RTS	;RETURN
		;		
0310	A2 FF	SDELAY:	LDX #$FF	;OUTER CONSTANT
0312	A0 FF	LOOP2:	LDY #$FF	;INNER CONSTANT
0314	88	LOOP3:	DEY	;INNER COUNTDOWN
0315	D0 FD		BNE LOOP3	;LOOP UNTIL ZERO
0317	CA		DEX	;OUTER COUNTDOWN
0318	D0 F8		BNE LOOP2	;LOOP UNTIL ZERO
031A	60		RTS	;RETURN
		;		
0320	00	COUNT:	[long delay count hold location]	
			END	

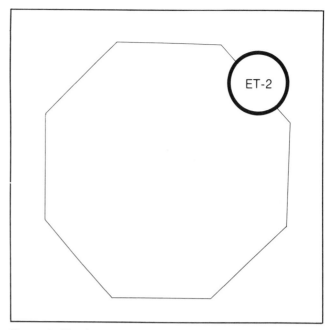

Figure 5. The (approximate) path resulting from the program.

setting it to a desired value "n" before calling LDELAY to give a total delay of n/4 sec. Finer control over the delay interval can be achieved by reducing the loop counts for the outer and inner loops within SDELAY (0311H and 0313H, respectively) from their original FF value.

The comments in the listing describe the action commands sent to the ET-2 at each step. This program makes use of the one-motor turn shown in Figure 4 (which may not be suitable for all surfaces). Since the outputs of the 6522 hold the values last set until the next output operation, the motor(s) will remain on (or off) during the call to LDELAY. The program has been simplified by using the default delay constant, 2, in the LDELAY loop. With just the right motor on, the robot will turn roughly 45° in the resulting interval, resulting in the approximate octagon pattern (Figure 5). Note that, as mentioned earlier, a power-off interval is commanded after each movement to minimize strain on the drive system (although it is not essential for a one-motor turn).

After the hex code in Table 4 is keyed into RAM at the locations give, the following steps should be followed to start the movement:

Step 1: Check the program carefully against the listing to verify each location. The single step (SST) button may be used to verify proper program execution (although stepping through the delay subroutines will prove tedious). Make sure that both the motor (1302H) and relay (1303H) output ports have been set to 03 (OFF). The motor switch may now be turned ON. Nothing should happen yet.

Step 2: Set the address to 0200, the start of the polygon program.

Step 3: Press the GO button. The robot will begin to traverse an octagon.

Step 4: To stop the program, press the ST key and turn the motor switch off.

Stopping the program is best done during one of the pauses, when both motors are off. If the ST key is pressed while a motor is running, it (they) will remain running, due to the latching action of the 6522/6530.

Conclusions and Future Work

A more elegant method of obtaining the program delay would be to make use of the interval timer in the 6522. The device may be set to count up to 256 prescaled clock pulses by writing to the counter address. Based on the write address used to load the counter, the system clock will be divided by 1, 8, 64, or 1024 to produce the prescaled clock pulses. The unit will begin to count down at the prescaled rate as soon as a value is loaded. The register may be read by a program at any time to obtain the current count, and it may optionally be told to generate an interrupt upon reaching zero. Also, each 6522 has two 16-bit programmable counters, but these lack the ability to scale the count rate.

The SUPERKIM controlled ET-2 robot is an excellent, moderately priced system to which the robotics experimenter can easily add more sensors and other equipment. More elaborate systems may make use of the computer's versatile interrupt handling capabilities to design an event-driven real-time control system for the robot. Programs can also be written to use the 6522/6530 I/O ports for A-to-D conversion and interfacing the ET-2's contact sensors.

In the configuration described here, the computer controlled ET-2 falls short of the definition of a true robot, since all of its movements are "open loop." It has no sensors to tell it that a successful 45° turn has been made or even if it is travelling straight. Until the contact sensors furnished with the ET-2 are interfaced, even simple obstacle avoidance behavior is impossible. Part 2 of this article will describe the addition of sensors and interfaces to enable much more interesting behavior. A good source of additional 6502 machine language programs can be found in Tod Loofburrow's book. [4] These programs, with minor modifications, can be used for controlling the ET-2 with the SUPERKIM.

The ET-2 robot shell is well-built and reliable. The only problem I could find is that it has a tendency to tip over when being driven backwards at full speed with the rear caster in certain positions. Lour points this out and recommends that backing be avoided by doing a 180° on-axis turn instead. The unit can be driven over thick pile carpets without loss of traction, a task which many home robots find troublesome. Each motor draws around nine amps at full speed, so that the system needs recharging after an hour or so of continued use. Lour offers the unit in plan, kit, or assembled form.

References

[1] "KIM-1 User Manual," MOS Technology, 950 Rittishouse Road, Norristown, PA 19401 (August 1976).

[2] "Instructions for SUPERKIM," Microproducts, 2107 Artesia Blvd., Redondo Beach, CA 90278.

[3] "ET-2 Assembly Manual," Lour Control, 1822 Largo Court, Schaumberger, IL 60194.

[4] Tod Loofborrow, *How to Build a Computer Controlled Robot*, Hayden Books, Rochelle Park, New Jersey (1979).

SUPERKIM MEETS ET-2
PART II: SENSORS

Don McAllister
4709 Rockbluff Drive
Rolling Hills Estates, CA

In the preceding article, I described how I interfaced and programmed a SUPERKIM single board computer (SBC) to /control the Lour Control ET-2 robot shell.

Without sensors, though, the SUPERKIM/ET-2 combination described in that article is not a true robot, since all of its movements are "open loop," that is, without feedback. This article describes how to interface contact sensors and sensors that require A-to-D conversion (such as infrared scanners or temperature sensors) to the SUPERKIM/ET-2. Once you interface the contact sensors furnished with ET-2, you can program avoidance behavior. This permits the SUPERKIM/ET-2 to sense when it has contacted an obstacle, and take appropriate avoidance actions. I refer the reader to the previous article for details concerning motion control of the ET-2 by the SUPERKIM.

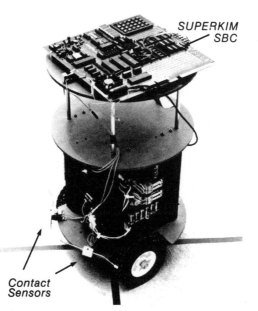

SUPERKIM
SBC

Contact
Sensors

Figure 1. Contact sensors can be mounted around the base of ET-2.

Interfacing ET-2 Contact Sensors to the SUPERKIM

ET-2 provides a number of contact sensor switches that can easily be interfaced to the SUPERKIM. These contact sensors, equipped with metal "feelers," can be mounted around the base of the ET-2 to sense contact with an obstacle by means of a switch closure.

Lour Control has provided four independent contact bumper assemblies, which are designed to ring around the base of ET-2 as shown in Figure 2. Whenever a guard rod, which projects out of either side of the assembly, comes in contact with an object during ET-2's motion, it is deflected laterally, activating a built-in momentary switch. Depending on which way the switch is toggled, and on the control program in SUPERKIM, the ET-2 can then perform an avoidance manuever.

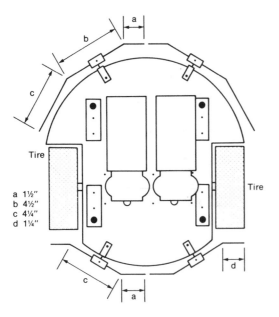

Figure 2. Location of contact switch/bumper assemblies.

a 1½"
b 4½"
c 4¼"
d 1¼"

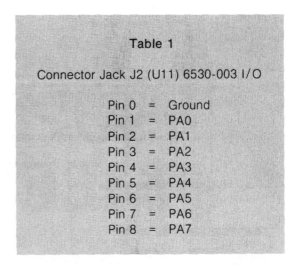

Table 1

Connector Jack J2 (U11) 6530-003 I/O

Pin 0	=	Ground
Pin 1	=	PA0
Pin 2	=	PA1
Pin 3	=	PA2
Pin 4	=	PA3
Pin 5	=	PA4
Pin 6	=	PA5
Pin 7	=	PA6
Pin 8	=	PA7

As figure 3 shows, each of the bumper switches have four basic parts—the guard rod, connecting block, switch, and mounting bracket. The guard is a 5/32 inch diameter rod that protrudes from both sides of the connecting block and acts as an extension of the switch's own toggle lever. You can easily distinguish the two bumper assemblies installed in the front section of the shell, since their guard rods are shorter than those mounted in the rear section. The switch's toggle lever and the guard rod are both attached to the connecting block by means of set screws. The switch itself is a momentary, on-off-on device that automatically returns to the center (off) position when released. A spring wire, wrapped around the switch's mounting stud, holds the connecting block in a horizontal position and aids in the resetting of the switch. The entire unit is attached to one of the four mounting holes on the tier of ET-2 by means of a corner angle mounting bracket.

We used SUPERKIM's 6522 to interface the SBC with ET-2's motor and relay controls. SUPERKIM also comes with two 6530 ROM/interface ICs, designated 002 and 003. To interface these sensors to SUPERKIM, we must first consider the operation of the I/O ports in a 6530. Each 6530 array provides 15 I/O pins. The microprocessor and operating program define whether a given pin is an input pin or output pin, determine what data are to appear on the output pins, and read the data appearing on the input pins. The I/O pins provided on 6530-002 are dedicated to interfacing with specific elements of the KIM-1 system, including the keyboard, display, TTY interface circuit, and cassette tape interface.

The I/O pins on the 6530-003 (U11) are brought out to connector jacks J2 and J3, and are available for user applications. Connector jack J2 has 8 pins constituting Port A, as shown in Table 1. Connector jack J3 has 5 pins constituting Port B, as shown in Table 2. Pin 0 on Port A is a ground line. Pins 1 through 8 on Port A and pins 1 through 5 on Port B are the programmable I/O lines. Figure 4 shows the location of the pins on the 6530-003 connectors J2 and J3 that are used as contact sensor input lines.

Figure 3. Contact sensor assembly (detail).

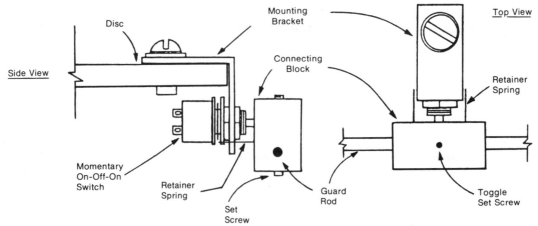

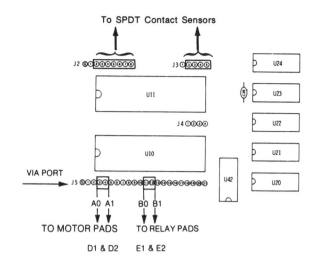

Figure 4. SUPERKIM/ET-2 contact sensor interface connections.

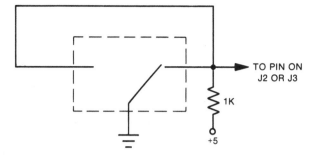

Figure 5. Contact sensor connection.

Motion Commands Based on Contact Sensor Data

The SUPERKIM can use contact sensor data to initiate a sequence of avoidance maneuvers any time the robot comes into contact with an obstacle. This behavior can be very complex, since a different avoidance maneuver routine can be triggered for every possible combination of contact sensor output. When all 8 contact sensors are mounted around the base of ET-2, the robot might use as many as 256 different avoidance maneuvers.

The principles behind this can be illustrated by considering two touch sensors on the front of ET-2, both wired to PA1 of Port A. In this case, KIM gets data byte FE if either front sensor contacts an obstacle. Table 4 gives a simple program making use of this data in a closed-loop fashion.

Execution of the program in Table 4 allows the SUPERKIM/ET-2 combination to go exploring somewhat in the manner of a billiard ball. The ET-2 moves forward in a stop-and-go fashion until one of the two forward contact sensors touch an obstacle. When this happens, the avoidance routine is called, which rotates SUPERKIM/ET-2 until the touch sensors are no longer in contact. Then the robot resumes its forward stop-and-go motion. Figure 6 shows the path of SUPERKIM/ET-2 under control of this program.

Each of the lines shown in Figure 4 go to one end of the desired (SPDT) contact sensor, as shown in Figure 5. Since the central pole of the switch is connected to ground, as the switch is opened and closed, the corresponding pin on jack J2 or J3 will be either an open circuit (corresponding to logic 1) or grounded (corresponding to logic 0). Read the data registers for Port A from memory location 1700H and the data registers for Port B from memory location 1702H.

You can interface the touch sensors by connecting one side of the SPDT switches mounted around the base of the ET-2 to signal ground and the other side to the appropriate pins of Port A and Port B. To understand how this connection works, consider the partial state diagram of the data register shown in Table 3.

If any of the pins PA1 through PA8 are connected to ground, then the corresponding state of the data line is set to zero, as shown in Table 3. The data byte stored in memory location 1700H—and read out by the KIM display—is the hexadecimal equivalent of the binary number represented by the states of the signals on PA1 through PA8, with PA1 being the least significant bit (LSB) and PA8 being the most significant bit (MSB). Thus, Port A alone can handle some $2^8=256$ on-off contact sensor states.

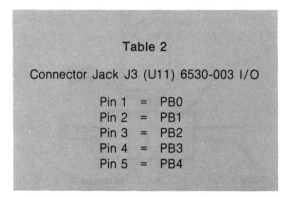

Table 2

Connector Jack J3 (U11) 6530-003 I/O

Pin 1 = PB0
Pin 2 = PB1
Pin 3 = PB2
Pin 4 = PB3
Pin 5 = PB4

Table 3

Data Byte Equivalent of Port A Sensor Signals

Address	PA1	PA2	PA3	PA4	PA5	PA6	PA7	PA8	DATA BYTE
1700H									
	1	1	1	1	1	1	1	1	FF
	0	1	1	1	1	1	1	1	FE
	1	0	1	1	1	1	1	1	FD
	1	1	0	1	1	1	1	1	FB
	1	1	1	0	1	1	1	1	F7
1 = Open	1	1	1	1	0	1	1	1	EF
0 = Grounded (closed)	1	1	1	1	1	0	1	1	DF
	1	1	1	1	1	1	0	1	BF
	1	1	1	1	1	1	1	0	7F

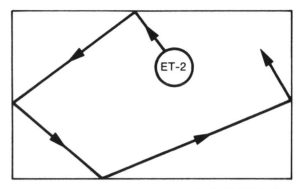

Figure 6. Path of ET-2 under control of the SUPERKIM "Billiard Ball" program.

As figure 6 shows, the path of ET-2 looks something like the trajectory of a billiard ball. By changing the program's delay constants at 0300H, 0311H and 0313H, you can change the angle of rotation of ET-2 during the avoidance manuever, as well as the duration of the start and stop motions.

Interfacing Analog Sensors to ET-2

Besides interfacing ET-2's contact (touch) sensors to the 6530 I/O parts you can also interface sensors that require analog to digital conversion (A/D). Sensors require A/D conversion when their output is a continuously variable signal or voltage as opposed to the 1 or 0 binary output of a touch sensor. Examples of such sensors useful in robotics are force/pressure transducers, temperature sensors, infrared sensors, or potentiometers used for shaft angle feedback in computerized servo control.

An LSI circuit, the ADC0817, is the primary IC in a 16 channel 8 bit A/D converter (ADC) system, which you can attach to the bus of the SUPERKIM 6502.* This ADC chip provides a relatively fast (100 microsecond) conversion time. Once the conversion has begun, the CPU can work on other tasks until the digital result is available.

The ADC0817 appears to the program as a block of memory starting at a base address, BASE, and extending through 16 locations to BASE + 15. (The actual circuit described occupies 4000 locations because of incomplete decoding which you can remedy if desired.) A conversion of a selected channel, say channel X, is started by writing to BASE + X. The 8 bit conversion result may then by read from any location in the block (eg. BASE) any time after the $100\mu s$ conversion time has elapsed. If you need multiple A/D conversions at the maximum speed, you can keep the 6502 busy with "housekeeping" during the conversion delay time. The system uses just five integrated circuits. The design, shown in Figure 7, occupies six square inches on the SUPERKIM prototype area, and draws only 60 mA of current from the 5 Volt DC power supply.

Operation of the circuit is simple because the ADC0817 performs all analog switching and A/D functions. The microprocessor R/W and ϕ1 lines, along with an inverted board select signal, are combined in two NOR gates, which 1) latch channel select bits A3-A0 and start A/D conversion during ϕ1 write cycles, and 2) enable the tri-state data bus drivers during ϕ1 read cycles.

You may want to take advantage of the SUPERKIM's interrupt circuitry to allow your program to go on to other tasks after starting the A/D conversion. The ADC0817 produces an end of conversion (EOC) signal when the most recent conversion has been completed. You can connect the EOC to a processor interrupt line (such as pin

Table 4
"Billiard Ball" Program Listing

Address	Contents	Label	Operation	Comments
0200	A9 03	Loop:	LDA #$03	;Polygon Program
0202	8D 03 13		STA $1303	;Turn Relays Off
0205	A9 00		LDA #$00	
0207	8D 02 13		STA $1302	;Both Motors On
020A	20 00 03		JSR LDELAY	;Wait
020D	A9 03		LDA #$ 03	
020F	8D 02 13		STA $1302	;Both Motors Off
0212	20 00 03		JSR LDELAY	;Wait
0215	AD 00 17		LDA $1700	;Check Contact Sensor
0218	49 FE		EOR	;Compare with FE
021A	FO 03		BEQ (Z=1)	;Avoidance if FE
021D	4C 00 20		JMP LOOP	;Keep On Going
0220	A9 01	Avoidance:	LDA #$01	
0222	8D 03 13		STA $1303	;Right Relay On
0225	A9 00		LDA #$00	
0227	8D 02 13		STA $1302	;Both Motors On
023A	20 00 03		JSR LDELAY	;Wait
023D	A9 03		LDA #$03	
023F	8D 03 13		STA $1303	;Turn Relays Off
0242	A9 03		LDA #$03	
0244	8D 02 13		STA $1302	;Both Motors Off
0247	20 00 03		JSR LDELAY	;Wait
024A	60		RTS	;Loop:
0300	A0 01	LDELAY:	LDY #$01	;Set Default Count
0302	8C 20 03	LOOP1:	STY COUNT	;Save It
0305	20 10 03		JSR SDELAY	;Call Short Delay
0308	AC 20 03		LDY COUNT	;Get Count
030B	88		DEY	;Count Down 1
030C	D0 F4		BNE LOOP1	;Continue Til Zero
030E	60		RTS	;Return
0310	A2 FF	SDELAY:	LDX #$FF	;Outer Constant
0312	A0 FF	LOOP2:	LDY #$FF	;Inner Constant
0314	88	LOOP3:	DEY	;Inner Countdown
0315	D0 FD		BNE LOOP3	;Loop Until Zero
0317	CA		DEX	;Outer Countdown
0318	D0 F8		BNE LOOP2	;Loop Until Zero
031A	60		RTS	;Return From Subroutine
0320	00		COUNT:	(Long Delay Count Hold Location)
			END	

*Both Texas Instruments and National Semiconductor produce the ADC0817.

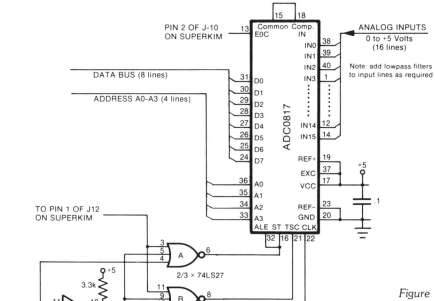

Figure 7. 16 channel analog-to-digital
converter system.

2 of connecter J-10) through one of the 74LS05 open collector inverters. These interrupts can only be cleared by starting another A/D conversion. To use the interrupt feature, you must write additional software to initialize the processor interrupt and to "handle" the interrupt when EOC occurs.

Wire-wrap construction is suitable for the circuit—and component layout is not critical. It is good practice, however, to orient the analog input area away from digital circuits. The ADC circuit has two limitations: 1) analog input voltages must be between 0 and +5 Volts, and 2) the signals being converted should not change appreciably during the 100 μs conversion period.

Table 5 depicts subroutine MCAD for multi-channel A-to-D conversion without using interrupts, along with an example of a calling routine for MCAD.

The program which calls the A-to-D conversion subroutine must initialize both the channel selection and storage-defining parameters before the JSR instruction is executed. In the program given, the channel selection information is contained in an index register for ease of use in starting a conversion.

Conclusions

The SUPERKIM controlled ET-2 robot is an excellent, moderately priced system to which the robotics experimenter can easily add more sensors and other equipment.

The contact sensors provided with the ET-2 leave something to be desired in that they do not make contact with overhanging obstacles such as tables and chairs. They do work adequately with vertical walls, and can be used to demonstrate obstacle avoidance behaviour in a suitably prepared environment. ®

References

[1]   D. F. McAllister, "SUPERKIM Meets ET-2," this book.

[2]   "Instructions for SUPERKIM," Lamar Instruments, 2107 Artesia Blvd., Redondo Beach, Calif. 90278.

[3]   "ET-2 Assembly Manual," Lour Control, 1822 Largo Crt., Schaumberger, Illinois 60194.

Table 5

A-to-D Conversion Routine

```
0200          BASE    *      $B000   ;BASE ADDRESS OF ADC0817
0200          STORE   *      $9000   ;START OF 16 BYTE STORAGE
                                             AREA
0200 9D 00 B0 MCAD    STAX   BASE    ;START CONVERSION ON
                                             CHANNEL X
0203 A0 0E            LDYIM  $0E     ;DELAY FOR CONVERSION
0205 88       DY      DEY            ;MINIMUM VALUE = $0E
0206 D0 FD            BNE    DY
0208 AD 00 80         LDA    BASE    ;GET CONVERTED DATA
020B 9D 00 90         STAX   STORE   ;STORE DATA
020E CA               DEX
020F 10 EF            BPL    MCAD    ;DO NEXT CHANNEL
0211 60               RTS            ;FINISHED
```

Example Calling Routine for MCAS

```
0212 A2 0F    MCMAIN  LDXIM  $0E     ;SELECT CONVERSION OF ALL
0214 20 00 02         JSR    MCAD    ;16 CHANNELS AND GO TO
                                     ;SUBROUTINE
0217 00               BRK            ;EXIT ** BE SURE TO INIT IRQ
                                     ;VECTOR**
```

TIMEL:
A Homebuilt Robot

PART 1

John Blankenship
Devry Institute of Technology
Atlanta, GA

The building of my robot, TIMEL, began with an assumption: hobbyists who want to contribute to the emergence of intelligent home robots should concentrate their efforts on software. It might be expensive, but we can build hardware theoretically capable of opening doors or washing windows. We can give robots the parts needed for voices, ears, and even eyes. But it is software—intelligent control programs—that will determine what these pieces can do and how useful home robots will be. Therefore, I built TIMEL for the purpose of developing intelligent software. If one had a budget like NASA or DoD, it might be best to start with a machine that had human strength and dexterity. Unfortunately, most of us do not have such resources. Most hobbyist systems are built with a low budget in mind. Therefore I tried to use inexpensive everyday items in building TIMEL's body. I wanted to save my major investment for interfaces and computing power. The hardware discussed in this article is only a vehicle for developing the software for Truly Intelligent Mechanical Electrical Life—TIMEL.

The Base and Body

A robot designed as a medium for software development does not have to be the six foot man-like structure we have come to expect. In fact, there are many reasons for going in the opposite direction. Since the system will be testing experimental software, its strength and size should prevent human intervention. The machine should not be able to do more damage than make small scratches on the furniture and walls. A hobbyist robot will also be expected to make appearances at club meetings or friends' houses and, therefore, should be easy to transport by car.

TIMEL is less than three feet tall, including a head that detaches to make moving him easier. A seat belt holds the robot snugly in the car.

As shown in Figure 1, TIMEL's base is made of plywood and measures 18 inches wide by 22 inches from front to back. The motorized wheels, shown in Figure 1, lift the base 7 inches off the floor. To keep a low center of gravity, the battery sits on an aluminum bracket that hangs from the base.

Both wheels can be independently driven in forward or reverse—and therefore provide both power and steering. With one wheel in reverse and the other moving forward, TIMEL gracefully pirouettes. (I purchased the wheels from Herbach & Rademan, 401 East Erie Ave., Philadelphia, PA 19134.)

To give the front caster a sturdy mount, the cradle for TIMEL's arm is also constructed of wood. Thin wooden skirts wrap around the base, making it seem lower than it is. These skirts will later serve as the place to mount fail-safe bumper switches.

To keep TIMEL's weight down, I built the entire body-shell out of cardboard. Its construction was fairly simple: First, I cut cardboard panels into the shapes shown in Figure 2, then taped and glued them to form TIMEL's front shell and removable rear panel. Next, I dipped six-inch squares of fiberglass cloth in resin and used them to laminate the entire body. The front shell was fiberglassed first. To make the removable rear panel fit exactly, I covered its edges with sandwich wrap (this prevents sticking), pressed it into place against the front shell, and fiberglassed it while in place.

A little automobile putty, a lot of sanding, and two coats of paint later, TIMEL started to take on some personality. I heartily recommend using cardboard and fiberglass if you are even a little handy. I had never worked with fiberglass, yet my first effort produced reasonable results.

TIMEL's head is an example of using common objects for robot construction. It is made of an inverted fishbowl, topped with a candy-dish lid. The surface is spraypainted with a light coat of metallic paint. Though normally opaque, the bowl glows nicely when lit from inside.

You can be sure that when you build a robot for software development, you will, at some time, need to modify or repair it. The robot should therefore be designed so you can easily access its internal workings. There are two ways to do this: make the robot shell removable or provide a way to pull the circuitry outside the robot. I chose a combination of the two. TIMEL's rear panel slips off, leaving very little covered. And all major circuitry is mounted on the battery cover, which you can pull out without disconnecting the ribbon cable. Figure 4 exposes TIMEL's insides.

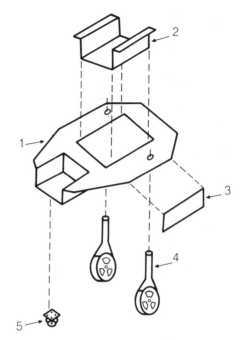

1 Plywood base
2 Aluminum battery support
3 Skirt (1 of many)
4 Motorized wheel
5 Caster

Figure 1. TIMEL's base.

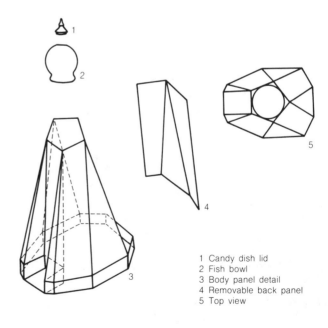

1 Candy dish lid
2 Fish bowl
3 Body panel detail
4 Removable back panel
5 Top view

Figure 2. The body-shell.

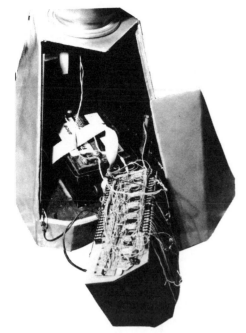

Figure 4. TIMEL's inner workings.

The Manipulator

In building TIMEL's manipulator, I decided to trade off a certain amount of strength and agility for low cost and simplicity. The manipulator has four degrees of freedom, rather than the traditional six: TIMEL's base rotates, its shoulder elevates the arm, its elbow flexes, and its wrist rotates. This simplicity allowed me to build TIMEL's arm out of humble materials which, though they fill their purpose well, may be overlooked by many hobbyists.

TIMEL's forearm, for example, needed to be lightweight and yet support three motors. The elbow consists of a section of tube from the center of a roll of carpet. Fitting snugly into the tube, a potato-chip can forms the wrist. The cardboard-to-cardboard bearing—with a little vaseline added for smoothness—may not pass an industrial inspection, but with lightweight objects, it works efficiently and never seems to wear down. Figure 5 shows TIMEL's forearm.

Figure 5. The rotating forearm/wrist assembly.

In Figure 6, you can see the arm in more detail. All the motors shown were found in junk stores. Almost any motor will do; and most can be fitted on the arm by simply cutting out the proper sized holes and epoxying the motors in place. I recommend gearhead motors—those with an appropriate gear train attached. They are usually lighter and easier to mount.

Referring to Figure 6, motor 1 rotates the elbow, motor 2 controls the shoulder (two coil springs help carry the weight of the arm) motor 3 rotates the hand, and motor 4 opens the fingers. The large gear that rotates the wrist—labeled 5 in the figure—was made from a wooden "donut" to which I glued a plastic strip of gear-like teeth (around the circumference of the annulus). Again, scavenging for parts, I found the strip of teeth with a toy airplane. (When the teeth are inserted and quickly withdrawn from the front of the plane, it spins the propeller.)

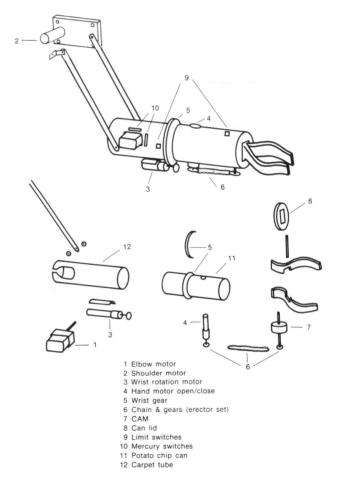

1 Elbow motor
2 Shoulder motor
3 Wrist rotation motor
4 Hand motor open/close
5 Wrist gear
6 Chain & gears (erector set)
7 CAM
8 Can lid
9 Limit switches
10 Mercury switches
11 Potato chip can
12 Carpet tube

Figure 6. The construction of TIMEL's arm.

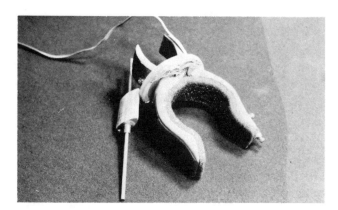

Figure 7. TIMEL's fingers.

After experimenting with many styles of grippers, I made TIMEL's hand from a short length of 2x4 pine board. I used a jig saw to cut out two S-shaped pieces and then fit them together much like a pair of pliers. As Figure 7 shows, a strip of foam rubber is glued inside the fingers to improve TIMEL's grip. A spring, not shown in that figure, holds the jaws in their normally closed position. Lying beside the gripper in Figure 7 is a wooden cam. Through a chain and sprocket drive, motor 4 turns the cam which in turn opens the robot's fingers. You may have noticed that two small switches are mounted on one of the fingers. They connect to TIMEL's onboard logic and allow for automatic override—which I'll say more about in Part II of this article.

Conclusion

A home robot does not have to bankrupt the hobbyist. TIMEL proves that by using common and inexpensive objects, you can build an adequate testbed for robot software development. This lets you put your money into computational power, rather than mechanics.

In Part II of this article, we'll take a look at TIMEL's control electronics.

Figure 8. TIMEL pours his master a stiff one.

TIMEL:

A Homebuilt Robot

PART 2

John Blankenship
Devry Institute of Technology
Atlanta, GA

In the first part of this article, I discussed the mechanical construction of TIMEL (Truly Intelligent Mechanical-Electrical Life). One of my major points in that article was that the real challenge in creating an intelligent home robot lies in the software that controls the robot's behavior. My design of TIMEL (Photo 1) was intended to provide the cheapest hardware to serve as a testbed for developing intelligent software.

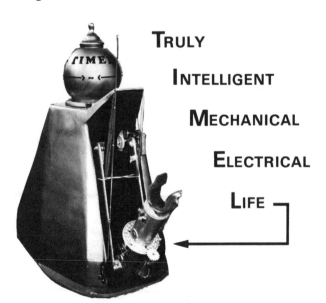

T<small>RULY</small>

I<small>NTELLIGENT</small>

M<small>ECHANICAL</small>

E<small>LECTRICAL</small>

L<small>IFE</small>

Photo 1. TIMEL: Truly Intelligent Mechanical-Electrical Life? The simple robot pictured here is one example of a system that will enable hobbyists to experiment with the software that make intelligent robots a reality.

This article will continue with the details of TIMEL'S construction—this time covering its internal electronics and its radio link to an external computer. Although I expect many of you to build TIMELs of your own, I don't expect you to build a duplicate of mine. Again, the idea of TIMEL is to use the simplest components available to you and to assemble them to suit your own interests.

Because of these expected variations (which I encourage) I'll be taking a somewhat different approach in explaining TIMEL's electronics. Instead of making this a step-by-step construction project, I'll take a more tutorial approach, designed to give you an understanding of the circuits I used so that you can adapt them to your own TIMEL.

Basic Design Requirements

I had several objectives in mind when I designed the control system for TIMEL. These were: an ability for direct manual control, control from an external computer, control from an on-board computer, communication between the two computers, and a failsafe, hard-wired "nervous system" to protect the robot.

Manual control is important for several reasons. One is convenience—there are many occasions when you want to move the robot without having to write a program or to carry it. Manual control is important for debugging the hardware, and it can give you personal satisfaction to see the robot move during the course of a long construction project. Even after the system is complete, manual control gives important clues to writing intelligent programs—if you can't do a task manually, how can you program it?

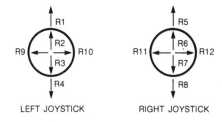

LEFT JOYSTICK   RIGHT JOYSTICK

| | HORIZONTAL POSITION | MODE | VERTICAL POSITION | |
			L. JOYSTICK	R. JOYSTICK
L. JOYSTICK	LEFT	R9: ARM	SHOULDER	ELBOW
	RIGHT	R10	COMMUNICATION MODE	
R. JOYSTICK	LEFT	R11: WHEELS	L. WHEEL	R. WHEEL
	RIGHT	R12: HAND	WRIST	FINGERS

Figure 1. The transmitter joysticks can activate twelve different switches, according to the chart shown above. The table shows which functions these switches control.

Most computers with a high-level language and enough compute power for sophisticated robot control programs are too bulky to fit inside a simple frame like TIMEL's—especially if you consider the keyboard, disks, monitors, etc. that are essential for program development. Battery power of such systems is also a problem. I decided to use an external APPLE II computer as TIMEL's controller and have it communicate with the robot on a radio link. The link can be adapted to any computer with a parallel output port.

The bandwidth of an inexpensive radio link (especially mine!) is usually too low for many of the real-time control functions the robot needs—monitoring sense switches, ultrasonic range measurement, counting wheel revolutions, etc. Using a small single board computer (SBC) on board solves this problem. The KIM I used can perform all these functions while still monitoring requests from the APPLE. The KIM can override external commands if necessary to protect the robot. This hard-wired "nervous system" provides failsafe protection—just in case software on the APPLE is faulty.

A Simple Radio Link

There are three ways of establishing a radio link from the external computer: buy a commercial unit that does just what you want, adapt some other type of unit, or build your own from scratch. In keeping with my design philosophy for TIMEL, I picked the method that required the least amount of effort and expense in hardware. Using a standard model airplane remote control (RC) unit provides the simplest manual control method and is easy to interface to the external computer. Photo 2 shows the RC transmitter (and also my KIM computer).

Rather than decode the signals from the on-board RC receiver for a direct electronic interface, I decided, again taking the simplest route, to use the servos that came with the RC unit to trip switches that control the robot. This approach leaves the RC receiver unmodified and eliminates any decoding circuitry.

Before you start to write letters cursing the inefficiency of such a system, consider the following: TIMEL is not meant to be state-of-the-art hardware, just a testbed for software. Depending on the brand of RC unit you use, the decoding may either be in the receiver, in the servo modules, or (most likely) distributed between them. To build your own decoder would require a custom design and additional hardware. Considering that the servos usually come with the RC unit, the mechanical connection

is the most economical, and it is entirely adequate for simple control of all the motors on the robot. A complete RC system costs about $200 new, and you can save a lot by checking with your local model airplane club for a used unit.*

Photo 2. A model airplane radio control unit transmits commands to TIMEL, either by manual joystick movement or external computer control.

My unit has four channels controlled by the horizontal and vertical movements of two joysticks. Since I needed to control several different functions (hand, arm, and

Editor's note: The availability of inexpensive computer-to-computer radio links may make an all electronic interface to your robot more attractive.

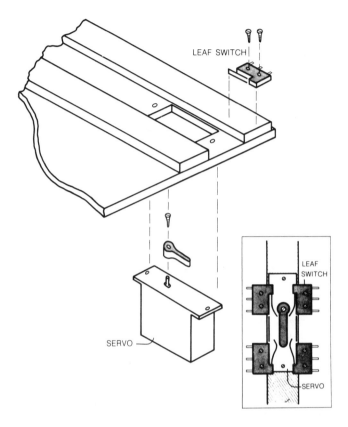

Figure 2. The mechanical action of the servos can cause small leaf switches to close. These switches are connected to the control electronics to activate TIMEL's motors or send signals to the onboard computer.

wheels), I decided to use the horizontal positions of the joysticks as a function selector and their vertical positions to control the rate and direction of movement. Figure 1 shows the mapping between the joystick positions and the receiver switches (R1-R12) that control the various functions.

Vertical movement of each joystick controls four switches. For example, if you move the right stick forward, R6 closes. If you move it further forward, R5 also closes. Thus, R5 and R6 correspond respectively to the slow forward and fast forward movement of the function(s) selected by the left/right positions of the joysticks.

As another example, consider moving the right stick left, which selects the wheel functions. Vertical movement of the left and right sticks control the left and right wheels, respectively. Moving the left stick forward moves the left wheel forward (backward moves it in reverse). Moving the

right stick controls causes the wheel to move more quickly. Moving the right stick controls the right wheel similarly. But you have to keep holding it to the left; this keeps the wheel functions selected.

Figure 2 shows how I mounted the servos to control the switch closures. In response to pulses sent by the RC transmitter and decoded by the receiver, the servo shaft rotates the lever arm left or right. The amount of movement is proportional to the joystick's movement. As the lever rotates, it trips one or more leaf switches. On the two vertical servos, I mounted four switches with the leaves bent so that one switch would activate before the other.

TIMEL's "Nervous System"

The receiver switches R1-R12 are only part of the inputs to the nervous system. There are also fifteen sense switches, S1-S15, that monitor various conditions. For the discussion that follows, refer to Figure 3, which shows where these switches are located.

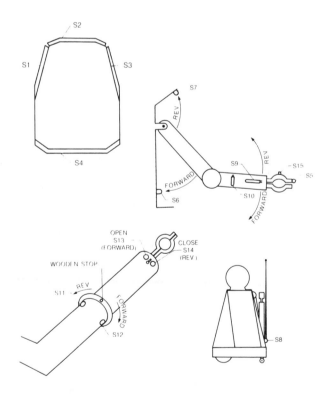

Figure 3. The diagrams above show the locations of fifteen sense switches. Each switch inhibits the motion of one of TIMEL's motors in the event of a collision.

The first four switches are made from pieces of ribbon switch. These are bonded with silicon glue to foam rubber around the base of the robot. They provide immediate shutdown of the drive motors in the event of a collision. (Eventually, ultrasonic ranging, monitored by the KIM, should prevent collisions in advance.) Photo 3 shows a close-up of a bumper switch.

To be most effective, the nervous system should have some built-in intelligence. For example, contact on S4 should void any attempt to move either wheel in reverse and all motors, including the wheels, to move forward. Similarly, contact on S1 should inhibit reverse motion of the left wheel and forward motion of the right.

To detect a collision with an overhanging obstacle, S8 is mounted at the base of the antenna (see photo 3). S5, on the end of one finger, works the same as S2; both can detect a collision while moving forward.

If the arm reaches its limits, S6 and S7 shut down forward and reverse motion of the shoulder. But using absolute limits on the elbow joint is not sufficient for safety. Instead, I mounted mercury switches S9 and S10 to keep the forearm from ever going beyond the vertical, either at the top or the bottom of its arc. To limit wrist movement, S11 and S12 are tripped by a small block glued to the rotating wrist gear. Similarly, S13 and S14 are tripped by a screw on the finger cam when the cam moves to its extreme positions.

Photo 3. Numerous sense switches serve as inputs to TIMEL's "nervous system."

One final switch, S15, is mounted on the side of the hand to stop elbow movement in the event of a collision. To make best use of S15, TIMEL's hand should be turned with S15 either up or down, depending on the direction of elbow movement. I mounted the wrist motion limiters so

that S11 and S12 will stop the hand with S15 either up or down. Should S15 close, only shoulder movement will move the arm to back off from the obstacle.

Receiver and sense switches are the inputs to TIMEL's nervous system. Of course, they are configured to suit the mechanical design of my robot. Should you choose a different design, you'll need a different switch configuration. In presenting the schematic of TIMEL's nervous system, I'll explain the design so that you can easily adapt the circuit to your own robot. I used only four types of logic gates in TIMEL's nervous system. The packages and their pin assignments are shown in Figure 4. All are standard TTL with +5V supplied by the circuit in Figure 5.

Photo 4. The servos from the RC unit, mounted next to leaf switches as shown, are the mechanical connection between TIMEL and the external world.

Consider the "meaning" of a logic signal. Although the line may be either low or high (0 or 1), either state may be the one that indicates the "active" condition of that signal. In TIMEL, for example, all the lines from the sense and receiver switches are normally high. When a switch closes, it grounds the line to indicate that the condition it monitors has occured. Thus, all these lines are "active low." This is important when it comes to reading the schematic.

The way you draw a gate can make the schematic easier to read. The output of a NAND gate in a 7400 package can be considered as an active low AND of two active high inputs (Fig. 6b). The same can be said of the other multi-input gates I use. In the schematic for TIMEL's logic,

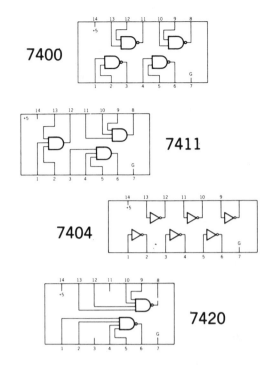

7400

7411

7404

7420

Figure 4. Logic packages used in TIMEL's control circuit. For low power consumption, I recommend using 74C—, 74L—, or 74LS— devices (in that order).

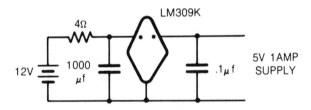

Figure 5. A simple 5V power supply that uses the onboard 12V battery as a power source.

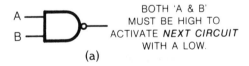

BOTH 'A & B' MUST BE HIGH TO ACTIVATE *NEXT CIRCUIT* WITH A LOW.

(a)

EITHER 'A' or 'B' CAN BE LOW TO ACTIVATE *NEXT CIRCUIT* WITH A HIGH.

(b)

Figure 6. Drawing a gate according to the function it performs makes reading the schematic easier. For a 7400 NAND gate, (a) shows it performing an AND function, and (b) shows its use to perform an OR. In Figure 7, I draw each type of gate according to how I use it.

shown in Figure 7, I give all the gates according to their intended *function*.

The schematic can be divided according to the functions performed by different sections. Transistors Q1-Q3 control the functions commanded by *vertical* movement of the left joystick. Q7-Q9 perform the same functions for the right joystick. When Q1 is turned on by a positive voltage at its input, it closes relay Y1, applying 12V to the diode chain. The four diodes drop about 2V, leaving 10V for "slow speed" output to the motors. If Q2 turns on, Y2 will bypass the diodes and pass 12V to the motors for "full speed."

You'll need to experiment with the number of the diodes to give a satisfactory "slow speed" with the motors you use in your robot, and choose their current rating according to your motors' requirements.

Q3, when active, reverses the polarity of the output voltage. Choose relays with ratings sufficient to carry the motor's drive current throughout its operating range, and then pick switching transistors hefty enough to carry the relay's drive current. For small relays, almost any NPN transistor will do. For larger ones, you may need to use Darlington transistors. Start with a 5K resistor on the transistor's base and experiment for the best results.

Transistors Q4-Q6 and their associated relays are energized by the horizontal mvement of the joystick to select one of the movement functions: arm, wheels, or hand. Even though vertical joystick motion has enabled output power, nothing will happen unless you also select a function, thereby passing the output voltage to the appropriate motor.

The next thing to consider is how to get a one (active high) on the inputs of the transistors. This is the function of the mass of gates on the left of the schematic. The inputs to this logic network come from the receiver switches, the sense switches, and the outputs of the internal KIM computer.

To use the computer interface for both monitoring and override control as I have done, you must use one whose lines can be programmed to be either inputs or outputs. Examples of such interfaces are the Programmable Interface Adaptor (PIA) and the Versatile Interface Adaptor (VIA). Using the PIA on the KIM, the control logic and can reprogram them as outputs to override the external commands.

Now let's look at the hard-wired logic functions. I will explain how two of them work—that should make the others clear, too.

Transistor Q2 controls the speed as commanded by the left joystick. It should close the relay whenever the stick is moved all the way forward or backward. These conditions correspond to closing R1 and R4, respectively. Referring to the circuit, note that R1 and R4 connect to a NAND gate, G2. G2 is drawn as an OR function to help indicate how we are using it to take the OR of the two active low signals. If either R1 or R4 is zero, the G2 will output a high (1), closing the "fast" relay Y2.

The on/off action of the left stick, controlled by transistor Q1, is more complicated. In general, Q1 should be on whenever R2 or R3 are closed (low). The AND gate G3 at the base of Q1 needs all three inputs high to activate the relay. These inputs are the basis for understanding how the "nervous system" works, so we'll look at each input in turn.

G1 takes the OR of the two receiver switch signals. If either switch closes, it outputs a one. If the other inputs to G3 are high, Y1 closes and gives power to whatever left-stick function you've selected. Note that a PIA I/O line is connected in parallel with each of the switches. Normally, these lines would be set as inputs so that the KIM could monitor the movement commands. If they were pro-grammed as outputs, setting a line to zero would cause the same results as closing the corresponding receiver switch.

The middle input line to G3 is the KIM override line. It is normally programmed to hold a one, but the KIM can disable the action of receiver switches R2 and R3 by outputting a zero.

The bottom input to G3 is the automatic override from the nervous system. The 3-input AND gate G12 is drawn to indicate that if any of its inputs goes low, it will output a low and disable Q1. The gates that produce these inputs, G9-G11, are used to monitor certain groups of sense switches, depending on which movement function you've selected. G9, for example, can only override when the left wheel function is requested (R11 closed or D0 low). In this case, it inhibits movement if G8 outputs a 1.

The inputs to G8 are provided by G6 and G7, which are active only during forward or reverse motion, respectively. If you move the left stick forward, R2 will be low and R3 will be high, placing a one on one input to G6 and disabling G7 with a low. Closing R3 by backward stick movement reverses the situation. With R3 high *any* of the sense switches S1, S2, S5, or S8 inhibit movement when closed. G4 takes the (active high) OR of these four (active low) signals. Refer to Figure 3 to see why these switches override forward motion of the left wheel.

Similarly, S3 or S4 override reverse motion of the left wheel through G5 and G7. This type of logic is applied through the rest of the circuit. After studying a few more examples, you should be able to thoroughly understand its operation and, more importantly, be able to design a nervous system for your own robot.

Remote Computer Control of TIMEL

For direct manual control of the robot, we can just move the joysticks on the RC transmitter. But to permit a stationary external computer to control it, we need a way to "move" the joysticks electronically.

Figure 8 shows a circuit that can accomplish just that. The 4016 ICs are CMOS analog switches, controlled from an output port on my APPLE (any parallel port will do). I wired the switches in parallel with the joystick potentiometers in the transmitter. (The transmitter circuitry is not shown.) A pair of switches control one direction (horizontal, vertical) of one joystick. When one switch of a pair is turned on, it connects a 6.8K resistor and its associated trim pot in parallel with one side of the joystick pot. This reduces the resistance of that side and, with the joystick centered, makes the transmitter think you've moved the stick in the corresponding direction.

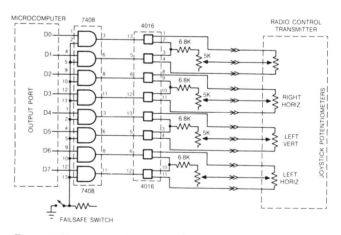

Figure 8. The 4016 analog switches allow any external computer with a parallel port to simulate movement of the joysticks. The resistance values will vary for different brands of RC units. The trim pots permit fine adjustment of how much the computer "moves" the joysticks.

With this circuit, the APPLE can only control TIMEL with one speed. Which speed that is depends on how much the resistance changes when the analog switch conducts. (Adjust the trim pot to get the speed you want— the values shown worked for my RC rig.) You can add

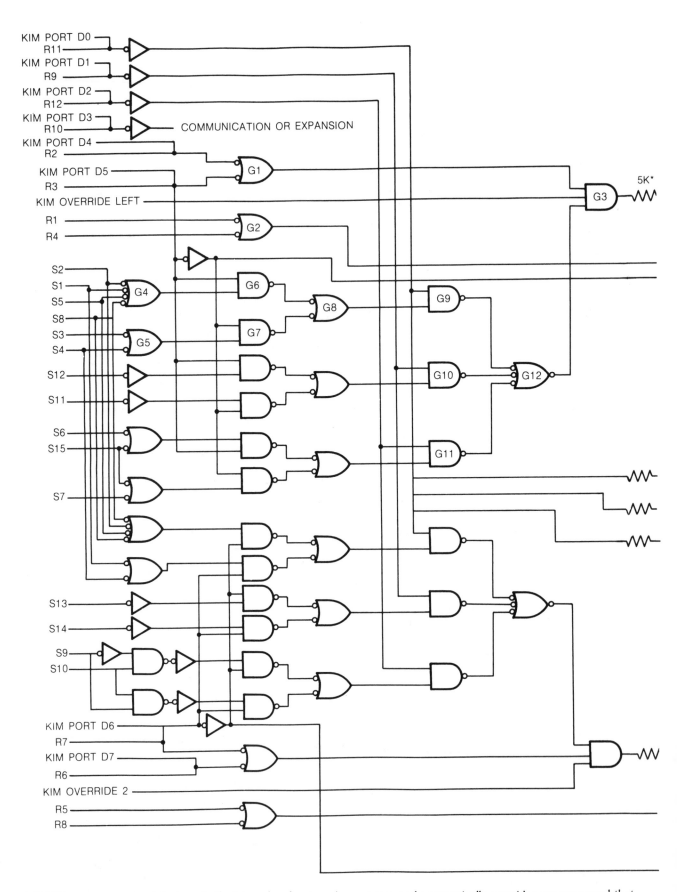

Figure 7. TIMEL's nervous system interfaces to the internal and external computers and automatically overrides any command that presents danger to TIMEL or its environment.

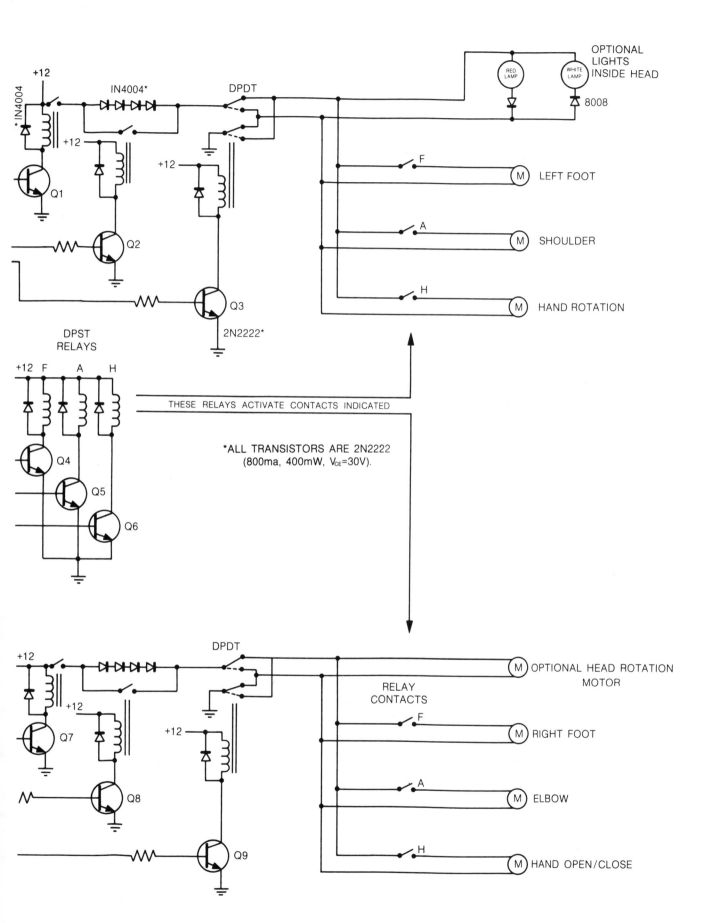

additional switches to get two-speed control if you wish. One warning: be sure not to turn on *both* switches of a pair at once—you'll short out the transmitter pot and possibly do some damage.

Using the System

Through the RC transmitter, either manual joystick motion or external computer commands can control each of TIMEL's motors. The internal computer can monitor or override the movement commands and perform whatever other sensory or control functions you choose to add. In the communications mode, the KIM can read the vertical stick movements (whether manually or electronically originated), but no motors will be activated. You could use this mode to request the KIM to perform some pre-programmed maneuver.

Of course, the sense switches will automatically override any erroneous command, regardless of its source. In such a case, it's up to you or your software to analyze the situation and take appropriate corrective action.

In the future, I'll be writing about TIMEL's control programs in the KIM and the APPLE. However, as I've said before, I view software development as an evolutionary task that I'll never finish. New ideas come much faster than I can add them to the programs. But then, that's what I think personal robotics is all about. ▣

A Homebuilt Computer Controlled Lathe

Jan Rowland
Visser-Rowland Associates, Inc.
2033 Johanna Drive, Building B
Houston, TX

Long before robots were accepted as valuable workers by industry, the use of automated machine tools had become an established practice. From the earliest *numerically controlled* (NC) machines, programmed by punched paper tape, to the latest software-programmed CNC (*computerized* NC) equipment, these tools could follow a complex cutting path, repeating it accurately and rapidly almost around the clock. In many manufacturing facilities, the use of CNC tools together with robots permit totally unmanned machining operations, where the robots load the CNC machines with raw stock and remove the finished pieces.

At first thought, it might seem that the Pipe Organ industry would be an unusual place to see automated manufacturing, since automated installations usually require a fairly high production volume to justify the large initial investment. In fact, though, since many of the parts used in organs are identical and require repetitive manufacturing steps, most of which are done by hand, opportunities for automation abound at many levels. Unfortunately, the "artistic bent" of most organ-builders is such that technology more advanced than an electric drill is ignored, for reasons ranging from simple ignorance to artistic arrogance. This appearance of *computer aided manufacturing* (CAM) in an organ-builder's shop is presently unique, although I hope this will not remain the

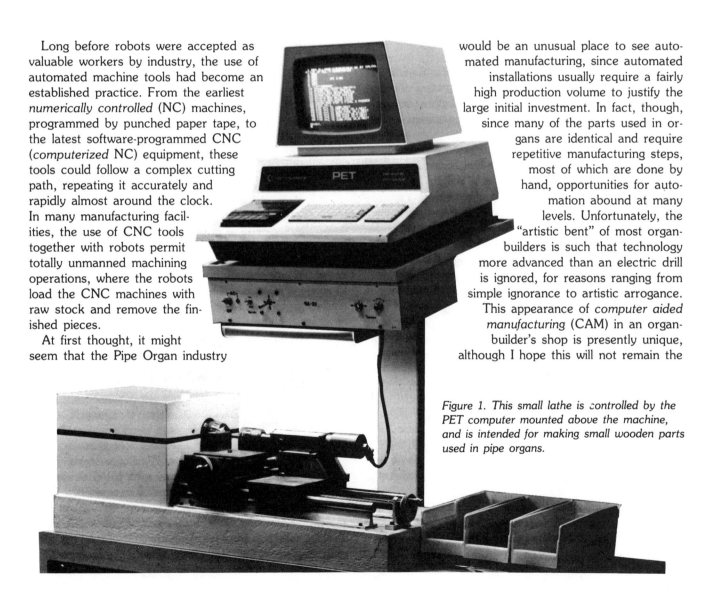

Figure 1. This small lathe is controlled by the PET computer mounted above the machine, and is intended for making small wooden parts used in pipe organs.

Reprinted from Robotics Age, Volume 3, Number 3, May/June 1981

case for long.

One of the many parts taken for granted, but one which must be made with reasonable precision and visual attractiveness, is the "drawknob," the handle which one pulls or pushes to turn an organ stop on or off. Low-end organs have molded plastic knobs, but the trend now is for a return to the finery of instruments of the Baroque period. These older instruments, and now again half or more of the contemporary organs will have hand-turned knobs of exotic woods (ebony, rosewood, etc.). There may be any number from two to nearly a hundred such knobs on a given instrument, some quite ornate, and at prices of up to $20.00 each for hand work, the cost for these parts alone can become significant.

After hand making a few hundred for our own work, my inherent laziness inspired the construction of a small lathe fitted with stepper motors to control the X and Y lead screws, controlled by a popular hobby-computer, in this case, a Commodore "PET." Few shops in this industry can afford the $30,000 and up for CNC chuckers or lathes, particularly mine. Besides, these machines are for metal working, usually with capacities for fairly large pieces, and making organ knobs on one would be something like using nuclear weapons to kill flies. With the availability of relatively cheap microcomputers of admirable flexibility and power, other items such as stepper motors on the surplus market, and Thomson linear ball-bearings, the construction of something as ambitious as a "home-brew" CNC lathe or, for that matter, other CNC tools, is not really as ridiculous as it may seem.

There are three basic components in this machine: the lathe mechanism, the electronics other than the computer, and the computer itself. One might rightly insist that software be included in the list, but as I have so little ability with programming other than in simple BASIC, I have managed to do all I ever intended (and all the machine itself is capable of!) without the use of any special software other than self-explanatory BASIC programs which control the machine directly in real-time, using the POKE command for communication to "the outside world" from the computer. Therefore, I leave "software" off the list, and leave sophisticated machine code programming to the more sophisticated reader.

The Lathe Mechanism

As can be seen from the photographs, the linear track, or "ways," of the lathe are standard case-hardened shafting (by Thomson Industries), mounted on a base of ordinary steel channel. Leveling the rough surface of this base or chassis would ideally be done by surface-grinding, but this would have cost at least a hundred dollars alone. Trick #1: Drill and tap the base for one of the shaft-support's mounting screws, smear a liberal coat of fast-hardening epoxy on the bottom of the support rail, then immediately screw it down to the base, making the screws just snug. Wipe away the excess epoxy which may squeeze out, and after it hardens, tighten the screws. What has happened is that the epoxy has filled in all the voids between the bottom of the shaft-support rail and the rough channel, making a very firm and level interface. Now, assuming you have already made the lathe saddle (or carriage or whatever, depending upon your specific application), it can be used as a "gauge" to locate the second shaft precisely on the base. Again use the epoxy trick, this time *before* any drilling. It might be necessary to level the second shaft to match the plane of the first by using shims. Run the carriage back and forth to insure that the two shafts are parallel, and let the epoxy harden. Now you can drill, tap and bolt the second shaft support down. While this procedure might make a professional machinist nauseous, it works. The adhesive qualities of the epoxy are unimportant since it is being used only for "grout" or filler.

The *carriage*, which supports the lateral axis of the cutting tool, and the *tail-stock*, which holds the free end of the workpiece, are both made with Thomson linear bearings, the basic idea being quite clear in the photographs. Since the precision is built into these ready-to-use pillow-blocks and shafts, it is possible to all but avoid the need for professional machine-shop operations. A drill

Figure 2. Closeup of the lathe mechanism. The tailstock center is operated pneumatically, maintaining continuous force against the workpiece, in this case by means of a cup-center.

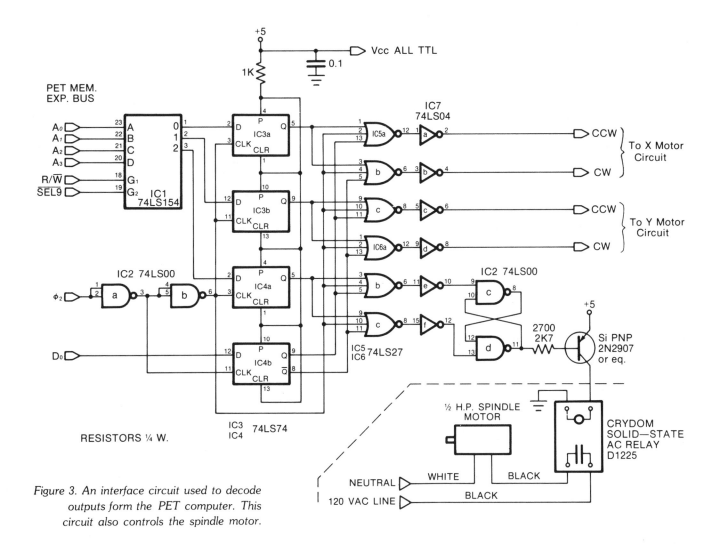

Figure 3. An interface circuit used to decode outputs form the PET computer. This circuit also controls the spindle motor.

press should be all one requires to do a passable job for all parts.

The lead screws, believe it or not, are replacement parts from a Whirlpool trash compactor. These screws in that particular make are ⅝″ diameter ACME threads, and the nuts which transfer the force to the mechanism in that machine are "split," consisting of two identical mating halves. This makes it possible to squeeze the two halves together by means of a spring in a suitable holder so that there is no play, ideal for the lathe application, or anywhere else where reasonably accurate position is needed, with fair forces involved. I merely sawed off appropriate lengths of the threaded portion of the screw shafts and performed very minor machining in an ordinary engine lathe to adapt the ends for my use. Alignment of a pair of bearings for a shaft such as these screws can be difficult, particularly when this shaft must be axial with a motor. But the stepper motors themselves have fine ball bearings, and I used the motor shaft as one support for the respective lead screw, leaving only one bearing to worry about. I used simple Oilite type bushings, as the speed is relatively slow, and

these can be reamed after installation for free operation—this is not possible with precision ball bearings.

The lathe spindle is an Atlas 12″ lathe spindle. This choice was made for several reasons. First, the cost of the part is far less than a custom-made spindle. Second, standard lathe chucks and collets can be fitted directly to it. This spindle was mounted in standard Boston Gear flange bearing blocks, the self-aligning type with spherical outer races. The spindle drive motor is a very ordinary and inexpensive 3450 RPM tool motor, Nr. 48 frame. It is connected to the spindle by sheaves (V-pulleys) and a V-belt. Ordinary cast zinc sheaves are not made with adequate precision for this speed, so I made both from chunks of scrap in the engine lathe. This was particularly necessary on the spindle, since the I.D. was 1½″, and no small sheaves have such a large bore, and cannot be reamed due to the shape of the castings. I should stress that at speeds over 2000 RPM, the precision of the V-belt drive sheaves must be very good, or considerable vibration will result. In fact, I was able to later reduce the vibration I encountered (in spite of all steps to prevent it initially) by trimming the motor sheave *after* its installation on the motor shaft. Placing the motor in the engine lathe between centers (with the lathe turned off) made the V-groove in the sheave perfectly concentric with the motor axis, running the motor itself provided the turning power for the cutting.

I can now see that another approach would be preferable, though it would require some expensive professional machining. If I had to do it over again, I would modify a standard motor by replacing its shaft and bearings with the lathe spindle itself, and suitably larger bearings. This would mean considerable work on the motor end plates, if not replacement with specially made ones. But the elimination of the V-belt drive and separate bearings would have made it worth it. This way, the motor itself would become the lathe's headstock. In any case, the machine is built and intended to run at only one spindle speed: 3450 RPM, ideal for woodworking and some plastics turning.

Controlling the Lathe

The lead screw drive system used on the lathe has more in common with general robotic methods. There are basically two ways to move a mechanism to a specific position upon command from a computer or other controller:, a servo system with feedback (closed loop) and stepper motors with no feedback (open loop). In the first system, the computer compares the desired position with that measured by the feedback system, typically potentiometers or shaft encoders of some sort. The direction and speed of the servo motor, whether electrical, pneumatic, or hydraulic, is determined by the computer, based upon this comparison, and information is sent to the servo motor to control it to perform the desired movement. For machine-tool precision, this system is usually far more expensive and complex than the second method, the stepper motor system. A stepper motor is a specialized DC motor which rotates in discrete angular increments (steps) upon command from its associated drive circuitry, usually manufactured and supplied by the same manufacturer as the motor itself.

All stepper motors are poly-phase devices. That is, the several windings in the motor are controlled by separate switches which sequence the current in these windings in some specific order of phase-related pulses. If certain windings are switched on and others off, the motor will step. If the sequence is repeated, the motor will step again in the same direction, and if the order of the pulse sequence is reversed, the motor will step in the reverse direction. If the "rules" are obeyed—the proper rate and sequence observed—the motor shaft will always rotate a specified amount. However, if the stepper motor is stalled by some outside force, such as an obstruction in the mechanism connected to it, it will not "catch up" after removal of the problem, since the pulses sent to it during the stalled condition are lost forever. Here, the servo system with feedback has an advantage, since the feedback element will continuously inform the computer or other controller what movement is actually occurring (or *not* occurring). Thus, once any problem is removed, the servo system will immediately attempt to catch up to where it should have been had there been no obstruction. In the present application, however, since the movement of the lathe carriage is relatively small, and since obstructions and such problems can be readily controlled, nothing more sophisticated than the familiar stepper motor drives are necessary.

In this application, the stepper motors are Superior Electric type MO 92 or HS 50, the latter being the older version of the same type, which I was able to locate on the surplus market for less than half the price of the new version. These have considerable average torque and 200 steps per revolution. Thus, with a lead screw pitch of 10 to the inch, the travel per pulse is 0.0005", a relatively fine resolution for a woodworking machine! Superior Electric manufactures an extensive line of solid state translators and preset indexers for their several types and sizes of stepper motors. The simplest of these is capable of up to 1000 steps per second, which converts to about 300 RPM. This STM-101 circuit board with connector sells for $100, and seems to be based upon early '60's technology—the flip-flops are constructed with discrete transistors, diodes, and capacitors! Figure 4 is a schematic of a stepper driver circuit which will perform easily as well, and can be built for much less. A great deal of energy must necessarily be wasted as heat in the two ballast resistors in series with the return leads form the motor windings due to complex electrical inductive actions at various speeds. The full details of this behavior are beyond the scope of this discussion and can be studied in Superior Electric literature and engineering guides.

The drives for both the X and Y axes are of course identical, and each has two inputs, one for forward and one for reverse (CCW and CW or "UP" and "DOWN" count). In order for the computer to control the driver circuits, an interface must be constructed which will decode the outputs from the computer and route them to the appropriate axis and direction. This interface for the Commodore PET computer is shown in Figure 3. It is interesting to note that somewhat simpler interface circuitry is possible for use with either machine code or BASIC "POKE" commands, but not both. For some reason I do not quite understand, the machine code "STA" command, which puts a byte into a particular address, has different timing than whatever MPU operation does this job whenever the "POKE" command is used in the high-level language, BASIC. But the interface shown will accept either type of manipulation, as it latches the address and data at different points on the clock cycle. This detail is undoubtedly different for other makes of computers, and the logical sense of the address select lines (if any) may be positive instead of negative as it is with the PET. But this basic circuit should work with other popular types of personal microcomputers with little or no modification. Note that the decoding of the addresses is not really

Figure 4. A suggested stepper motor translator/driver circuit. This circuit will drive small and medium sized stepper motors to rates up to 1000 steps per second. The driver transistors should be mounted on insulated heat sinks of at least 8 sq. in. area per device. The 100 watt ballast resistors should be mounted on a large heatsink with strong forced-air cooling. Regulation on the 24 volt supply is not critical.

complete, and a number of individual addresses will be decoded identically, although all will lie within the $9000-9FFF range. I use decimal 36864 for the X axis, 36865 for the Y axis, and 36866 for the spindle motor switch. Any byte in which the LSB is "1" (an odd number) will cause a forward pulse, and any byte in which the LSB is "0" will cause a reverse pulse. For example, if you enter the command "POKE 36864,1" the X axis motor will move one step forward. In a BASIC program, the following line

100 FOR I=1 to 1000:POKE 36864,1:NEXT

will cause the X axis to move exactly ½". It is best to assign the values to legitimate variable names, e.g., "LET XA=36864," as this will run faster and result in less typing in a program.

Backing up a bit, the spindle motor is switched on and off by means of a commercial solid state AC relay, this one being a CRYDOM (International Rectifier) D1225, able to switch up to 25 amps of 120 V AC. The controlling input to this type of relay is any DC voltage from 3 to 32 V DC, so a normal TTL signal will control the motor. A short delay

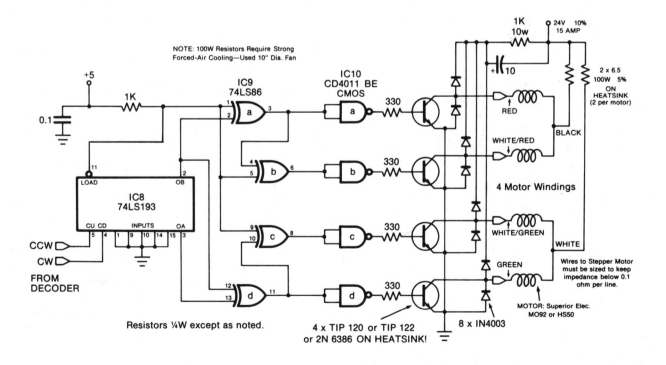

Some Sources for Hard-to-Find Mechanical Parts

It is amazing how quickly a tinkerer's imagination can dream up new and clever ideas once sources for weird and hard-to-find parts become available, particularly when the catalogs from the suppliers contain good drawings of the parts offered and suggestions for their use. There are many manufacturers of precision gears, sprockets, dials, bearings, pawls, ratchets, shafting, motors, and so on which are potentially useful in the type of mechanical movements found in robotic devices. But there are relatively few single sources for many of these parts. You'd be surprised how few professional designers are aware of some of the best sources.

Often, custom parts that are made at great cost could have been found "off-the-shelf" in a precision mechanical components catalog. The availability of some standard parts should inspire slight changes in designs on paper, allowing the use of standard parts instead of custom items.

The Winfred M. Berg, Inc. firm (449 Ocean Ave., East Rockaway, L.I., NY 11518) is one of the best sources for small precision components and assemblies. They offer by return mail a free catalog, that contains not only detailed drawings of the parts offered in thousands of sizes and types, but a number of design ideas as well. A competitor, the PIC Design division of the Wells Benrus Co., (6842 Van Nuys Blvd., Van Nuys, CA 91405, phone 213/782-6702, with additional offices in the Benrus Center, Box 335, Ridgefield, CT 06877, phone 203/431-1500) also has many small precision parts. Both companies offer parts in metric measure.

Stock Drive Products (55 South Denton Ave., New Hyde Park, NY 11040) offers more small mechanical components, but leans toward timing-belt parts. They also have a line of small DC motors with and without gearboxes.

It is fair to point out that the pricing of the precision parts offered by the above firms may shock the neophyte, but you must realize that the cost is in the precision—few of these parts are "stamped out." Most are machined on very expensive semi-automatic equipment that must be amortized. For precision movements, be prepared to pay a price!

Surplus sales firms offer some items at comfortable prices. But the disadvantage in dealing with a surplus firm is that they offer only what they have on hand at any given time, and often it is impossible to get replacement parts or any quantity. You can find some excellent companies handling surplus and new items in most major cities. Some good examples are C&H Sales Co. (2176 E. Colorado Blvd., Pasadena, CA 91107), AST/Servo Systems, Inc. (930 Broadway, Newark, NJ 07104), and H&R, Inc. (401 E. Erie Ave., Philadelphia, PA 19134). These firms offer both new items in unopened packages (from bankruptcies, etc.), and used items, ranging in condition from nearly worthless to virtually new. Items may be sold on an "as is" basis or with some sort of guarantee; and this is usually indicated in their catalogs. They offer such things as stepper motors, DC servo motors, encoders, gauges, tools, instruments, and many other types of miscellaneous items. They have nice catalogs with photographs of the items offered.

Heavier gears and bearings are available from Boston Gear Dvision of Rockwell International (14 Hayward St., Quincy, MA 02171). Boston also supplies standard sprocket chain and gears in a huge range of sizes. They publish an extensive catalog, again with drawings and explanations, engineering charts, etc.

W.W. Grainger, Inc. (5959 Howard St., Chicago, IL 60648) has distributors in most cities throughout the United States. They offer many different types of industrial motors in the fractional and integral horsepower sizes, and all sorts of industrial equipment—even some domestic items. For the robotics enthusiast, they also offer some small gear motors, which are very reasonably priced, and a few fractional horsepower DC drive motors. The latter might be useful for wheel-drives on larger robots since they can be controlled in much the same way as slightly more sophisticated DC servo motors. The best thing about Grainger, aside from their large line of products, is that their prices range from very low to reasonable, and their service and availability of offered items is good to excellent. "Check Grainger's first" is a good motto for any shop, whether for hobby or heavy industry!

There are a number of makers of stepper motors and controls. Some specialize in very small types—of the kind used in computer printers, others in military versions. Superior Electric Co. (383 Middle St., Bristol, CT 06010) is one of the best known firms making stepper motors, and the line they build appears in many types of commercial robots, machine tools, and office machines. Superior's line of stepper motors range from roughly 25 oz-in of torque up to one or two horsepower. They also manufacture solid state translators and indexers for their motors. Their catalogs contain torque-vs.-rate charts for all types with all controllers which might be compatible, as well as some engineering data. They also publish a small handbook about stepper motors and the AC Slo-Syn motors they also make. This booklet tells how to match or select a motor to a given load and whether the drive system uses belt, pulley, or lead screw. Formulas are given for each case, and step-by-step examples are included.

While it is true that large population centers are the best areas in which to find the type of items mentioned above, it may surprise you how much can be found in isolated places. The Yellow Pages is one of the best "catalogs" a designer can possess, and should be one of the first places to look—in conjunction with the telephone, of course—before giving up.

For those interested in the source of the spindle used in the lathe, the best and most direct source is, believe it or not, Sears. The spindle is Sears part number 10-31T for the model 101.28980 lathe. You can order it as a replacement part. It is actually manufactured by Clausing, a large and well-known machine tool manufacturer. Atlas is a subsidiary of that firm. Sears offers the Atlas lathe with the Craftsman brand pasted over the Atlas name. The spindle nose is internally a standard No. 3 Morse taper, and the outer diameter is 1.5-8 threads. Therefore, you can fit it with standard chucks and collet devices of many brands, available from any machine tool supplier.

Jan Rowland

loop in the program must be entered after the motor is turned on via "POKE MA,1" (MA is the Motor Address of 36866 decimal) to allow the spindle to come up to speed.

Describing the Cutting Path

Since memory is limited, the storing of points along the shape of the desired part to be turned is not a good idea for the drawknob turning application, since there are often many curves, difficult to express in terms of a series of Cartesian coordinates. Instead, I write the programs as a series of one or more mathematical functions such as expressions for circles ($R*R=X*X+Y*Y$, for example), ellipses, etc. Straight line paths are, of course, easiest. The individual Y values along the X axis are computed separately, and the new step count is the algebraic difference between the old Y and the newly computed Y value. This value is used in a FOR-NEXT loop, POKEing a 1 or 0 as many times as appropriate. But this calculation requires quite a lot of time in BASIC, and if the Y value for every single step in X were calculated, a typical organ drawknob could take as much as an hour to make! By taking 10 steps in X between Y movements, the cutting rate is speeded up tenfold, and the reduced resolution is

no problem due to the nature of the parts being made—they are usually sanded right in the lathe and later polished with compound. Clearly, clever machine language coding would speed things up considerably, but stepper motors are cantankerous and must be *accelerated* to step rates over 400 steps per second (this depends upon many mechanical variables, so one must usually make actual performance tests to determine acceleration and deceleration requirements). If any faster computer output is to be contemplated, special self-decrementing delay loops to insure correct acceleration must be provided. This can become quite involved, and I feel very lucky that the CBM BASIC running in my machine is just the right speed to permit direct, acceleration-independent running of the stepper motors without loss of position. In other words, I can write lines such as this

```
100 FOR I=1 TO 1000:POKE XA,1:NEXT
200 FOR I=1 TO 1000:POKE XA,0:NEXT
```

and the carriage will move ½" to the right, instantly reverse, move ½" to the left and stop, with a dial-indicator showing a very precise return to the starting point. In a robotic mechanism using stepper motor drives with no form of feedback, careful consideration must be given to this acceleration problem. In some computer-controlled stepper motor devices with software provided to run on popular computers, the acceleration details are embedded in the software, and are "transparent" to the user.

In my original design, I included a simple joy-stick (operating only contacts, not pots) to make it easier to set the carriage at a desired starting location during initial set-up. This control invokes local oscillators which ramp up to speed and run much faster than the computer can output its pulses. I have found that this device is not as useful as I thought it might be, and I have deliberately omitted the circuit details from the schematics. Simple is better, and one can easily enter direct-mode commands on the CRT and use the screen cursor control keys to affect direct movement of the carriage. Once set up for a run, no further human intervention is necessary in the operation of the lathe other than replacing a finished part with a new blank, and pressing the RETURN key, since the tool bit always returns to the starting point after finishing work— that is, if the program is written correctly!

It now remains for me to build a true robot to stand before this lathe and exchange the finished parts with prepared blanks, as is done in heavy industry. Speaking of pipe-dreams...

THE INDUSTRIAL ROBOT AS TRANSFER DEVICE

Vern Mangold
Kohol Systems
P.O. Box 1185
Dayton, OH

Since nearly every U.S. industry demands more productivity, industrial robots are proliferating. However, as robots make their way into our factories, we find that most of them are only used in *tool handling* applications—spot welding or spray painting, for example. Yet there exists another class of applications, one that has a tremendous untapped potential. This class uses *work piece handling* robots, which transfer products from one work station to another. Although tool handling tasks, such as spot welding, are readily performed by robots, I believe we should also consider applications that involve the transfer of products within the plant.

The concept of transferring a product from one work station to another is relatively new. Centuries ago, products were manufactured in roughly the same location. A single craftsman or mechanic simply changed tools when he started a new manufacturing task. The modern assembly line changed that arrangement. With the assembly line, factories had to transport products to dedicated areas, each of which was assigned to one specific subtask.

In many cases, robots can dramatically reduce the need to invest in parts transfer equipment designed for one particular part or process. Fixed automation equipment can be supplanted by a combination of programmable robots and *generalized* parts transport equipment—conveyors or pallets, for instance. Since you can often produce different product styles by simply reprogramming the same robot, the same manufacturing line can produce different product batches. This brings the benefits of automation to factories whose production volumes would not justify customized "hard" automation.

How is Part Transfer Done Today?

Though industry uses many types of transfer devices, we can group these into two main categories: *bulk transfer devices*, which transfer randomly placed parts, and *orientation transfer devices*, which hold the workpiece in position. Of the two, industry more frequently uses bulk transfer devices. Conveyor belts, sheet metal chutes or guides, and simple part slide mechanisms are all devices of this kind. The familiar tote bin—often assuming different names, such as gond, wire basket, or part box—is the most basic form of bulk transfer device.

If you use machines that can only be fed with parts precisely positioned, then you need an orientation transfer device. The actual device you choose typically depends on how accurately you need to locate parts.

The overhead conveyor is a simple type of orientation transfer device which you can configure in one of three ways: indexing from position to position, continuously moving, or *powered and free*. In a powered and free conveyor system—the most sophisticated of the three types—hanging hooks or racks move continuously around a closed loop. When a work station needs a hook at a given point, it can divert the hook from the main conveyor system and rigidly lock it into position. After the hook is loaded with a product, it returns to the main loop, from which it can move to a different work station. In this type of system, as well as in other overhead chain conveyor systems, gravity orients the part.

Another type of orientation transfer device is the *pallet conveyor*. This device is simply a normal conveyor equipped with pallets for individual parts. Parts are placed into the fixture of a part pallet and then transferred to another work station, either by continuous motion or by indexing in controlled steps. This type of transfer device can be extremely accurate. By using shot pins and other mechanical devices, a robot can locate a palletized part within 5 mils (.005").

A *lift and carry device*—which uses a mechanical linkage system to index parts from one work station to the next—is similar to the pallet conveyor. Like the pallet conveyor, it has high accuracy and good part repeatability.

Other types of orientation transfer devices include the *bulk hopper feeder with orienter* and *iron man* type devices. The bulk hopper feeder lets you automatically feed loose, randomly oriented parts from a bowl-like vessel by elevating individual parts to a drop-off position. Using the right fixture at the drop-off point, you can accurately deliver the part to a conveying device. An iron man operates by closing around a part and orienting it by the pressure of its grasping surfaces.

The question naturally arises: How much do these devices cost? Table 1 compares the costs of some of the transfer devices you might use to move a part about twenty feet, with accurate positioning in some cases. (Note that the prices given represent approximate data, gathered from industrial suppliers in the midwest region of the U.S. Prices, of course, will vary with geographical area and industrial activity.)

When evaluating the cost effectiveness of a transfer device, remember that you need a significant amount of manual labor when you use the most "economical" methods listed in Table 1. Consider the typical use of a tote bin, for example. After a work station is finished with a part, a worker manually loads the part into a bin for transport. When the tote bin arrives at its destination—usually by fork lift truck—another human worker unloads it. Transport by tote bin is labor intensive.

With the rising cost of labor, the relatively inexpensive transfer devices—such as tote bins, baskets, parts pallets,

and so on—become less attractive. In the U.S. automotive industry, for example, the average first shift worker costs around $30,000 a year (if we include benefits and overhead). Clearly, we need ways to transfer parts with as little human intervention as possible.

Why a Robot?

Strangely enough, there has been no stampede of managers eager to use robots as transfer devices. Yet robotic technology has been proven in this country since 1961. Industrial robots have given millions of hours of productive service with availability (uptime) records of better than 98%. Industrial robots are physically able to transfer a wide variety of parts. Computerized servo-controlled robots can produce a broad range of complex motions in space and, with few exceptions, can be easily equipped with pneumatic clamps for handling parts.

Before we consider the advantages of using robots to transfer parts, we should examine the manufacturing process in more detail. In general, a finished product is the result of a number of sequential operations: fabricating, machining, assembling, testing, coding, packaging, and shipping, for example. But regardless of the number and type of operations a product requires, there are really only two basic ways to configure station-to-station parts flow: for *serial* or for *parallel* manufacturing operations.

The goal, when you design a manufacturing process, is to combine operations in a way that optimizes the

TABLE 1

Item 1	Tote bin—$125 to $250 each
Item 2	Gravity roller conveyor—$2,200 to $4,000
Item 3	Power roller conveyor—$6,500 to $9,000
Item 4	Gravity roller conveyor with fixture and escapement—$7,200 to $11,000
Item 5	Power roller conveyor with fixture and escapement—$14,000 to $35,000
Item 6	Part pallet—$100 to $175
Item 7	Part pallet with donnage—(ten layers) $400 to $550 each
Item 8	Indexing chain conveyor—$12,500 to $14,000
Item 9	Continuous chain conveyor—$9,000 to $11,000
Item 10	Powered and free accumulating conveyor—12,000 to $16,000
Item 11	Indexing pallet conveyor—$25,000 and up
Item 12	Iron man carry system—$20,000 and up
Item 13	Lift and carry device—$30,000 and up
	Note: all prices include internal installation cost.

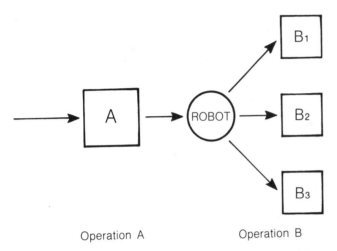

Operation A   Operation B

Figure 1. Serial and parallel processes are combined to achieve maximum throughput. The combination depends on the cycle times of each operation. In this case, operation B has three times the cycle time of operation A.

throughput of the process. To do this, you often have to process a product in parallel through an operation that has a relatively long cycle time—using several identical machines to perform the same function. For operations with relatively short cycle times, you may be able to use a single machine and still balance the process flow. This is serial operation. The processing mode you choose depends entirely on the individual cycle times of the operations involved.

The industrial robot can be effective in both serial and parallel operations. In a serial operation, momentary imbalances in the cycle times of two or more operations will interrupt smooth product flow. You can use the flexibility of an industrial robot to compensate for this by programming the robot to offload the product into a buffer storage area. In this way, even when unequal cycle times and the downtimes of capital equipment interrupt the flow of product through a process, manufacturing operations do not have to come to a standstill.

In a parallel operation, you can program an industrial robot to service on demand two or more identical devices. You can maintain your throughput on operations with longer cycle times by letting the robot load processing elements as they become available.

For the purpose of discussion, consider a hypothetical part transfer task. Suppose the part must be transferred from a prior operation to a process device called A. After process operation A has been completed, the part must be

transferred to operation B. Assume that operation B has a cycle time that is three times longer than operation A. To maintain a constant throughput, three identical work stations, each performing operation B, will be clustered with the work station for operation A. After completing operation A, we transfer the part to operation B1, B2, or B3. Figure 1 diagrams this process. In this way, we can process the parts through a work area with maximum efficiency and balanced flow.

In both the serial and the parallel cases, industrial robots have an advantage over dedicated transfer devices. Industrial robots are not limited to one fixed mode of operation. You can program them to perform both serial and parallel operations in one work area. Robots can place parts much more accurately than simple conveying mechanisms: there are commercially available industrial robots whose repeatability is within 4 mils. By judiciously selecting a robot and effectively designing its tooling, you can get performance accurate enough to do nearly any kind of parts transfer.

Industrial robots are extremely reliable. No robot could survive in an operation that has a greater uptime than the robot itself. Manufacturing managers cannot tolerate any automation device that becomes, through its own downtime, the limiting factor in the operation. For this reason, quality control in manufacturing the robots themselves has become perhaps the most stringent of any industry. This strict control also helps reduce operating costs, since it lessens the robot's need for maintenance.

The majority of industrial robots in the U.S. can boast of an operating cost of less than five dollars an hour. This figure includes the base price of the equipment and is calculated over a two-shift, twelve month year. Since first shift labor costs average about seventeen dollars an hour, the industrial robot can easily cost less in the long run than manual transfer.

Robots also save labor costs by eliminating the equipment needed to protect the safety of human workers. Using industrial robots for parts handling can sizably reduce costs in shielding, guarding, ventilating and lighting. It also reduces the insurance costs and liability risks associated with human labor.

When you calculate how much manpower is replaced with a robot, remember that since a robot works more than one shift, it can replace more than one worker. It can also replace more than one man per shift—because two or more workers are often needed to move a heavy workpiece. In addition, a normal production worker will grow

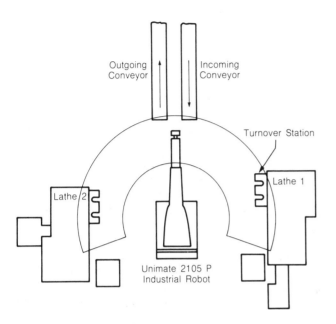

Figure 2. A Unimate 2000 Series robot transfers steel forgings in a serial process.

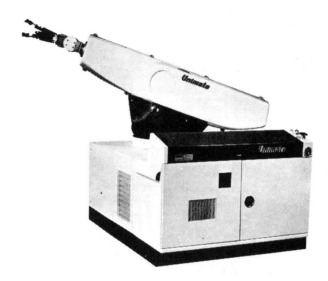

Unimation's UNIMATE Robot—widely used for parts transfer applications.

tired during the course of a typical work shift. It is a general rule in industry that, in 95% of the applications, industrial robots take *more* time than manual labor to perform a given task. However, averaged over the course of an entire work shift—in nearly *every* case reported—robots produced a higher *total* number of finished parts. This is entirely due to fatigue and "downtime" for human needs. It is not uncommon for managers to justify a robot installation solely on the direct replacement of first shift manpower.

In many cases, parts handling critically affects the quality of a product. Consider an extreme example—steel forgings. A billet of steel must be kept within a limited range of temperatures. Therefore, any delay in handling the product will cause incorrect forging and, in some cases, extensively damage the dies. With steel billets, the consistency of the transfer process is as critical as the consistency of the steel and the force of the forging press. A robot can insure that consistency.

O.K. So How Can I Use My Robot?

Rather than discuss the "general case," let's look at some examples in which robots are used as transfer

devices. The following five cases offer a good cross-section of this kind of application:

Serial Operation is shown in Figure 2. The problem is to transfer steel forgings through two successive machining operations. The parts are six kinds of forged steel housings that weigh from 2 to 40 pounds each. The outer diameters of the parts are about 14″, and their thicknesses are about 3″. This application uses a Unimate 2000 Series robot. The Unimate accepts raw parts from an incoming conveyor and preloads the part on a positioning table in front of a Jones and Lamson (J&L) turning machine. When the J&L machine has completed its cycle, it signals the Unimate. The robot extracts the semi-finished part, places it on a second rest stand, and loads another raw part into the first J&L. When the raw part is correctly positioned in the first J&L, the robot retracts and signals the machine to begin its cycle. The robot then transfers the semi-finished part to a third rest stand for preload into the second J&L. Upon receiving a signal that the second J&L has completed its cycle, the robot extracts a finished part and places it on a fourth rest station. As in the first machine operation, the robot loads the semi-finished part into the second J&L, retracts, initiates the machining cycle, and deposits the finished part on an outgoing conveyor. The robot costs $80,000, and its payback period is about 9 to 11 months of two shift operation.

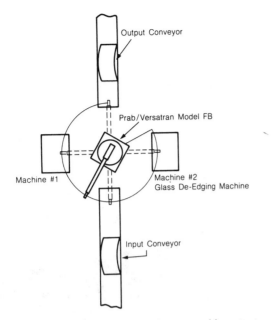

Figure 3. A Versatran FA Series robot is used here to transfer automobile windshield and window glass in a parallel process.

Parallel Configuration is shown in Figure 3. The task is to transfer glass windshield and window panels from a washing conveyor to a glass de-edging station. The panels weigh from 6 to 48 pounds and range in size from 6" by 4" to 54" by 22", each .25" thick. The tooling on the glass de-edging station has three stops that must be positioned within 5 mils. In this application, a Versatran FA Series robot with 660 control transfers the panels from the washer conveyor to the de-edging station on demand. The robot selects the appropriate de-edging machine based on input signals from either machine. To optimize cycle time, the robot is equipped with a dual hand. The robot's hand tooling is adjustable to handle the full range of the 22 different parts required in this application. The transfer's

cycle time is 5.5 seconds. The Versatran costs about $65,000, and its payback takes around 13 to 14 months of two shift operation.

A process that combines serial and parallel operation is shown in Figure 4. Steering knuckles for large off-the-road earth-moving equipment must be processed through a numerically controlled (NC) turning machine, a gauge station, and a dedicated transfer line for drilling and counter-boring. Because of the NC machine's cycle time, the factory needs three of them to maintain throughput. The steering knuckles weigh 78 pounds, are 7" high, 8" wide, and 3" thick. A Series 4000 Unimate robot moves the steering knuckles through available NC turning machines, then takes the finished parts through the gauge and subsequent transfer machining operations. The Unimate is equipped with a dual hand to optimize cycle time. Ten seconds before the completion of a machine cycle on the NC turning machine, it sends a pre-initiate signal. This allows the robot to position itself in a "ready" position just outside the NC machine. Cycle time for this operation is 8 seconds for each part transfer to the turning machine and 6 seconds for the gauge station cycle. The Unimate costs $120,000, and its payback takes about 15 to 18 months.

Random transfer is shown in Figure 5. Various appliance components require transfer from one moving conveyor to another. In this case, the parts are transferred from an unpalletized part conveyor to a conveyor that transports the parts to a dipping tank. After the parts have been dipped, they must go back to the original conveyor. The parts are appliance components, weighing from 10 to 15 pounds and coming in assorted sizes. A Cincinnati T3 robot, with a tracking option, performs the transfer. It is

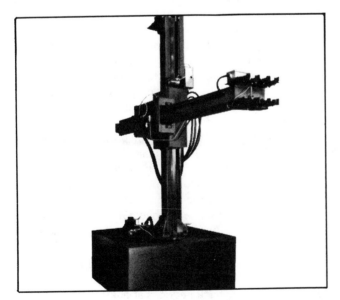

Versatran Model FA Robot

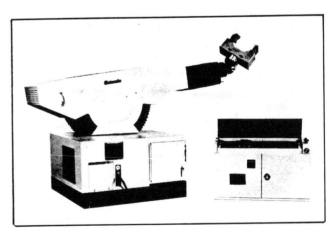

Unimate 4000 Series

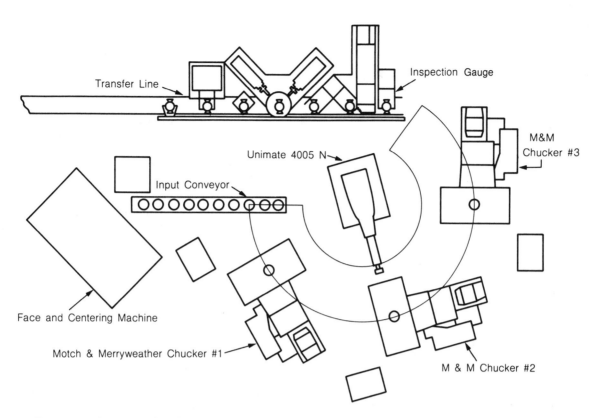

Figure 4. *To transfer the steering knuckles of large earth-moving equipment, a Series 4000 Unimate robot is configured to combine serial and parallel process.*

equipped with special tooling that allows it to pick up a wide variety of parts. Both conveyors are continuously moving—at different rates. Steel hooks carry the parts on the conveyor. The robot has a cycle time of from 4 to 7 seconds to transfer parts from conveyor to conveyor, depending on the part. The Cincinnati costs $96,000, and its payback takes approximately 10 to 11 months of three shift operation.

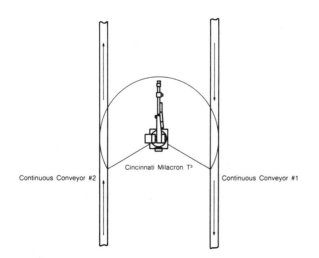

Figure 5. *A Cincinnati Milacron T³ robot transfers appliance components in a random select process.*

Cincinnati Milacron's T³—able to lift a hundred pound part.

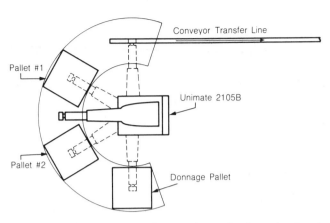

Figure 6. A Unimate 2000 Series robot unloads engine heads from parts pallets.

Palletizing is shown in Figure 6. Broached engine heads for four cylinder engines need to be loaded into pallets then transferred to another plant for machining. The parts must be unloaded from the pallets onto a conveyor that feeds a transfer machine. The engine head weighs 42 pounds, and has the dimensions of 18″ high by 4″ wide by .75″ thick. In this case, a 2000 Series Unimate robot works in conjunction with three pallet locaters. The locaters are fixed to the floor adjacent to the robot. Pallets loaded with 96 parts (6 layers with 16 parts per layer) are placed into the locaters for accurate positioning. The robot automatically selects the proper program for de-palletizing a given pallet. When the robot finishes a layer of parts, it removes the donnage* separators and deposits them in another pallet location. The tooling is designed to indicate whether a part has been properly loaded into the donnage. If a part has been loaded incorrectly, the robot hand tooling will signal the robot to select a different program in order to properly orient the part. After the robot has finished unloading one pallet, it will automatically select a program to unload the second pallet while giving a signal to an outside operator, who places a full pallet of parts in the empty parts locater. The entire engine manufacturing cycle time is 18 seconds, of which the robot takes 6 seconds. Since the whole plant is paced on an 18 second cycle, the robot has a larger percentage of dead time in this application. Still, the robot costs $45,000, and its payback period is only about 5 or 6 months of two shift operation.

Conclusion

Although I've spoken favorably about using robots as transfer devices, I should point out that robots do have some limitations. Currently, robots have very limited

---

Donnage is a device, often a plastic mold, placed on top of a pallet to precisely locate the palletized object for a robot.—ed.

TABLE 2

Robot Manufacturer	Type	Approximate Weight Carrying Capacity	Price Range
ASEA	DC electric servo, microprocessor	15 to 150 lbs.	$96,000 to $125,000
Auto-Place	Non servo pneumatic, iterative processor	20 lbs.	$4,000 to $11,000
Cincinnati Milacron	Hydraulic servo, microprocessor	100 to 250 lbs.	$67,000 to $125,000
Prab	Non servo hydraulic, iterative processor	50 to 125 lbs.	$25,000 to $35,000
Prab-Versatran	Hydraulic servo, microprocessor control	80 to 2,000 lbs.	$45,000 to $95,000
Seiko	Non servo pneumatic, iterative process	1 to 10 lbs.	$6,000 to $9,000
Unimation	Hydraulic servo, microprocessor	50 to 450 lbs.	$27,000 to $60,000
Unimation-PUMA	DC electric servo, microprocessor	3 to 7 lbs.	$37,000 to $42,000

Note: Actual weight carrying capacity figures are estimates only. The load carrying capacity of any industrial robot is dependent upon part configuration factors and application parameters.

sensory perception. They have to rely on relatively crude sensing techniques to orient and align parts. This means that you have to engineer a significant amount of structure into the manufacturing environment before a robot can function there successfully. Today's robots can't just reach into a tote bin and pull out the right part!

Giving the robot a properly oriented and accurately located part can be expensive. Methods as economical as a bulk flow feeder are often as expensive as the robot itself. This lengthens the robot's payback period.

The robot itself, especially for the first-time user, is not inexpensive. Table 2 gives the prices and capabilities of some industrial robots. Added to the base price of a standard robot, is the cost of hand tooling and accessories. A hand can cost from $4,000 to $25,000, depending on the application. The price of the engineering needed to expedite a project can run from $4,500 to $15,000, depending on the application. Add to this the cost of one-time accessory items—programming tools, program storage options, and spare parts kits, for example. All these factors inflate the bottom line cost of a robotic project.

Still, I believe that none of the disadvantages mentioned are significant enough to stop the widespread use of robots in transfer applications. The major restraint, at this time, is the lack of education on the part of plant, manufacturing, and industrial engineers and management. As these groups learn how to effectively use robots, the robot as transfer device will come into its own.

TIG Welding with Robots

New techniques are leading to the automation of Tungsten Inert Gas (TIG) arc welding.

John Meacham
Vice President, Engineering
System Technika

Introduction

In arc welding, heat from an electric arc fuses metals together. In gas tungsten arc welding (GTAW), the arc emanates from a nonconsumable tungsten electrode held above the workpiece. A blanket of inert gas (usually helium, argon, or a mixture of the two) protects both the tungsten electrode and the weld metal, since exposure to air would quickly oxidize them. People often refer to the GTAW process by its popular nickname, TIG (tungsten inert gas) welding.

Let's take a look at how a human performs TIG welding. Before welding, he cleans the workpiece of all contaminants, such as rust, dirt, oil, grease, and paint. The welder then strikes an electric arc—perhaps by quickly tapping the electrode on the work. Once the arc is lit, the welder moves his torch in a small circle until the heat creates a small pool of molten metal. When the welder gets adequate fusion at one point, he moves the torch slowly along the seam between the parts to be welded, melting their adjoining surfaces, and feeding the welding rod into the pool of molten metal, just ahead of the arc.

The welder must painstakingly control his welding speed, the speed he feeds the welding filler, and his welding current. He moves the welding rod and torch smoothly forward, making sure that the hot end of the welding rod and the hot solidified weld are unexposed to contaminating air. With foot controls, the welder adjusts the current to get proper fusion and penetration in the weld. He judges how much current to apply by the size of the molten metal puddle.

Manual TIG welding is a discipline that demands skill and experience. There are not enough TIG welders to meet industry's need for them, and, naturally, the welders are paid highly for their time. The power-generating, chemical, petroleum, and aerospace industries—all of which use high quality TIG welding—would all benefit from the automation of the TIG process.

TIG in Fixed Automation

Before we can understand how TIG is automated, we must understand the four process variables that determine the quality of the weld. These are: welding current, arc voltage, travel speed, and filler-feeding speed. The changes in these variables, taken over time, defines the *weld profile* for the TIG welding process.

To a large extent, welding current determines the quality of the weld. When we reduce the current, we reduce both the penetration and the width of the weld. For a given workpiece geometry, we usually want the current to be held constant. When we need more than one current level during a weld, the current should "slope" up or down to the new level, rather than discontinuously jump from one level to the next.

If current is constant, we can measure the distance from the electrode to the workpiece (arc gap) as a voltage potential called *arc voltage*. In some welding systems, arc voltage is used for feedback to control the arc gap.

For a given current and arc voltage (arc gap), travel speed determines the amount of energy delivered per unit length of weld. In this way, travel speed affects the quality of the weld. Increasing the speed, while keeping current constant, reduces both the penetration and the width of the weld.

Finally, the welder needs to control filler-feeding speed—how quickly he adds welding wire to the molten weld. One method, commonly used in surfacing and in making large welds, is to feed filler metal continuously into the molten weld pool by oscillating the welding rod and the arc from side to side. The welder always keeps the welding rod near the arc, feeding into the molten pool.

In the simplest kind of automation, an operator sets the welding parameters to fixed values. The welding machine follows the preset values, with feedback on travel speeds and current. This kind of operation is known as fixed automation, and in TIG welding it is commonly used to join

two cylindrical tubes together. A fixture clamps the two tubes together. The electrode moves over the seam between the tubes, welding in an "orbital" path around the tube, keeping the welding parameters constant. (It did not take much imagination to call this "orbital tube welding.")

TIG Welding with Robots

It is the tube's simple geometry that makes it so amenable to fixed automation. For a tube, we can keep the travel speed constant. We need only weld around the circumference. Since the thickness of the tube stays constant, we need just one current level, sloping up to the current level at the start of the weld, and sloping down at the end. To keep arc voltage constant, we need only keep the electrode at a fixed distance from the weld.

With geometries more complicated than a tube's, we need greater positioning precision. We also need a programmable weld profile—one that describes the complicated changes in current, arc-voltage, travel speed, and wire-feed rate at all times during the weld. To obtain positioning precision, we can use a robot arm, perhaps with an extra control for moving the electrode. To obtain a programmable weld profile, we need a programmable welding power supply.

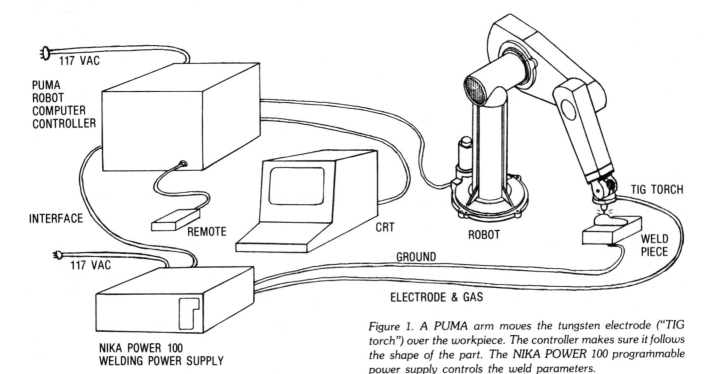

Figure 1. A PUMA arm moves the tungsten electrode ("TIG torch") over the workpiece. The controller makes sure it follows the shape of the part. The NIKA POWER 100 programmable power supply controls the weld parameters.

TIG welding with the PUMA.

At System Technika, we developed a system that interfaces a robot arm with a programmable welding power supply. Figure 1 diagrams the major components of System Technika's welding system. A robot arm (PUMA and modified Seiko arms have been used) runs the TIG torch over the workpiece, as a controller makes sure it follows the shape of the part, and as our NIKA POWER 100 power supply controls the weld parameters. Let's examine how this system performs a typical TIG weld.

During a set-up test run, the "welding engineer" programs the weld profile. In simplified terms, he sets the current (75 amperes would be a typical value), arc voltage (8–10.5 arc volts is typical with argon as the shielding gas), welding wire-feed rate (20 inches per minute is typical), and travel speed (5 inches per minute is typical). The welding engineer can program a different set of process variables for every time interval during the welding process. He can program more than a hundred points to hold different weld parameters. At this time, he also "teaches" the robot arm the shape of the workpiece.

Once programmed (taught), the power supply keeps welding parameters at their correct values for the appropriate portion of the welding sequence. The robot arm produces continuous path motion, traveling over the length of the seam.

A human welder would adjust the current to get the suitable molten weld puddle size. Since the system cannot measure puddle size, it measures arc voltage, which (given constant current) is a function of the *z-axis* distance between the electrode and the workpiece. The further the electrode travels from the workpiece, the greater the arc voltage, the wider the arc gap, and the less heat penetrates the weld. The potential voltage drop across the arc is fed back to the NIKA power source, which then controls a z-axis motor on the end of the robot's arm. In this way, the system maintains the required arc voltage.

The system also monitors itself for faults detected during the weld. If gas flow is interrupted at any time, the system immediately signals the operator, stops all motions, extinguishes the arc, and terminates the welding sequence. The system takes similar measures if it discovers that the workpiece is improperly connected to the power source's ground connector, if the liquid coolant for the torch head circulates too slowly, or if the torch accidentally touches or strays too far from the workpiece.

We have the option of programming the system for special welding techniques, such as oscillation and weaving. The system allows us to include weaving and oscillation

A PUMA arm positions the TIG welding torch over a workpiece.

A modified SEIKO arm, set up for TIG welding. Notice that the z-axis distance between the electrode and the workpiece is controlled by a special z-axis motor.

speeds in the "travel speed" parameter. The weld engineer can program multipass welding by storing more than one set of welding parameters. While the system repositions the electrode for a second pass, it sets a "pilot current," as opposed to the normal welding current.

Conclusion

As we have seen, our current system has adaptive feedback in one direction—along the z-axis, perpendicular to the workpiece. In the near future, we hope to add feedback in the x-axis and y-axis directions as well, to keep the electrode tracking along the seam.

Seam tracking requires more sophisticated sensing. Eddy current sensors, sonic sensors, infrared sensors, and computer vision are all being developed. In addition, vision systems are being developed that can monitor the size and geometry of the weld puddle—information available, up to now, only to a human welder.

TIG welding was once possible only by using highly skilled, hard-to-find welders. Now, precise computer-controlled robots, working interactively with computer-controlled welding power supplies, have given industry the ability to automate the TIG welding process. Robots repeat their programs reliably, time and time again. With adaptive feedback correcting for variations in part fit-up, the robotic welding system welds as well as the best welder on his best day, day after day. Robotic welding amplifies the productivity of each *human* welder—who can now teach and supervise robots instead of doing all the *repetitive* welding by hand. ⬛

Industrial Editor Jerry Saveriano displays the hand held teach pendants for the welding power supply (left) and the PUMA robot (right).

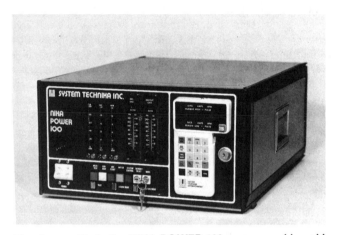

The System Technika NIKA POWER 100 programmable welding power supply.

Rehabilitative Robots

Dr. Larry Leifer
Director, Rehabilitative Engineering
Research and Development Center
Veterans Administration Medical Center
Palo Alto, CA

This article examines the objectives and evolution of Rehabilitative Robotics and defines specifications for an interactive robotic aid.

Man is a tool-using species. Our ability to manipulate our environment has given us the status we enjoy in the world today. We have created tools to extend this ability, culminating in programmable industrial automation—robots. Such robots are machines designed to manipulate other machines. We might think of robots as the ultimate result of our need to manipulate—they have the ability to take our place in many manual tasks.

To understand the importance of manipulation, just consider how much of the cerebral cortex is dedicated to hand and arm functions. Figure 1 caricatures a human body, drawn along the medial-lateral cross-section of the brain. Each part of the body is drawn in proportion to the number of neurons which respond to sensory stimulation or evoke motor output at that part. This representation—referred to as a *homunculus*—clearly shows the disproportionate amount of attention the human brain gives to our power to manipulate.

A severely disabled person has lost the power to manipulate. He is cut off from the direct control of his own personal space, impaired in performing personal maintenance functions, and cut off from the vast array of gadgets that most of us use in the course of our day.

Quadriplegia, paralysis of four limbs, is one disability that impairs the power to manipulate. It is typically the result of a traumatic spinal cord injury that disrupts neural tissue in the spinal column. When this happens at the level of the neck, the individual will have neither sensation nor muscle control in his trunk and legs. Depending on the details of the injury, the quadriplegic will have little muscle control or sensation in his arms. Grasp function is almost always lost. Respiratory function, because it is controlled from higher in the brain stem, is usually left intact. This injury occurs with increasing frequency, is seven times more likely to happen to males than females, and is most likely to happen to young people from eighteen to twenty-two years old.

In spite of an overwhelming loss of body function, a quadriplegic has no impairment of his mental faculties, and, with good medical handling, can have a normal life-expectancy. Though statistical data are sparse, I have estimated that the total cost of medical care and living support functions—for a twenty year old quadriplegic who lives the expected sixty-seven years—will be about two-and-a-half million (1980) dollars.

At the Rehabilitative Engineering Research and Development (RER&D) Center, in the Palo Alto Veteran's Administration hospital, we are working toward this goal: to give persons who are severely disabled, but mentally alert, physical control of their environment, without continuous assistance from other people.

We are applying advanced robotic technology toward this end, and are studying four areas in which we believe a disabled person can benefit from rehabilitative robotics: in activities of daily living, such as preparing food, eating, and personal hygiene; in personal clerical tasks, such as operating a calculator, computer, or telephone; in vocational tasks, such as computer programming or secretarial work; and in recreation—controlling electronic games, playing chess, or painting, for example. We believe that when a severely disabled person has the ability to manipulate he will become vocationally independent. Dignity and a higher quality of life should result from such independence.

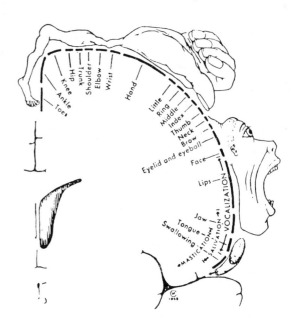

Figure 1. The allocation of human cortical capacity to sensory motor integration is visualized as a homunculus. The body is scaled so that the size of various segments is proportional to the number of neurons responding to or driving that portion of the body. It is clear that the arm and hand receive a disproportionate amount of attention. (Reproduced, with permission, from Penfield & Rasmussen: The Cerebral Cortex of Man: A Clinical Study of Localization of Function. Macmillan, 1950.)

Design Philosophy

Before reviewing the evolution of rehabilitative robots, let us examine the concept of rehabilitation that, so far, has guided the development of assistive devices. The dominant philosophy, implicit and explicit, has been to replace lost or damaged anatomy. Most designers of rehabilitative aids assume that a missing limb must be replaced. They try, in effect, to rebuild the disabled individual.

Unfortunately, when we try to rebuild anatomy we place severe constraints on the size, weight, power, and geometry of potential solutions. At present, when we improve the anatomical fidelity of a manipulator we lessen its functional performance. This fact led me to consider using industrial manipulators, rather than human-like arms, to help disabled persons satisfy their need to manipulate.

At the RER&D Center, we try to replace the missing limb's function, not its anatomy. The manipulation system should not pretend to be part of the user's body and need not be in physical contact with the user. However, the robotic aid should be handsome and worth including in the user's home. If we could not build a good-looking manipulator, then cosmetic criteria would fully justify replacing anatomy.

Our design philosophy is based upon the development of a single general-purpose system rather than a collection of special-purpose devices. Surrounding a disabled person with a clutter of rehabilitative gadgets will only further isolate him from people and public places.

A rehabilitative robotic aid must be interactive because it serves the constantly shifting attention of its user. It does not operate in the industrial context of repeated tasks. With an emphasis on flexibility, it must operate in largely unstructured environments and must be programmed in real-time. It represents the cutting-edge of robotics in the generic area of *Interactive Robotics*.

We have to consider two global value questions at each step in the evolution of Rehabilitative Robotics: how does one clinically evaluate a rehabilitative robotic aid to assess its benefit to the client and patient, and how much will it cost?

Clinical evaluation of the robotic aid is important and should be an integral part of the development plan. Assessment variables must be quantitative and objective. There is a long line of rehabilitative devices which have been developed, delivered to the patient or clinic and promptly closeted. Often the developer could have easily dealt with the problem if he had taken the time to consult the prospective users. In other circumstances, the device is not used because of psychological or social reasons, again not known to the developers. Had the designer explored these issues *before* technical development, he would have realized that the proposed device was irrelevant to the situation. Machines have no needs; only people have needs. It is imperative that robot designers establish and maintain a constructively critical dialogue with prospective users of their product.

At this time it is difficult to predict how the economics of robotic aids will evolve. However, the forces which propel development of industrial manipulators—especially labor costs—are also experienced by severely disabled individuals who must pay for up to twenty-four hour attendant care. Cost, reliability and service factors will depend upon progress in the industrial application of robotics. However, we can observe several important trends: First, the cost of computation is declining. In every other year one can obtain equivalent computational power for half the price (Turn, 1974). This factor affects the cost of voice recognition, voice synthesis, manipulator control, and sensory information processing. Second, as manipulators are produced in larger numbers, their cost per unit will fall. Currently the number of industrial manipulators produced yearly is in the low hundreds. It is expected to reach the tens of thousands by 1985. Finally, rising labor costs and declining productivity are contributing to increased investment in automation. A parallel phenomena in medical care makes the use of robotic aids increasingly attractive in rehabilitation.

There are three conditions a robotic aid must fulfill before it can economically substitute for human caretaking:

1. It must manipulate well enough to replace some classes of human assistance;
2. It must be reliable enough to encourage the user and his attendants to use it;
3. It must save (or earn) enough money to pay for itself.

To my knowledge, all previous projects have failed on one or more of these criteria (see reference Table 1). Yet I believe that the time is correct to develop powerful and economic robotic aids for the physically disabled. The necessary technology is evolving rapidly, industrial robots are now commercially viable, and the "independent-living movement" has become widely accepted among disabled individuals. Universities and industry have the technology. Technology transfer should now be carried out by the individuals who created that technology—guided by the people who need it.

Figure 2. *Veteran Asher Williams damaged his spinal cord at the neck level while diving. Mr. Williams was wearing a temporary head brace at the time of this photograph when he was participating in a project evaluation interview. He has since recovered considerable use of his arms. Dr. Inder Perkash (standing), Chief of the VA Spinal Cord Injury Service in Palo Alto, California, is a co-investigator on the VA sponsored Rehabilitative Robotics Project.*

Technical Considerations

There are three classical technical questions that we must consider at each step in the evolution of Rehabilitative Robotics:

1. How does the human user input command and control information to the robotic aid?
2. How does the robot process new inputs in the context of its current operation to produce a smooth manipulative sequence?
3. How is the robot's performance output communicated back to the human user?

The robotic aid must be able to accept both *command* and *control* input from the disabled person. Command, usually accomplished with a keyboard, refers to a discrete item of information which specifies a task, operating mode,

Table 1 (a thru d)

Table 1a Critical review of projects in which manipulators were adapted for the physically disabled; approximate chronological order; prosthesis projects not included; DF = degree-of-freedom; R = rotational; T = translational.

Project	Arm	Discrete Inputs	Continuous Inputs
1 CASE version.1 floor mounted	laboratory 4DF, 4R, 0T 1kg lift	photo switches brow switches	none
2 RANCHO chair mounted	orthosis 6DF, 6R, 0T 1kg lift	6 tongue switches	2DF manipulandum
3 HEIDELBERG floor mounted	industrial 5DF, 5R, 0T 20kg lift	2 tongue contacts 1 suck&puff	3DF mouth manipulandum
4 VAPC chair mounted	rehabilitative 4DF, 3R, 1T 2kg lift	5 position switch	2DF chin manipulandum
5 JPL (VAPC derived) chair mounted	rehabilitative 6DF, 5R, 1T 1kg lift	36 words	none
6 UCSB table mounted	same as JPL 6DF, 5R, 1T 1kg lift	10 words	none
7 DENVER table mounted	same as JPL 6DF, 5R, 1T 1kg lift	1 tooth click	2DF eye tracking
8 JOHNS HOPKINS table mounted	prosthesis 5DF, 5R, 1T 2kg lift	1 switch keyboard	1DF chin manipulandum
9 SPARTACUS floor mounted	industrial 6DF, 6R, 0T 3kg lift	16 vox patterns 1 switch keyboard	6DF, head, elbow, hand, manipulandum
10 SPAR chair mounted	rehabilitative 5DF, 3R, 2T 2kg lift	2 switches	3DF manipulandum
11 STANFORD (PROPOSED) table mounted self-powered	industrial 6DF, 6R, 0T 1kg lift	100 words head motion keyboard	3DF head motion manipulandum

Table 1b Critical review of projects in which manipulators were adapted for the physically disabled; approximate chronological order; prosthesis projects not included; W = world coordinates; T = tool coordinates.

Project	Control Modes	Automatic Routines Capability	User Programming
1 CASE version.1	joint velocity	position replay	none
2 RANCHO	joint on/off velocity	none	none
3 HEIDELBERG	cylindrical velocity	none	none
4 VAPC	Cartesian (W) velocity	none	none
5 JPL	joint on/off velocity	none	none
6 UCSB	joint on/off	none	none
7 DENVER	Cartesian (W) position	none	none
8 JOHNS HOPKINS	joint velocity	limited	limited
9 SPARTACUS	Cartesian (W,T) position velocity	limited	none
10 SPAR	joint velocity	none	none
11 STANFORD (PROPOSED)	joint Cartesian (W,T) position velocity	locate/avoid grasp surface follow	unlimited sequence stack, run stored programs

Table 1c Critical review of projects in which manipulators were adapted for the physically disabled; approximate chronological order; prosthesis projects not included; modified 3 level rating scale (low, medium, high).

Project	Sensory Functions	Hand	Learning Difficulty
1 CASE version.1	none	2 finger vise	very high
2 RANCHO	none	braced hand 2 finger hook	very high
3 HEIDELBERG	none	pneumatic 2 finger vise + poker	medium
4 VAPC	none	2 finger hook	low medium
5 JPL	none	2 finger hook	high
6 UCSB	none	2 finger hook	medium
7 DENVER	none	2 finger hook	low medium
8 JOHNS HOPKINS	none	2 finger hook + suction	medium
9 SPARTACUS	10cm optical grasp force vertical force	2 finger 4 bar linkage removable	low very flexible
10 SPAR	none	2 finger hook	high
11 STANFORD (PROPOSED)	4mm optical 100mm optical grasp force	2 finger 4 bar linkage modular	low very flexible

Table 1d Critical review of projects in which manipulators were adapted for the physically disabled; approximate chronological order; prosthesis projects not included; modified 3 level rating scale (low, medium, high); low price <$10,000; medium <$20,000; high <$50,000.

Project	Manipulation Capability	Reliability	Arm Price Control Price
1 CASE version.1	medium	medium	high very high
2 RANCHO	very low	high	low very low
3 HEIDELBERG	medium	medium	very high very high
4 VAPC	low	medium	medium medium
5 JPL	low	low	medium very high
6 UCSB	low	low	medium high
7 DENVER	medium	low	medium high
8 JOHNS HOPKINS	low	medium	low low
9 SPARTACUS	high	high	very high very high
10 SPAR	very low	medium	very high very low
11 STANFORD (PROPOSED)	high	high	

ROBOTICS AGE: IN THE BEGINNING

datum, or label. Control, typically done with a joystick or steering wheel, refers to continuous input functions, such as those required to pilot a vehicle. Possible control inputs include head displacements, jaw displacements, eye displacements, tongue displacements, respiratory pressure patterns, vocalization patterns, electromyographic activation patterns and electroencephalographic activity patterns.

In most man-machine interfaces, the human hand performs both command and control functions. For the quadriplegic this is impossible. Unfortunately, alternative channels are poorly understood. Head displacements are limited in range and speed. Their precision has not been evaluated quantitatively. While the human voice is capable of making thousands of utterances, computers can recognize only one or two per second (a high information bandwidth with a low temporal bandwidth). The physiological signal channels (electromyogram and electroencephalogram) are conceptually elegant but unreliable. In the case of eye movement command/control, we must be careful not to impair the normal function of an already vital information channel.

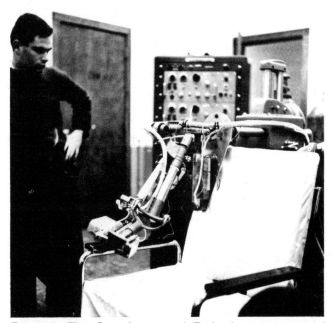

Figure 3. The Case Institute of Technology computerized orthosis was the first application of robotics technology to a rehabilitative manipulator. This 4 degree-of-freedom device was designed to carry the user's paralyzed arm while performing prerecorded manipulation tasks. Motion sequences were taught to the system by an able–bodied assistant during training sessions.

We believe that an interactive manipulator interface should use a variety of command/control channels tailored to the task at hand. The interface could include automatic computer control to coordinate individual joint motions, manage grasp/release reflexes, and control precise trajectories; voice command to enter data, select modes, identify objects, and call stored programs; synthesized voice feedback to appraise the user of system status information, express error conditions and provide usage prompts; and analog head displacements to pilot the manipulator in real-time.

We should remember that fatigue is usually proportional to the amount of conscious attention needed by the control task. Accordingly, input modes should be amenable to subconscious performance.

As the robot processes human input, the information transfer between man and machine must be carefully managed. Every physical disability reduces our capacity to transmit, receive, or process information. Therefore, we should have the user specify goals, the system attempt to acheive those goals, and the user supervise the system's operations. Of course, the exact division of labor between man and computer will vary according to the task. The following examples delineate the system control modes we believe are necessary at this time:

1. *Open Loop* performance of predetermined trajectories and trajectory sequences;
2. *Closed Loop* tracking of a target in six degrees of freedom (6-DF:X, Y, Z, roll, pitch, yaw);
3. *Manual*, "teach" mode operation;
4. *Background* command entry and program editing:
5. *Foreground* performance of the manipulation task;
6. *Reflex* control of grasp, obstacle avoidance, and unplanned human contact;
7. *Hierarchical Command* and *Control* structures which place the user at different positions in the control network:
    a. *Man-in-the-Loop*, direct control;
    b. *Supervisory Command*, auto-pilot mode;
    c. *Mixed Mode*, 6-DF command with 2-DF control.

The usefulness of artificial limbs had been limited mainly by inadequate sensory feedback. Movements performed open loop, without sensory feedback, are awkward. Visual feedback of gross limb displacements is not as easy to manage as kinesthetic feedback. Unlike muscle, tendon, and joint sensation, visual feedback requires constant attention. Constant attention fatigues the user and diminishes his ability to attend to other stimuli.

We are faced with the same problem at the performance feedback interface that we had at the command input interface: human channel bandwidths are limited and poorly understood. At this time we can best proceed by minimizing the amount of information the user is required to process. If possible, the system should process sensory data to the point of making manipulation decisions. The user can then supervise the outcomes of those decisions. The human role is one of task selection, command generation, and performance supervision.

The user needs some performance feedback. The robotic aid designer should consider using vibro-tactile stimulators, electro-tactile stimulators, visual displays, auditory codes, or voice response as output channels.

Using the design considerations discussed above, we can now envision a typical rehabilitative robotic aid. The aid should include one or more electro-mechanical arms which the user can move about the environment to perform useful work; one or more microcomputers to control the arm(s); one or more input channels to give the user complete command and control of the robotic system; and, finally, one or more feedback channels to make the user conscious of the arm's performance.

The Evolution of Rehabilitative Robotics

Since the dawn of pre-history, man has tried to extend his power of manipulation beyond the limits of his flesh. *Telemanipulators*, extensions of man's arms and hands, were the first fruits of this drive. Telemanipulation was first used for rehabilitation in the form of prosthetics—anatomical replacements for lost arms or legs.

Rehabilitative engineers have often tried to build externally powered prosthetic arms, only to be severely hampered by weight and power constraints. Most designers prefer body-powered artificial arms because the user then has some sensory feedback on limb performance. Attempts to control the prosthesis with electromyographic (EMG) signals from residual muscles have been frustrated by the user's need to consciously maintain visual attention to the terminal device. Though efforts to do adaptive EMG signal processing (Graupe et al, 1977) are promising, the lack of sensory feedback remains a problem. Some designers have attempted to build tactile displays for joint and grasp feedback; but these displays are not included in production prostheses (Solomonow and Lyman, 1977). Even the most sophisticated prostheses do not incorporates any computational capability.

While engineers have built prostheses for persons with

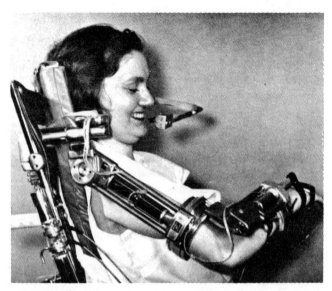

Figure 4. The Rancho Los Amigos Orthosis had 7 degrees-of-freedom. It was the first electric orthosis ammenable to external control. Joint specific control was accomplished with mechanical switch arrays. One of these arms was interfaced to the Stanford Artificial Intelligence Laboratory DEC-10 as part of a proof-of-concept demonstration.

missing limbs, they have built *orthoses* for those with paralyzed arms. An *orthosis* is an exoskeletal structure that supports and moves the user's arm.

This line of development produced the first computer-manipulator system, at Case Institute of Technology during the early 1960's (Figure 3). The Case four degree-of-freedom (4-DF) externally powered exoskeleton carried the paralyzed user's arm through a variety of manipulation sequences (Reswick and Mergler, 1962; Corell and Wijinshenk, 1964). In the first of two versions, the system performed preprogrammed motion. The user initiated the motion by pointing a head-mounted light beam at photo-receptors mounted in a structured environment. An able-bodied assistant, moving the orthosis manually, taught arm-path sequences to the system. While stored digitally, the data were effectively analog. By using numerically controlled pneumatic actuators with feedback from an incremental encoder, the system achieved closed-loop position control.

In an upgraded second version, a minicomputer performed coordinate transformation along X, Y, and Z axes. Case employed electromyographic (EMG) signals to specify endpoint velocity within this coordinate space. Photo-receptors, mounted on each arm segment, could be used to control individual joint displacements. In the sense of having a stored operating code, neither version was programmable. Yet this was a milestone project in many respects. For more than ten years, no other project employed the technology or concept of computer-augmented manipulation with as much sophistication.

The Rancho Los Amigos Manipulator (Figure 4) was designed as an orthosis with seven degrees-of-freedom. It

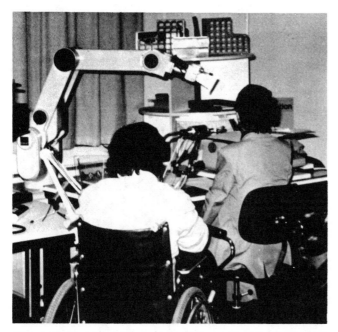

Figure 5. The Heidelberg manipulator and structured work station was the first rehabilitative application of an industrial manipulator. The system used a minicomputer for end-point control in cylindrical coordinates. The user controlled the manipulator at all times. They decided against computer augmentation of the task.

followed the design philosophy of the Case system but did not augment manipulation with computer control. It used direct current servo motors at each joint and controlled each motor with a variety of ingenious switch arrays. Several similar versions of the "Golden Arm" were built. At least one version was wheelchair-mounted and battery-powered. General Teleoperators (Jim Allen, president and principal designer) still offers manipulators descended from this line of evolution.

Extensive clinical trials confirmed the impracticality of joint specific control. These trials confirmed results from the Case group and underlined the need for computer augmention. Moe and Schwartz (1969) computerized the Rancho Arm to provide coordinated joint displacement and proportional control. In 1971, Freedy, Hull and Lyman studied the feasibility of using a computer to adaptively help the user control the manipulator.

These efforts, though, could not overcome limitations inherent in the orthosis. In 1979, Corker et al, evaluated remote medical manipulators. They observed that fitting a manipulator to the specifics of an individual's anatomy and range of motion makes construction and control very difficult. Furthermore, there is no functional reason for the manipulator to carry the user's arm, which has neither grasp nor sensation. In fact, there is a danger of injury because the user's arm could be driven beyond its physiological range without any warning sensation. The orthotic approach is a clear case of anatomical replacement thinking. This line of evolution in rehabilitative telemanipulation is effectively extinct.

As an "evolution of the species" footnote, I should mention that Victor Scheinman and I purchased one of the Rancho orthoses in 1964 for the then-budding robotics project in the Stanford University Artificial Intelligence Laboratory (SAIL). We instrumented the arm for joint position feedback and interfaced it to a DEC PDP-10 computer. Preliminary experience with computer control of that arm helped establish reliability as the most important performance criteria.

In the late sixties and early seventies, work began with manipulators which were physically isolated from the user

Figure 6. The Johns Hopkins University Applied Physics Laboratory manipulator and work station evolved from the goal of developing a prosthesis. Like the Heidelberg system, the applications environment is highly structured. Their experience with semi-automated feeding suggests that computer augmented manipulation may be well received by users and their attendants.

and made no pretext of being limb replacements. The designer was now free to follow the dictates of efficient machine design for computer control. In 1969, working with remote master-slave manipulators, Whitney formalized the concept of resolved motion rate control. Whitney's work helped lead to the use of end-point control in robotics and to computer augmented remote manipulation. In 1974, Roth demonstrated that for certain geometries one can efficiently and explicitly solve all of the manipulator configuration alternatives (for 6-DF) to minimize a variety of parameters (such as net joint displacement for a specified point-to-point path length).

Roesler and Paeslack, in 1974, at the University of Heidelberg, were the first to use an industrial manipulator and a highly structured work environment (Figure 5). The electromechanical manipulator had five degrees-of-freedom plus grasp. The floor-mounted manipulator was about 1.3 times human scale and occupied a fixed location within the work station. The Heidelberg manipulator used a minicomputer to transform coordinates for end-point control in cylindrical coordinates. They used computer augmentation to store intermediate points of special interest and to integrate the status of many special purpose devices within the environment. Among the devices were a modified telephone, special typewriter, custom mouth-stick keyboard, motorized parts bin, and a three degree-of-freedom mouth manipulandum.

In this system, the user controlled all movement. However, the system did not allow the user to program the manipulator, nor did it execute preprogrammed manipulation sequences. This appears to have been the result of cultural or investigator bias, since the developers report that prospective users stated a clear preference for direct control at all times. Social and cultural factors are crucial in determining system specifications. In this case, they may have prevented the Heidelberg group from fully using the technology and expertise vested in the project.

A conceptually similar manipulation work station was developed by investigators at the John Hopkins University Applied Physics Laboratory (Seamone, et al., 1978; Schneider, et al., 1981). Their four degree-of-freedom manipulator (plus grasp) was human scale, having evolved from a prosthesis for shoulder-level amputees (Figure 6).

Though overweight and underpowered, the prothesis was successfully adapted to limited microprocessor control. One DC torque motor drove elbow flexion-extension, wrist pronation-supination and shoulder flexion-extension. Cable-driven spring-return mechanisms with solenoid-actuated locks isolated the shared degrees-of-freedom. A second motor drove shoulder rotation and grasp. The arm was mounted on a translation table that

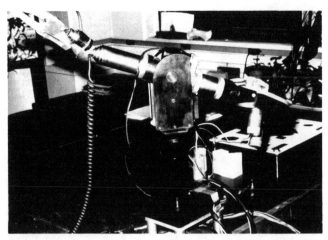

Figure 7. A telescoping 4 degree-of-freedom powered reacher developed at the VA Prosthetics Center in New York (now the VA Rehabilitation Engineering Center) used analog electronics to effect end-point velocity control. A derivative of this design by General Teleoperators added two additional degrees-of-freedom at the wrist. In this form the manipulator has been adapted to voice command, eye-motion control and wheelchair mobility. Reliability problems have severely hampered the further evolution of the system.

provided one additional degree-of-freedom, enlarging the manipulator's working volume. Simultaneous motion on all axes was precluded by the motor drive design.

The John Hopkins system used preprogrammed manipulation sequences for standard tasks such as book retrieval from fixed storage bins and food service from specified serving bowl locations. The user could define, enter by keyboard, and store ten motion sequences for later execution. John Hopkins' experience with semi-automated self-feeding supports the idea that users and their attendants respond favorably to some degree of automation and appreciate the independence it makes possible.

Carl Mason of the Veterans Administration Rehabilitation Engineering Center (formerly the VA Prosthetics Center) was the first to use a prismatic shoulder joint in a rehabilitative remote manipulator. The manipulator was designed to retrieve objects from the floor and overhead shelves. The earliest version had four degrees-of-freedom plus grasp and might be better referred to as a "powered reacher" (Figure 7). The device was used in both table and wheelchair-mounted settings. It was controlled by mechanical mode switches and a 2-DF chin manipulandum. Mason used analog electronics to obtain Cartesian velocity control. The device was not programmable.

Later, General Teleoperators used its design as a basis for several second-generation telescoping manipulators. Three additional degrees of rotational freedom were added to the wrist of these models (5-DF rotation, 1-DF translation). Researchers have used several such devices to study alternative approaches to handling the man-machine interface.

The NASA Jet Propulsion Laboratory (JPL) in Pasadena mounted one of these manipulators on an electrically powered wheelchair (Figure 8) and used a minicomputer-based voice recognition system to control the velocity of individual joints. JPL also used the voice system to command and control the wheelchair (Heer, et al., 1975). While the 36 word vocabulary should have been adequate, low recognition reliability and slowness discouraged further use of voice command.

Robert Roemer, at the University of California at Santa Barbara, adapted one of the 6-DF General Teleoperator manipulators to command-control by a microcomputer-based voice recognition unit. John Lyman's laboratory at the University of California at Los Angeles evaluated the system, and Corker, et al., reported preliminary results in 1979. In summary, the manipulator was not sufficiently reliable for clinical tests. The average recognition rate was 68.9%, too low to permit real-time interactive command-control of a manipulation task.

The Spartacus Project in France was the first to use a computer-augmented nuclear industry master-slave manipulator for rehabilitative purposes (Figure 9). The CEA-LaCalhene MA-23 manipulator had six degrees-of-freedom plus grasp (Guittet, et al., 1979). Control modes included discrete mechanical switches, a 3-DF manipulandum, mechanical head motion instrumentation, and laryngophone input. Theirs was the first robotic aid with optical proximity detectors on the terminal device. They accomplished computer augmentation with a minicomputer (SOLAR 16/65 with 48-k main memory) and included end-point control in both world and tool reference Cartesian coordinates. Position, velocity and motor current derived force control modes were available. These sensory capabilities provided the system with "reflexes" for precise object localization, and grasp. The manipulator was programmable in a high level language and capable of repeating preprogrammed motion sequences. At the time of their most recent publication, the system had no provision for user programming.

The manipulator itself, designed by Vertut, et al., (1976), uses DC torque motors, steel ribbon reversible transmissions, and analog position servo systems. In spite of its apparent size, the manipulator has relatively low inertia, high natural compliance, and limited reversible force output. The entire system weighs over 1,000 pounds and stands over two meters high—quite a formidable aid! The project has done pioneering work with sensors, control augmenting reflexes, and heirarchical control software. The Spartacus Project had all the ingredients of an interactive robotic aid, thought the MA-23 was too big for a domestic environment.

The VA–supported Stanford Robotic Aid is the first system to incorporate a human scale industrial manipulator (Unimation PUMA-250), a standard microprocessor

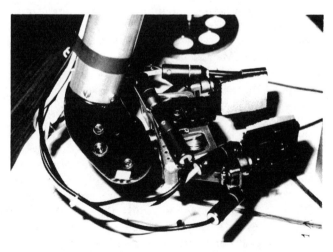

Figure 9. The French Spartacus project developed the first truly robotic aid. The system used a successful nuclear industry master-slave manipulator with six degrees-of-freedom and grasp (the CEA-LaCalhene MA-23 developed by Jean Vertut). A 16-bit minicomputer (SOLAR 16/65) was used for several levels of computer augmented manipulation. Proximity sensors on the two fingered hand were used to provide basic reflexes for grasp and object localization. Weighing over 500 kg (1200 pounds) and standing over 2 meters high (7 feet) without the computer rack, the system was simply too large for the domestic environment.

Figure 8. Researchers at JPL outfitted the General Teleoperators manipulator with a minicomputer based voice recognition system and mounted it on an electrically powered wheelchair. They attempted to control both manipulation and the wheelchair motion with voice commands. Manipulation was not computer augmented.

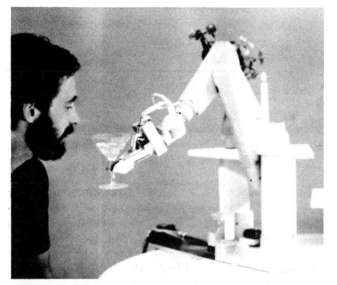

Figure 10. Stanford University's Robotic Aid is the most recent attempt to demonstrate the feasibility of using robotic techniques to satisfy the manipulative needs of severely disabled individuals. The system contains 12 microcomputers and is built around Unimation's successful human scale PUMA-250 programmable manipulator. Voice command and synthesized voice output with additive manipulandum input promise to give the user effective real-time interactive command and control of a wide variety of manipulation tasks. A smart sensate hand supports reflex control of grasp, object localization and object avoidance.

based voiced command unit, and mixed mode heirarchical control software running in five independent microcomputers (Figure 10). As reported [in Leifer 1980ab], the first version of this system is about to enter clinical pretesting. The six degree-of-freedom PUMA-250 is DC torque motor driven with incremental optical encoders in a digital position servo. Manipulation is controlled in the background while user interactions are monitored in the foreground of a BASIC-like operating sytem (VAL). The PUMA controller is a 16-bit DEC LSI 11/2 with 24K bytes of main memory. The VAL operating system occupies about 12K of PROM leaving 12K of RAM for user programs and coordinate transform data.

In the Stanford Robotic Aid, the PUMA controller is driven by a Z-80 based Zilog microcomputer system executive that integrates voice and sensor inputs with arm and voice outputs. Programs and data are stored on the executive system's dual 8″ floppy disk drives. Voice recognition and hand control units are both Z-80 based. The two-fingered hand incorporates twelve optical proximity sensors, half of which are short range (4mm), high resolution detectors (0.05mm), and the other half are long range (100mm), low resolution detectors (4.0mm).

In response to voice commands the system can do preprogrammed manipulation tasks, immediate mode movement control and deferred command sequences. The robotic aid has world, tool and joint coordinate systems available to it, and it can intermix coordinate systems in

the command syntax. The system can mix voice initiated motion with joy-stick (head control unit) inputs during real-time manipulation. This project appears to have most of the ingredients needed to decisively test our hypothesis that robotic aids can help rehabilitate severely disabled individuals. I will discuss it in more detail in future articles in this series.

Conclusion

There are two dominant themes in the evolution of robotics. Along one path we observe the development of analog extensions to the human arm. I have referred to this approach as *Telemanipulation*. It includes the most primitive stick and the most advanced master-slave manipulator. The second evolutionary path features devices which have no physical connection to the human user. *Robotic Manipulation* is intended to augment function, not anatomy. Robotic devices typically follow commands, telemanipulators are controlled. I believe, as Figure 11 shows, that these two paths will give rise to a third approach, which I refer to as *Interactive Robotics*.

Rehabilitative Robotics is one application within the generic field of Interactive Robotics. Neither telemanipulation nor robotics in their pure form can satisfy rehabilitive needs. Telemanipulation fails because of man-machine

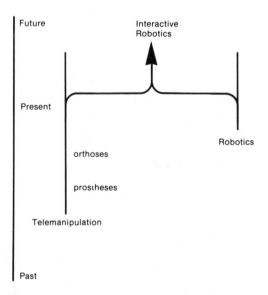

Figure 11. Interactive Robotics is a natural result of the evolution of telemanipulation and robotics. It is a consequence of human needs and the availability of necessary resources.

interface constraints encountered with disabled users. Robotics based solely upon computer control of electro-mechanical manipulators fails because the human living environment is much less structured than the industrial assembly environment, and because human needs change more rapidly than manufacturing tasks. We expect inter-active robotics to become increasingly important within the industrial environment as robotic applications prolif-erate, users with less "computer tolerance" are accommo-dated, and task assignments become more complex and varied.

Industrial robotics is now strongly influencing the design of rehabilitative aids. I think, though, that the lines of influence may yet run in both directions. Though designed to aid the physically disabled, interactive robotic systems will have important consequences within the field of robotics. ®

Acknowledgements

The Stanford Robotic Aid Project has been supported by the Veterans Administration (contract V640-P-1434), and generous technical assistance from Brian Carlisle, general manager of Unimation's Advanced Development Laboratory. The following project members have all made valued contributions: Bernard Roth, Ph.D., ME, co-investigator; Inder Perkash, M.D., VA Spinal Cord Injury Service Chief, co-investigator; Urs Elsasser, Ph.D., ME; Charles Buckley, MSME, Ph.D. candidate; Charles Wam-pler, MSME, Ph.D. candidate; Mitchel Weintraub, MSEE, Ph.D. candidate; John Jameson, MSME, Ph.D. candidate; Michael Van der Loos, MSME, Ph.D. candidate; and Jeff Kamler, BA, training specialist. Recognition is due to persons who made their contribution during the particu-larly difficult days prior to receipt of project funding: James Sachs, MSME; David Kelley, MSME; Rickson Sun, MSME; Michel Parent, Ph.D. visiting scholar from France; and especially Victor Scheinman, long time associate and most valued hallway discussion contributor. Vernon Nickle, M.D., long time advocate of rehabilitative engineer-ing, has provided the spiritual support needed to get people going on a project of this magnitude and complex-ity. Photographs were contributed by individual designers and project leaders. They are sincerely appreciated.

Bibliography

[1]   Graupe, D., Beex, A., Monlux, W., and Magnussen I., A Multifunctional Prosthesis Control System Based on Time Series Identification of EMG Signals Using Micro-processors, Bulletin of Prosthetics Research, (Spring 1977), pp. 10-27.

[2]   Solomonow, M., and [illegible] J. Artificial Sensory Com-munications via the Ta[illegible] (Space and Frequency Optimal Displays), [illegible] edical Engineering, Vol. 5, (1977) pp. 27[illegible]

[3]   Reswick, J., and M[illegible] dical Engineering and Progress Report [illegible] arch Arm Aid, Case Institute of Technology, Report No. EDC 4-64-3, (1962).

[4]   Corell, R., and Wijnschenk, M., Design and Development of the Case Research Arm Aid, Report EDC 4-64-4, Engineering Design Center, Case Institute of Technol-ogy, (1964).

[5]   Moe, M., Schwartz, J., A Coordinated, Proportional Motion Controller for an Upper-Extremity Orthotic Device, Proceedings of the Third International Sympo-sium on External Control of Human Extremities, Dubrov-nik Yugoslavia, (August 1969), pp. 295-305.

[6]   Freedy, A., Hull, L. Lucaccini, L., and Lyman, J., A Computer Based Learning System for Remote Manipu-lator Control, IEEE Transactions on Systems, Man and Cybernetics, Vol. SMC-1, (October 1971), pp. 356-363.

[7]   Corker, K., Lyman, J., and Sheredos, S., A Preliminary Evaluation of Remote Medical Manipulators, Bulletin of Prosthetics Research, (Fall 1979), pp. 107-134.

[8]   Roth, B., Performance Evaluation of Manipulators from a Kinematic Viewpoint, in Sheridan, T., (ed.), Performance Evaluation of Programmable Robots and Manipulators, N.B.S. SP-459, (October 1976).

[9]   Roesler, H., and Paeslack, V., Medical Manipulators, First CISM-IFToMM Symposium of Theory and Practice of Robots and Manipulators, Udine, (1974), pp. 158-169.

[10]   Seamone, W., Schmeisser, G., and Schneider, W., A Microprocessor Controlled Robotic Arm/Worktable Sys-tem with Wheelchair Chin-Controller Compatibility, Pro-ceedings of the IRIA International Conference on Tele-manipulators for the Physically Handicapped, Rocquen-court, France, (September 1978), pp. 51-62.

[11]   Schneider, W., Schmeisser, G., and Seamone, W., A Computer-Aided Robotic Arm/Worktable System for the

High-Level Quadriplegic, IEEE Computer, (January 1981), pp. 41-47.

[12] Mason, C., Medical Manipulator for Quadriplegics, Proceedings of the Midcon Professional Program, Biomedical Applica... ...ystems Theory, (1980), pp. 1-6.

[13] Heer, E., Wi... ...and Karchak, A., Voice Controlled Adaptive ...or and Mobility System for the Severely Hand... ...roceedings of the Second Conference on Re... ...ned Systems Technology and Applications, (June 1979), pp. 85-86.

[14] Guittet, J., Charles, J., Coiffet, P., and Petit, M., Advance of the New MA 23 Force Reflecting Manipulator System, Second CISM-IFToMM Symposium on Theory and Practice of Robots and Manipulators, Warsaw, (1976), pp. 307-322.

[15] Vertut, J., Charles, J., Coiffet, P., and Petit, M., Advance of the New MA 23 Force Reflecting Manipulator System, Second CISM-IFToMM Symposium on Theory and Practice of Robots and Manipulators, Warsaw, (1976), pp. 307-322.

[16] Leifer, L., Sun, R., and Van der Loos, H.F.M., A Smart Sensate Hand for Terminal Device Centered Control of a Rehabilitative Manipulator, 1980 Advances in Bioengineering, Mow, V.C. editor, ASME, New York, 1980, pp. 69-72.

[17] Leifer, L., Sun, R., and Van der Loos, H.F.M., The Need for a Smart Sensate Hand in Rehabilitative Robotics, 1980 Advances in Bioengineering, Mow, V.C. editor, ASME, New York, 1980, pp. 285-288.

The Robot Builder's Bookshelf

David Smith and
Mike Schoonmaker III

Many of our readers have asked us to print a reading list of books that will provide a broad and thorough background in robotics and artificial intelligence (AI). Readers Smith and Schoonmaker have given us a good start with this annotated listing. We will be supplementing the list from time to time.

We start our list with two books that offer constructive criticism of AI:

What Computers Can't Do: The Limits of Artificial Intelligence, by Hubert L. Dreyfus, Harper Collophon Books, Harper and Row, Publishers, New York, NY (1972).
This book is not about robots but covers the problems and progress of artificial intelligence in general.

Computer Power and Human Reason (From Judgement to Calculation), by Joseph Weizenbaum. W. H. Freeman.
Weizenbaum argues that computers will never achieve true intelligence. They will only be able to mimic behavior that humans will interpret as intelligent. He also argues that building such systems could be dangerous.

The following books are about building your own robot:

How to Build Your Own Self-Programming Robot, by David L. Heiserman, Tab Books, Blue Ridge Summit, PA 17214 (1979).
The robot in this book uses an 8085 microprocessor in a wire-wrapped, on-board system. It has no hands or arms. It senses obstacles by checking if the motors have stalled. Not recommended for a room full of breakables!

Build Your Own Working Robot, by David L. Heiserman, Tab Books.
This robot does not use a microprocessor. The "brain" is built up out of separate TTL logic chips, and it has no hands or arms. The robot uses bumpers with switches for object contact sensing.

Robot Intelligence...With Experiments, by David L. Heiserman, Tab Books (1981).
This is the book for those who want to experiment with robotic software without spending money on a robot body. David Heiserman wrote a series of programs in Level II basic for a TRS-80 with 4K of memory. The "robot" is a graphics figure that wanders around on the screen. No disk drive is required and the standard screen is used.

How to Build Your Own Working Robot Pet, by Frank DaCosta, Tab Books.
DaCosta shows how to make a robot "dog." It has no hands or arms and features an on board wire-wrap system using the 8085 CPU. Obstacle detection is done with ultrasonics and bumper contact switches.

Android Design, by Martin Bradley Winston, Hayden Book Company, Inc., 50 Essex St., Rochelle Park, NJ 07662 (1981).
There is no other book on the market that has as much solid information as this one. It has over 90 addresses of parts suppliers. It shows how to make a very flexible finger out of bike chain. It shows how to make a 64 by 64 solid-state TV camera out of a dynamic ram chip. It includes information on VMOS FET speed and direction controls for DC motors. It covers a triangular wheel subsystem that can climb stairs.

Computers That Talk, by George Papcun and Lloyd Rice, Byte/Mc-Graw-Hill, $18.95 (available winter 1982).
Get your robot to talk. If only it was as easy to get it to hear.

How to build a Computer-Controlled Robot, by Tod Loofbourrow, Hayden Book Company, Inc. (1978).
Ultrasonics and bumper contact switches do obstacle detection. This is the book to read if you don't want to build the CPU "brain." The robot, "Mike," uses an on-board KIM (6502) with a memory expansion unit. One problem: the KIM is a little out of date. Take a look at the AIM, SYM, VIM, or SuperKIM. Even so, Loofbourrow gives solid technical information in a lucid presentation.

The next books provide general information related to robotics and AI:

The Brains of Men and Machines, by Ernest W. Kent, Byte/McGraw-Hill, $15.95.
Kent gives computer-like models for human intelligence and argues how human intelligence may influence computer design.

Brains, Behavior, and Robotics, by James S. Albus, Byte/McGraw-Hill, $15.95, (1981).
Albus takes a look at how the human brain works and gives design considerations in building a robot brain.

Handbook of Remote Control and Automation Techniques, by John E. Cunningham, Tab Books (1978).

A general look at making objects move. It contains some information on DC motors, hydraulics, electronics and control systems.

Human Factors Applications in Teleoperator Design and Operation, by Edwin G. Johnsen and William R. Corliss, Wiley-Interscience, A division of John Wiley & Sons, Inc., New York, (1971).
Johnsen and Corliss discuss robot arms, hands, and the senses. The bibliography is a gold mine of information.

Computers and Thought, by Edward A. Feigenbaum and Julian Feldman, McGraw-Hill, (1963) $12.50.
This is a well-organized collection of 28 papers by experts in the field. It gives a good background in artificial intelligence.

Introduction to Artificial Intelligence, by Philip C. Jackson, Petrocelli/Charter (1974).
This book emphasizes the GPS system and heuristic search theory. It also includes game trees, pattern perception, theorem proving, and semantic information processing.

The Thinking Computer, Mind Inside Matter, by Bertram Raphael, W. H. Freeman (1976), paperback.
This is probably one of the most readable introductions to AI, and a bargain at $6.95.

A.I.—The Heuristic Programming

Approach, by James R. Slagle, McGraw-Hill (1971).

The Relevance of General Systems Theory, by Ervin Laszlo, George Braziller, N. Y. (1972).
Many people think systems theory is important to robotics. I have my own system. If it doesn't work, use a hammer. If it does work, use a hammer. I found this book interesting.

Cybernetics, by F. H. George, Dover (1971).
Cybernetics is dated as a word, but this is still a reasonable introductory text.

The Complete Strategist, by J. D. Williams, McGraw-Hill, (Revised Copyright by the Rand Corp. 1966).
This is recommended reading for anyone interested in games. It is very readable, illustrating all techniques covered in the text. Economic game theory is not covered.

Cybernetic Frontiers, by Stewart Brand, Random House (1974).
Though entertaining, this book is non-technical and rather dated.

Great Ideas in Information Theory (Language and Cybernetics), by Jagjit Singh, Dover (1966).
Singh gives a good history of the early work in AI. It is very readable.

Man and the Computer, by John G. Kemeny, Chas. Scribner and Sons (1972).
This book explains the symbiosis of

man and the computer. Next he should do a book on the symbiosis of man and the A.I. device.

Great Ideas in Operations Research, by Jagjit Singh, Dover (1968).
Operations Research is problem solving by mathematics to seek optimal solutions.

The Psychology of Thinking, by Robert Thomson, Pelican (1957-1967).
A well documented book that admits it knows nothing. Reprinted again and again.

IEEE Transactions on Pattern Analysis and Machine Intelligence, IEEE, New York (1979).
Recommended reading.

Essentials of Statistics for Scientists and Technologists, by C. Mack, Plenum Publishing Co., 227 West 17th St., NY 10011.
A book on statistics for someone who wants to build robots? Yes, I think you'll find statistics useful in evaluating the performance of your robot. This book is very readable and is the best one I have found so far.

Electronics books for robot builders:

TTL Cookbook, by Don Lancaster, Howard W. Sams & Co., 4300 West 62nd St., Indianapolis, Ind. 46268 (1964).
Lancaster describes the operation and function of the most common

chips. He also gives some valuable hints on construction.

CMOS Cookbook, by Don Lancaster, Howard W. Sams & Co. (1977).
CMOS uses less power than TTL, which is important when you see the drain that the CPU puts on a robot's batteries. The TTL Cookbook and the CMOS Cookbook both cover about the same material, but using the different logic families.

Engineer's Notebook, Integrated Circuit Applications, by Forrest M. Mims, III, Radio Shack (1979).

Mims presents a collection of circuits that use Radio Shack parts. It would be helpful to someone learning electronics.

Understanding IC Operational Amplifiers, by Roger Melen and Harry Garland, Howard W. Sams & Co., Inc.

IC Timer Cookbook, by Walter G. Jung, Howard W. Sams & Co. (1977).

The 555 Timer Applications Sourcebook, With Experiments, by Howard M. Berlin, Howard W. Sams & Co. (1976).

Digital Interfacing with an Analog World, by Joseph J. Carr, Tab Books (1978).